Casting and Forming of Light Alloys

Casting and Forming of Light Alloys

Editor

Wenming Jiang

Basel • Beijing • Wuhan • Barcelona • Belgrade • Novi Sad • Cluj • Manchester

Editor
Wenming Jiang
Huazhong University of
Science and Technology
Wuhan, China

Editorial Office
MDPI
St. Alban-Anlage 66
4052 Basel, Switzerland

This is a reprint of articles from the Special Issue published online in the open access journal *Metals* (ISSN 2075-4701) (available at: https://www.mdpi.com/journal/metals/special_issues/Casting_Forming).

For citation purposes, cite each article independently as indicated on the article page online and as indicated below:

Lastname, A.A.; Lastname, B.B. Article Title. *Journal Name* **Year**, *Volume Number*, Page Range.

ISBN 978-3-0365-8918-3 (Hbk)
ISBN 978-3-0365-8919-0 (PDF)
doi.org/10.3390/books978-3-0365-8919-0

Cover image courtesy of Wenming Jiang

© 2023 by the authors. Articles in this book are Open Access and distributed under the Creative Commons Attribution (CC BY) license. The book as a whole is distributed by MDPI under the terms and conditions of the Creative Commons Attribution-NonCommercial-NoDerivs (CC BY-NC-ND) license.

Contents

About the Editor vii

Wenming Jiang and Yuancai Xu
Casting and Forming of Light Alloys
Reprinted from: *Metals* 2023, 13, 1598, doi:10.3390/met13091598 1

Feng Guan, Suo Fan, Junlong Wang, Guangyu Li, Zheng Zhang and Wenming Jiang
Effect of Vibration Acceleration on Interface Microstructure and Bonding Strength of Mg–Al Bimetal Produced by Compound Casting
Reprinted from: *Metals* 2022, 12, 766, doi:10.3390/met12050766 5

Bei Yuan, Dunming Liao, Wenming Jiang, Han Deng and Guangyu Li
Study on Friction and Wear Properties and Mechanism at Different Temperatures of Friction Stir Lap Welding Joint of SiCp/ZL101 and ZL101
Reprinted from: *Metals* 2023, 13, 3, doi:10.3390/met13010003 23

Igor A. Petrov, Anastasiya D. Shlyaptseva, Alexandr P. Ryakhovsky, Elena V. Medvedeva and Victor V. Tcherdyntsev
Effect of Rubidium on Solidification Parameters, Structure and Operational Characteristics of Eutectic Al-Si Alloy
Reprinted from: *Metals* 2023, 13, 1398, doi:10.3390/met13081398 41

Wenchang Zhang, Kun Xu, Wei Long and Xiaoping Zhou
Microstructure and Compressive Properties of Porous 2024Al-Al$_3$Zr Composites
Reprinted from: *Metals* 2022, 12, 2017, doi:10.3390/met12122017 55

Shuhui Huang, Baohong Zhu, Yongan Zhang, Hongwei Liu, Shuaishuai Wu and Haofeng Xie
Microstructure Comparison for AlSn20Cu Antifriction Alloys Prepared by Semi-Continuous Casting, Semi-Solid Die Casting, and Spray Forming
Reprinted from: *Metals* 2022, 12, 1552, doi:10.3390/met12101552 71

Xu Zheng, Jianguo Tang, Li Wan, Yan Zhao, Chuanrong Jiao and Yong Zhang
Hot Deformation Behavior of Alloy AA7003 with Different Zn/Mg Ratios
Reprinted from: *Metals* 2022, 12, 1452, doi:10.3390/met12091452 89

Gengyan Feng, Hisaki Watari and Toshio Haga
Fabrication of Mg/Al Clad Strips by Direct Cladding from Molten Metals
Reprinted from: *Metals* 2022, 12, 1408, doi:10.3390/met12091408 105

Vadlamudi Srinivasa Chandra, Koorella S. V. B. R. Krishna, Manickam Ravi, Katakam Sivaprasad, Subramaniam Dhanasekaran and Konda Gokuldoss Prashanth
Mechanical and Tribological Behavior of Gravity and Squeeze Cast Novel Al-Si Alloy
Reprinted from: *Metals* 2022, 12, 194, doi:10.3390/met12020194 119

Mahmoud Ahmed El-Sayed, Khamis Essa and Hany Hassanin
Influence of Bifilm Defects Generated during Mould Filling on the Tensile Properties of Al–Si–Mg Cast Alloys
Reprinted from: *Metals* 2022, 12, 160, doi:10.3390/met12010160 131

Anastasiya D. Shlyaptseva, Igor A. Petrov, Alexandr P. Ryakhovsky, Elena V. Medvedeva and Victor V. Tcherdyntsev
Complex Structure Modification and Improvement of Properties of Aluminium Casting Alloys with Various Silicon Content
Reprinted from: *Metals* **2021**, *11*, 1946, doi:10.3390/met11121946 . . . . **147**

Ming Sun, Depeng Yang, Yu Zhang, Lin Mao, Xikuo Li and Song Pang
Recent Advances in the Grain Refinement Effects of Zr on Mg Alloys: A Review
Reprinted from: *Metals* **2022**, *12*, 1388, doi:10.3390/met12081388 . . . . **159**

About the Editor

Wenming Jiang

Wenming Jiang, Ph.D., professor and doctoral supervisor for the State Key Laboratory of Materials Processing and Die & Mould Technology, School of Materials Science and Engineering, Huazhong University of Science and Technology, Wuhan, China, was selected as the most beautiful scientific and technological worker in the national Foundry Industry in 2021 and the scientific and Technological Innovation (double innovation) talent in Jiangsu Province. He served as the director of the National Magnesium Alloy Society, the director of the National Casting Society, and the Executive Deputy Secretary general of the National lost mold Casting Committee. He has presided over more than 20 scientific research projects, such as the National Natural Science Foundation, the National Key Research and Development Plan, and the National Key Research Plan for Basic Strengthening. He has published more than 150 academic papers, applied for/authorized 35 invention patents, and participated in the compilation of five books/textbooks. He is an Editorial Board Member of journals such as *Materials*, *Metals*, *China Foundry*, *Casting*, and *China Foundry Equipment and Technology*. He has mainly carried out research on magnesium, aluminum alloys, and their precision casting forming theory and technology; 3D printing rapid casting technology; and green casting theory and technology research.

Editorial

Casting and Forming of Light Alloys

Wenming Jiang * and Yuancai Xu

State Key Laboratory of Materials Processing and Die & Mould Technology, School of Materials Science and Engineering, Huazhong University of Science and Technology, Wuhan 430074, China; mutouxyc@163.com
* Correspondence: wmjiang@hust.edu.cn

1. Introduction and Scope

With the rapid development of aviation, aerospace, navigation, automotive, electronics and other fields, the demand for light alloys components is increasing, and the performance requirements are becoming higher and higher, especially for large complex light alloys components. Therefore, high performance light alloys will have a great application potential in the future. The casting and forming of light alloys is an important step to obtain large and complex light alloy components with a high performance. Together with the compositions of light alloys, they determine the formability, defects, microstructure and mechanical properties of light alloys.

This Special Issue aims to present the latest developments in the casting and forming of light alloys. The preparation process and performance enhancement of Al, Mg or their composite materials are mainly studied. Particular attention has been paid to the relationship between process conditions, microstructural features and mechanical properties.

2. Contributions

This Special Issue contains a total of 11 articles covering the topic of the casting and forming of light alloys such as Al and Mg. Among them, six papers are about Al alloys, four papers are about light metal composite materials, and one paper is about Mg alloys.

Huang et al. [1] compared the microstructure of a AlSn20Cu wear-resistant alloy prepared using semi-continuous casting, semi-solid die casting and spray forming. The results showed that the tin phase particles of the alloy prepared using semi-continuous casting had a prolate particle shape, the tin phase of the alloy prepared using semi-solid die casting was nearly spherical and strip shaped, and the tin phase in the alloy prepared using spray forming and hot extrusion was nearly equilateral shaped. Among the three preparation methods, the semi-solid die casting had the shortest process time, and the spray molding process could obtain a finer and more uniform tin phase structure.

Shlyaptseva et al. [2] developed a new modifier with complex effects on the structure of Al-Si alloys. The modifier was composed of TiO_2, BaF_2 and KF. Under the role of the composite modifier, the α-Al dendrites, Al-Si eutectic and primary Si were all refined to different degrees. The SDAS and the average area of eutectic silicon in aluminum alloys with different Si contents were all reduced. The composite modifier could increase the strength of the hypoeutectic and eutectic silumins by 10–32% and the plasticity by 24–54%.

Zheng et al. [3] studied the effect of the Zn/Mg ratio on the hot deformation behavior of an AA7003 alloy. The optimum hot working temperature of the ternary alloy AA7703 was in the range of 653 K to 813 K, and the strain rate was lower than $0.3~S^{-1}$. Materials with a low Zn/Mg ratio could cause problems with hot deformability. Alloys with higher ratios had better machinability. The Al_3Zr dispersoid in the alloy could effectively inhibit the recrystallization of the AA7003 alloy, and the Zn/Mg ratio could potentially affect the drag force of the dispersoids.

El-Sayed et al. [4] discussed the influence of casting process parameters on bifilm defects in aluminum alloy casting. The bifilm defects produced during the filling process of

Citation: Jiang, W.; Xu, Y. Casting and Forming of Light Alloys. *Metals* **2023**, *13*, 1598. https://doi.org/10.3390/met13091598

Received: 29 August 2023
Accepted: 8 September 2023
Published: 15 September 2023

Copyright: © 2023 by the authors. Licensee MDPI, Basel, Switzerland. This article is an open access article distributed under the terms and conditions of the Creative Commons Attribution (CC BY) license (https:// creativecommons.org/licenses/by/ 4.0/).

the aluminum alloy had adverse effects on the mechanical properties. Adding filters to the gating system and reducing the hydrogen content of the molten metal could minimize the possibility of bifilm defects and significantly increase the tensile strength and elongation of the casting.

Petrov et al. [5] reported the effect of rubidium on the solidification parameters, structure and operational characteristics of a eutectic Al-Si alloy. The rubidium was relatively distributed in the silicon phase, effectively refining the eutectic silicon and changing its morphology. Rubidium modification changed the solidification parameters of the alloy. The solidus temperature and eutectic solidification onset temperature were significantly lowered, leading to an expansion of the solidification range.

Chandra et al. [6] compared the mechanical properties and wear resistance of an Al-Si alloy prepared using squeeze casting and gravity casting. The results showed that compared with gravity die-casting, the Al-Si alloy prepared using high pressure squeeze casting was refined due to the increased cooling rate and the destruction of primary dendrites during solidification via extrusion pressure. The grains were refined and the dendrite arm spacing was reduced. The reduction in casting defects in high-pressure squeeze casting alloys resulted in a lower coefficient of friction and an improved alloy wear resistance.

Yuan et al. [7] studied the friction and wear properties of friction stir welding CiCp/ZL101 and Zl101 composites at different temperatures. The results showed that the sliding friction process at each temperature was relatively stable, and the average friction coefficient was stable at about 0.4. The wear forms at room temperature were mainly oxidative wear and abrasive wear. As the temperature increased, the main wear form became fatigue wear. When the temperature reached 200 °C, the characteristics of adhesive wear appeared. After 250 °C, the composites had high-temperature lubricating properties. The composite materials had good high-temperature friction and wear properties.

Zhang et al. [8] prepared porous 2024Al-Al_3Zr composites using in situ and spatial scaffolding methods, and studied the effects of Zr content and space scaffold (NaCl) content on the properties of the composites. Studies showed that with the increase in Zr content, the powder cohesion was enhanced and the defects were significantly reduced. The increase in Al_3Zr reduced the stress concentration and hindered the crack growth. However, too much Al_3Zr increased the brittleness and reduced the performance. The increase in space scaffold content led to a gradual decrease in the compressive properties and energy absorption performance of the material.

Feng et al. [9] studied the effects of roll speed, pouring sequence and solidification length on AZ91D/A5052 clad strips prepared using direct cladding from molten metals. The results showed that the rolling speed had an influence on the average thickness of the solidified layer. The thickness of solidified layer decreased with the increase in rolling speed. The high-melting-point A5052 alloy, when poured into the lower nozzle, could solve the remelting problem of the low-melting-point AZ91D. Extending the solidification length could reduce the generation of intermetallic compounds.

Guan et al. [10] studied the effect of vibration acceleration on the microstructure and properties of a composite-casted Mg-Al bimetal interface. With the increase in vibrational acceleration, the cooling rate of the bimetal increased, leading to reductions in the reaction duration to form the intermetallic compound and its thickness. And the Mg_2Si phase in the IMC's layer was refined and distributed more uniformly.

Sun et al. [11] reported the latest progress on the effect of Zr in the grain refinement of magnesium alloys. The Mg-Zr master alloy ensured a clean interface between the Zr particles and Mg solution, which was beneficial to the diffusion of Zr elements and improved the utilization rate of nucleation. It was an efficient way to introduce Zr elements. The mechanism of grain refinement using Zr was attributed to the heterogeneous nucleation and constitutional supercooling effect. Pretreatment of the Mg-Zr master alloy or treatment of the solution could improve the utilization rate of Zr and obtain a better refining effect.

Acknowledgments: As Guest Editors, we would like to express our sincere gratitude to all the contributing authors and reviewers for their outstanding work, which has made this Special Issue possible. We are also deeply grateful to the staff at the Metals Editorial Office and MDPI for their invaluable support and active involvement in the publication process. Last but not least, we extend our heartfelt appreciation again to all the contributing authors and reviewers whose exceptional contributions have played a crucial role in the success of this Special Issue. We hope that it will serve as an informative and valuable reference for readers.

Conflicts of Interest: The authors declare no conflict of interest.

References

1. Huang, S.; Zhu, B.; Zhang, Y.; Liu, H.; Wu, S.; Xie, H. Microstructure Comparison for AlSn20Cu Antifriction Alloys Prepared by Semi-Continuous Casting, Semi-Solid Die Casting, and Spray Forming. *Metals* **2022**, *12*, 1552. [CrossRef]
2. Shlyaptseva, A.D.; Petrov, I.A.; Ryakhovsky, A.P.; Medvedeva, E.V.; Tcherdyntsev, V.V. Complex Structure Modification and Improvement of Properties of Aluminium Casting Alloys with Various Silicon Content. *Metals* **2021**, *11*, 1946. [CrossRef]
3. Zheng, X.; Tang, J.; Wan, L.; Zhao, Y.; Jiao, C.; Zhang, Y. Hot Deformation Behavior of Alloy AA7003 with Different Zn/Mg Ratios. *Metals* **2022**, *12*, 1452. [CrossRef]
4. El-Sayed, M.A.; Essa, K.; Hassanin, H. Influence of Bifilm Defects Generated during Mould Filling on the Tensile Properties of Al-Si-Mg Cast Alloys. *Metals* **2022**, *12*, 160. [CrossRef]
5. Petrov, I.A.; Shlyaptseva, A.D.; Ryakhovsky, A.P.; Medvedeva, E.V.; Tcherdyntsev, V.V. Effect of Rubidium on Solidification Parameters, Structure and Operational Characteristics of Eutectic Al-Si Alloy. *Metals* **2023**, *13*, 1398. [CrossRef]
6. Chandra, V.S.; Krishna, K.S.V.B.R.; Ravi, M.; Sivaprasad, K.; Dhanasekaran, S.; Prashanth, K.G. Mechanical and Tribological Behavior of Gravity and Squeeze Cast Novel Al-Si Alloy. *Metals* **2022**, *12*, 194. [CrossRef]
7. Yuan, B.; Liao, D.; Jiang, W.; Deng, H.; Li, G. Study on Friction and Wear Properties and Mechanism at Different Temperatures of Friction Stir Lap Welding Joint of SiCp/ZL101 and ZL101. *Metals* **2022**, *13*, 3. [CrossRef]
8. Zhang, W.; Xu, K.; Long, W.; Zhou, X. Microstructure and Compressive Properties of Porous 2024Al-Al$_3$Zr Composites. *Metals* **2022**, *12*, 2017. [CrossRef]
9. Feng, G.; Watari, H.; Haga, T. Fabrication of Mg/Al Clad Strips by Direct Cladding from Molten Metals. *Metals* **2022**, *12*, 1408. [CrossRef]
10. Guan, F.; Fan, S.; Wang, J.; Li, G.; Zhang, Z.; Jiang, W. Effect of Vibration Acceleration on Interface Microstructure and Bonding Strength of Mg-Al Bimetal Produced by Compound Casting. *Metals* **2022**, *12*, 766. [CrossRef]
11. Sun, M.; Yang, D.; Zhang, Y.; Mao, L.; Li, X.; Pang, S. Recent Advances in the Grain Refinement Effects of Zr on Mg Alloys: A Review. *Metals* **2022**, *12*, 1388. [CrossRef]

Disclaimer/Publisher's Note: The statements, opinions and data contained in all publications are solely those of the individual author(s) and contributor(s) and not of MDPI and/or the editor(s). MDPI and/or the editor(s) disclaim responsibility for any injury to people or property resulting from any ideas, methods, instructions or products referred to in the content.

Article

Effect of Vibration Acceleration on Interface Microstructure and Bonding Strength of Mg–Al Bimetal Produced by Compound Casting

Feng Guan [1], Suo Fan [2,*], Junlong Wang [1], Guangyu Li [1], Zheng Zhang [1] and Wenming Jiang [1,*]

1 State Key Laboratory of Materials Processing and Die & Mould Technology, School of Materials Science and Engineering, Huazhong University of Science and Technology, Wuhan 430074, China; guanfeng@hust.edu.cn (F.G.); u201615655@hust.edu.cn (J.W.); 2020509030@hust.edu.cn (G.L.); u201611197@hust.edu.cn (Z.Z.)
2 School of Mechanical & Electrical Engineering, Wuhan Institute of Technology, Wuhan 430073, China
* Correspondence: fan_suo@wit.edu.cn (S.F.); wmjiang@hust.edu.cn (W.J.)

Abstract: Vibration was adopted to enhance the interface bonding of Mg–Al bimetal prepared by the lost foam compound casting (LFCC) technique. The Mg–Al bimetallic interface was composed of three layers: layer I (Al_3Mg_2 and Mg_2Si phases), layer II ($Al_{12}Mg_{17}$ and Mg_2Si phases), and layer III ($Al_{12}Mg_{17}$ + δ-Mg eutectic structure). With the increase in vibration acceleration, the cooling rate of the Mg–Al bimetal increased, resulting in the decrease in the reaction duration that generates the intermetallic compounds (IMCs) layer (including layers I and II) and its thickness. On the other hand, the Mg_2Si phase in the IMCs layer was refined, and its distribution became more uniform with the increase in the vibration acceleration. Finally, the shear strength of the Mg–Al bimetal continued to increase to 45.1 MPa when the vibration acceleration increased to 0.9, which was 40% higher than that of the Mg–Al bimetal without vibration.

Keywords: Mg–Al bimetal; vibration; interface; microstructure; bonding strength

1. Introduction

Bimetallic materials have received close attention in recent years because of their unique superiority in combining the advantages of both different materials. Aluminum and magnesium alloys are two commonly used structure materials for lightweight applications for economic savings and ecological protection [1,2]. The former has the characteristics of good formability, excellent corrosion resistance, high specific strength, and stiffness [3]. The latter has low density, good castability, and high damping capacity [4]. Mg–Al bimetal, which consists of these two components, combines these advantages, and is widely applied in the automobile, aviation, and aerospace fields [5,6].

At present, a large number of technologies have been used to fabricate Mg–Al bimetal, such as welding [7], rolling [8], and casting [9]. The rolling process is an effective way to produce laminated or rodlike material, but it is challenging to manufacture complex parts. The welding methods are usually used to prepare the bimetallic products with a simple shape with high efficiency. Compared to the rolling and welding processes, compound casting has the advantage of being suitable for the production of bimetals with large and complex geometry. Moreover, the production procedure of the compound casting process is simple and low cost. Recently, much attention has been attracted to compound casting technology. Mg–Al bimetals have been prepared by various compound casting methods. Zhu et al. [9] fabricated AM50–Al6061 bimetallic products by a compound die casting method, followed by a low-temperature (200 °C) annealing schedule. The shear strength of the AM50–Al6061 bimetal reached the maximum value of 8.09 MPa, after a three-hour annealing. Their research found that the thickness of the Mg–Al interface

greatly affected its shear strength. He et al. [10] used a solid–liquid compound casting process to manufacture the arc-sprayed Al–AZ91D bimetal via casting AZ91D melt into the molds deposited by arc-sprayed aluminum coating. Hajjari et al. [11] used compound casting to produce a pure Al–pure Mg bimetal joint. The Mg–Al bimetal obtained in this experiment has a thick interface of about 1 mm, and the interface consists of multiple layers of different microstructures. The shear strength of the Mg–Al interface is about 23 MPa. Emami et al. [12] produced Mg–Al bimetal by both conventional and lost foam compound casting. The research also found a thick Mg–Al interface, achieving a millimeter level in the contact area of the two kinds of metals. In our previous research, lost foam compound casting was used to fabricate the A356–AZ91D bimetal, and key parameters such as pouring temperature and liquid–solid volume ratio were systematically studied. The results show that the improvement of the bonding strength of Mg–Al bimetal can be achieved by adjusting the processing parameters [13–16]. The properties of the Mg–Al bimetal are still poor due to the massive brittle and hard Al–Mg intermetallic compounds (IMCs) phase in the Mg–Al interface [10–12], and the Al_2O_3 film on the surface of the solid Al alloy, which hinders the direct bonding of the solid and liquid alloy and brings down the wettability [17,18].

In the recent research, the metal coating is usually prepared on the solid matrix to strengthen the mechanical properties of the bimetal prepared by compound casting [19,20]. Mola et al. [21] prepared a Zn coating with a thickness of 0.1 mm on the surface of the 6060 inserts by diffusion bonding. Then, the coated insert was used for the compound casting. After introducing the Zn element in the interface, the microstructure of the Mg–Al interface changed from Al–Mg IMCs to Mg–Al–Zn ternary IMCs, and the shear strength of the Mg–Al joint increased significantly, from about 8 MPa to about 42 MPa. Liu et al. [22] fabricated the Mg–Al bimetallic composites by a compound casting process. Moreover, a Ni interlayer was coated on the aluminum insert to hinder the direct reaction between the solid aluminum and liquid magnesium. As a result, the bonding strength of the Mg–Al bimetal was improved from 17.3 MPa to 25.4 MPa after adopting the Ni interlayer. However, the preparation process of the coating will make the preparation process of the bimetal more complicated and greatly increase the energy consumption and production cost, thus producing more pollution.

In addition, the researchers also tried a variety of other methods to improve the bonding properties of bimetals produced by the compound casting process. Chen Yiqing et al. [23] added La element to the magnesium alloy melt when preparing Mg–A390 bimetal by casting liquid Mg alloy onto the solid Al alloy panel. Experimental results show after adding the rare earth La into the magnesium alloys, the $Al_{12}Mg_{17}$ phase at the interface becomes lesser and thinner, and the cast grain is gradually refined. When the magnesium alloys contain 1% rare earth La, the maximal shearing strength of the interface can achieve 88.5 MPa. Wu Li et al. [24] proposed a modified horizontal continuous casting process under the electromagnetic field for preparing AA3003–AA4045 clad composite hollow billets. When rotating electromagnetic stirring was applied, the flow pattern of fluid melt was greatly modified; the temperature field in the interface region became more uniform. As a result, the microstructure of the clad composite hollow billet was refined, and the diffusion of the elements at the interface was promoted. Tayal et al. [25,26] used sandpaper with different mesh sizes to grind the surface of the aluminum alloy to research the influence of surface roughness on the shear strength of the A356–Mg bimetal produced by vacuum-assisted sand mold compound casting. The results show that it is more appropriate to use 800-grit sandpaper to polish the surface of the A356 insert. Babaee et al. [27] machined a special concentric groove pattern on the surface of the Al insert to improve the bonding properties of Al–Al-4.5%Cu bimetal prepared by squeeze casting. The tensile strength of the bimetal increased from 17 MPa to 54 MPa, after applying this method.

Although many studies have been conducted on the preparation of Mg–Al bimetal, few studies have been conducted to improve the performance of the Mg–Al interface by improving its microstructure through external assistance. The dependence of the mechanical properties on the microstructure of the Mg–Al interface is still insufficient. Vibration-assisted solidification is a technology of applying vibration in metal casting. It can effectively improve the microstructure solidification and the properties of the materials. Meanwhile, it has the advantages of low cost and no pollution. On the one hand, the molten metal was compressed and stretched by periodic force under the vibration, which was beneficial to the degassing and crystallization process [18,19,28,29]. Moreover, mechanical vibrations can form forced convection in the melt. The flow of the melt causes the temperature equalization of the melt, promotes the heat exchange between the molten metal and the mold, and increases the cooling rate of the melt [30]. At the same time, the flow caused by the melt also leads to the transport process of the solid phase and an increase in crystal nucleation. Therefore, it promotes the formation of more uniform fine-grained solidification microstructures [31]. Moreover, the vibration can also affect the distribution and shape of precipitates by promoting the element diffusion and solute exchange in the bimetallic interface [32]. It has excellent application prospects for the bimetals produced by compound casting.

In our previous study, we have investigated some key process parameters of a lost foam composite casting (LFCC) [13–16]. The microstructure of the interface of the Mg–Al bimetal prepared by the lost foam solid–liquid composite casting process was also reported. However, there are still many unsolved problems in our previous studies. For example, the bonding strength of the Mg–Al interface was not strong, the composition of the interface was relatively complex, and the formation process of the interface has not been fully clarified. In this work, the AZ91D–A356 bimetal was produced using the LFCC process. Vibration with different accelerations was applied during the casting and solidification process to enhance the bonding strength of the Mg–Al bimetal, by the vibration table used in the original lost foam casting process. The effect of the vibration acceleration on the microstructures and the bonding strength of the Mg–Al bimetal was investigated.

2. Experimental Procedure

2.1. Materials and Fabrication Process

The commercial A356 aluminum alloy (Al-6.81Si-0.44Mg-0.21Fe-0.02Ti wt.%) and AZ91D magnesium alloy (Mg-9.08Al-0.62Zn-0.23Mn wt.%) were used to fabricate the Mg–Al bimetal. The aluminum rods with a diameter of 10 mm and a height of 110 mm were obtained from the commercial aluminum ingot by wire-electrode cutting. Figure 1a shows the original microstructures of the A356 rods. They were mainly composed of the Al substrate and the Si phase dispersed in the substrate. Figure 1b,c shows the microstructure zoom in areas b and c in Figure 1a, indicating that the original shape of the Si phase in the A356 rods is mostly needle-like or slate-like.

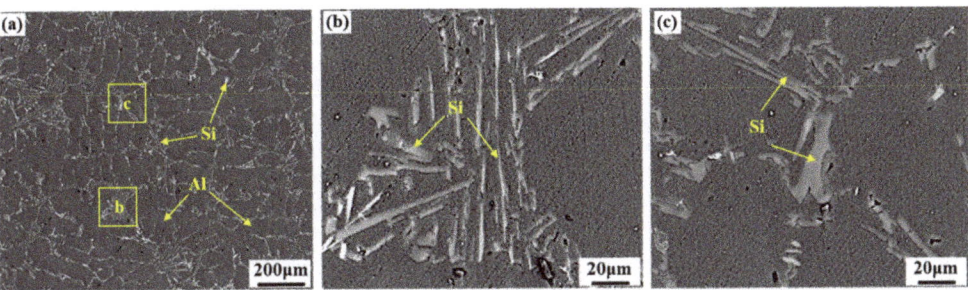

Figure 1. (**a**) Original microstructures of the A356 insert; (**b**) microstructures zoom in area b in (**a**); (**c**) microstructures zoom in area c in (**a**).

Before being used in the experiment, the as-cast A356 rods were polished with silicon carbide sandpapers, followed by cleaning with acid (50% HNO$_3$ + 48% HF + 2% water) and lye (20 g/L NaOH, 5 g/L ZnO) to remove the oxide film on the surface. The treated A356 rods were assembled with the foam pattern. Then, the foam patterns were used for the LFCC process, of which the schematic diagram is illustrated in Figure 2. The foam pattern was placed in the sand flask, which was vibrated through the vibration table, while the loose sand was added to the sand flask. Under vibration, the loose sand in the sand flask was compacted. Then, a plastic film was placed over the surface of the loose sand. Finally, the vacuum pump was launched, and the sand flask was vacuumized and maintained at a vacuum of −0.03 MPa during the experiment. Under atmospheric pressure, the sand mold became tough and could tolerate the applied vibration. During the LFCC process, the pouring temperature of the AZ91D magnesium alloy was 720 °C. The 35 Hz vibration with different peak accelerations of 0.3 g (with the peak–peak displacement of 0.2 mm) and 0.9 g (with the peak–peak displacement of 0.6 mm) was applied to the manufacturing process of the Mg–Al bimetal by the vibration table under the sand flask, as shown in Figure 2, to investigate the effect of the vibration acceleration on the microstructure and bonding strength of the Mg–Al bimetal. During the casting process, the solidification temperature curve of the Mg–Al bimetal was measured at a sampling frequency of 75 Hz by the thermocouple, placed against the surface of the A356 insert, as shown in Figure 2.

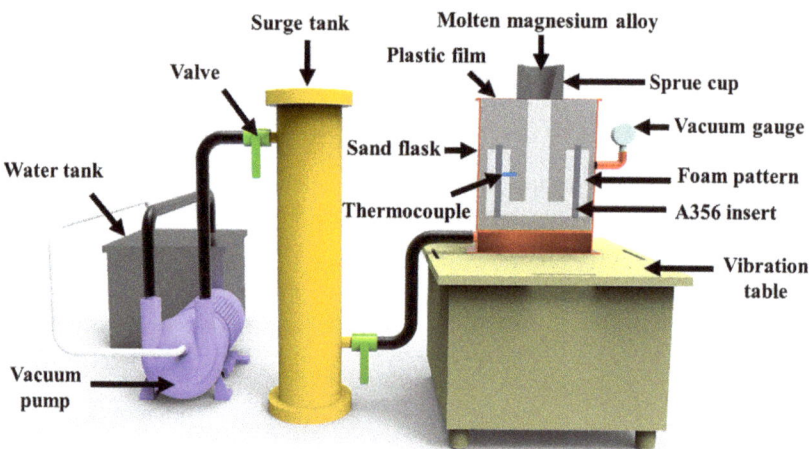

Figure 2. The schematic diagram of the experiment.

2.2. Characterization

The Mg–Al bimetal specimens were cut along their cross section. A Quanta 200 scanning electron microscope (SEM, FEI, Eindhoven, The Netherlands) equipped with energy-dispersive X-ray spectroscopy (EDS) was used to investigate the microstructure and chemical composition of the Mg–Al bimetallic interface. The thickness of the Mg–Al interface and the size of the precipitated phase were measured using the image-pro software, and the measure method and process are illustrated in supplementary materials, as shown in Figures S1–S11. The element distribution of the interface was tested by WDS equipped with EPMA (EPMA-8050G (Shimadzu, Tokyo, Japan)). An XRD-6100 X-ray diffractometer (XRD, Shimadzu, Tokyo, Japan) was employed to identify the phase compositions at the Mg–Al interface. Further investigation of the constitutive phases at the interface was performed using transmission electron microscopy (TEM; JEOL2100, Tokyo, Japan). The bonding properties of the Mg–Al bimetal were tested by a push-out experiment in the ZwickZ100 universal testing machine (Zwick, Roell, Germany) with the compression rate of 0.5 mm/min. The schematic diagram of the push-out experiment is shown in Figure 3. The bonding property of the Mg–Al bimetal was evaluated according to Equation (1) [33–35]:

$$S = F/(\pi d h), \tag{1}$$

where S is the shear strength of the Mg–Al bimetal, F is the maximum force loaded obtained from the testing machine during the compression process, d is the diameter of the aluminum rod, and h is the height of the specimen. The fracture behavior of the Mg–Al bimetal was analyzed using SEM equipped with EDS.

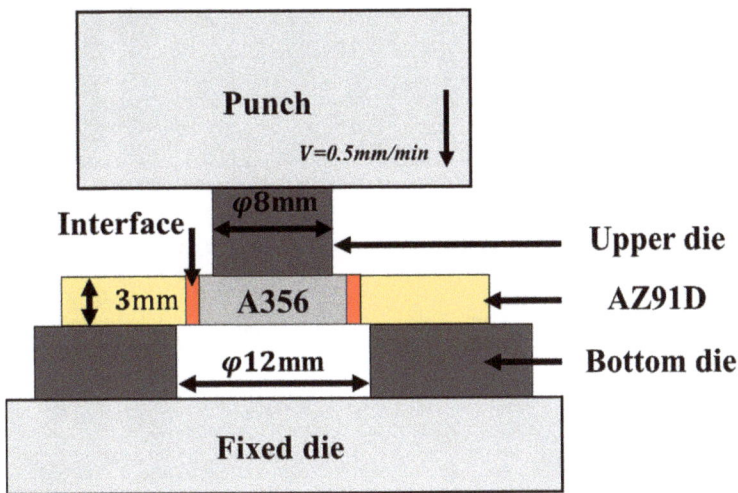

Figure 3. Schematic diagram of the push-out test for the shear strength testing.

3. Results

3.1. Effects of the Vibration Acceleration on Microstructure of the Mg–Al Bimetal

Figure 4a–c shows the interface morphology of the Mg–Al bimetal obtained with different conditions. Without the vibration, the outline of the interface is relatively flat. After applying the vibration, the thickness of the interface changes, and its morphology becomes irregular. When the vibration is not applied, there is a boundary in the Mg–Al interface at the location marked in Figure 4a. In the Mg–Al interface, many long dendrites grow towards the AZ91D matrix on the left side of the boundary, as shown in Figure 4a. On the right side of the boundary, the distribution of the grey precipitates in that area is not uniform. Near the A356 matrix, fewer precipitates can be observed.

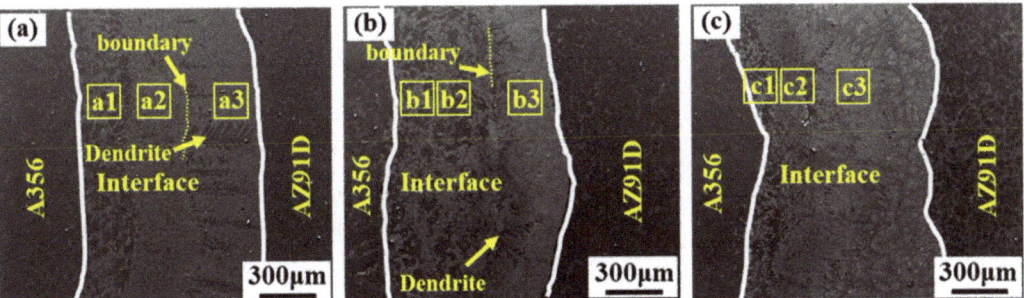

Figure 4. *Cont.*

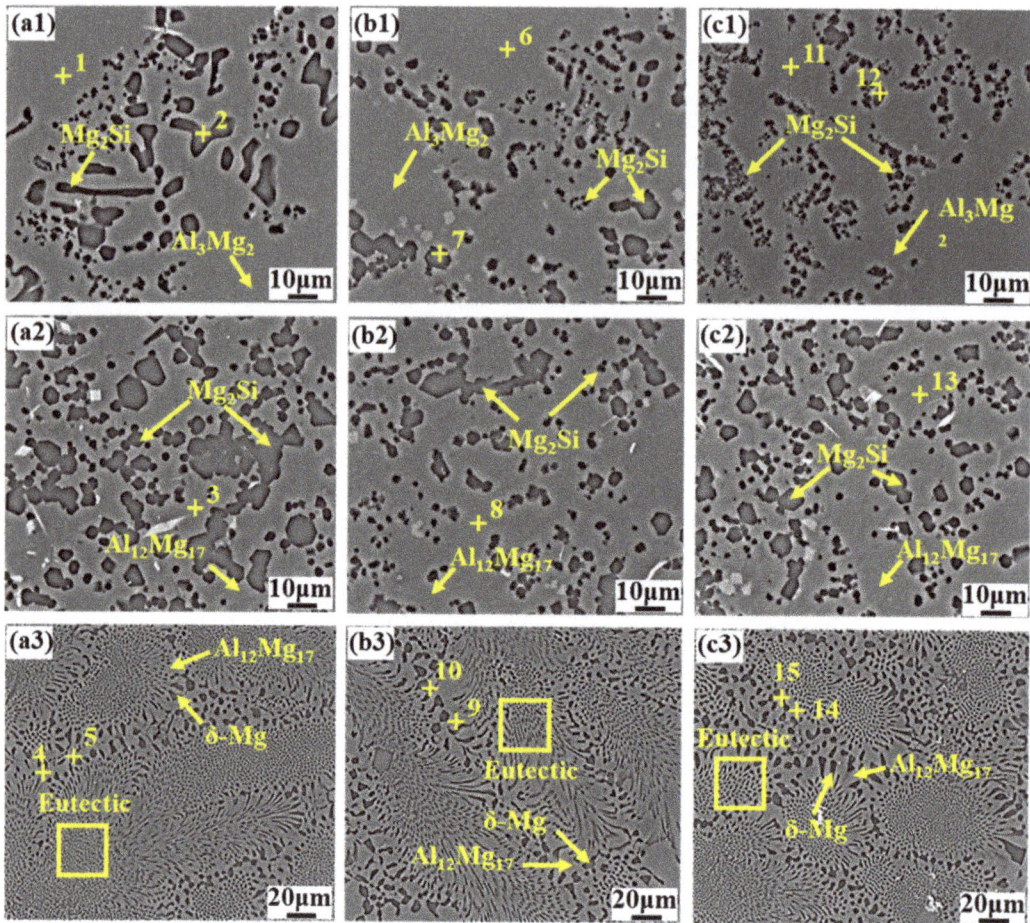

Figure 4. Microstructures of the Mg–Al bimetals obtained with different conditions: (**a**) without vibration; (**b**) with the vibration acceleration of 0.3 g; (**c**) with the vibration acceleration of 0.9 g; (**a1–a3**) microstructures under high magnification in the regions a1–a3 in (**a**), respectively; (**b1–b3**) microstructures under high magnification in the regions b1–b3 in (**b**), respectively; (**c1–c3**) microstructures under high magnification in the regions c1–c3 in (**c**), respectively.

When the vibration with 0.3 g is adopted, a clear boundary can still be observed in the Mg–Al interface at the location marked in Figure 4b, and there are obvious dendrites on the right side of it. However, the dendrites in the interface zone decrease compared with that without vibration. As the vibration acceleration increases to 0.9 g, no obvious boundary can be observed in the Mg–Al interface, and the dendrites in the Mg–Al interface decrease further. According to the above observations, three regions are selected for further observation to analyze the effect of vibration acceleration on the microstructure of the Mg–Al interface. Figure 4(a1–c3) shows the microstructures under high magnification in the regions marked in Figure 4a–c, respectively. EDS point analysis is used to analyze the composition of the phase existing in the Mg–Al interface, and the results of the corresponding points are shown in Table 1. The possible phases of the analyzed points are identified and indicated in Table 1, combing the EDS point analysis results with the Al–Mg [36] and Mg–Si [37] binary phase diagrams. According to these results, the Mg–Al interface can be divided into three regions: layer I (composed of Al_3Mg_2 and Mg_2Si phases), layer II (composed of $Al_{12}Mg_{17}$, and Mg$_2$Si phases), and layer III (Al$_{12}$Mg$_{17}$ + δ-Mg eutectic). Layers I and II can also be collectively called the IMCs layer because their substrates are the Al–Mg IMCs, and layer III can be named the eutectic layer. Layer I is mainly composed of the Mg$_2$Si precipitates and Al$_3$Mg$_2$ substrate. There are many Mg$_2$Si bulks and bars when the vibration is not applied, as shown in Figure 4(a1). After applying the vibration with the acceleration of 0.3 g, the Mg$_2$Si phase is dispersed into granular form, as shown in Figure 4(b1). When the acceleration increases to 0.9 g, a large number of fine Mg$_2$Si particles can be observed in the Al$_3$Mg$_2$ substrate, and the size of the Mg$_2$Si phase is refined, as shown in Figure 4(c1).

Table 1. Results of EDS analysis at different locations in Figure 4.

Area No.	Element Compositions (At.%)			Possible Phase
	Mg	Al	Si	
1	38.73	61.27	-	Al$_3$Mg$_2$
2	58.02	12.56	29.43	Mg$_2$Si
3	50.82	49.18	-	Al$_{12}$Mg$_{17}$
4	63.94	36.06	-	Al$_{12}$Mg$_{17}$
5	83.66	16.34	-	δ-Mg
6	38.73	61.27	-	Al$_3$Mg$_2$
7	61.23	14.84	23.93	Mg$_2$Si
8	49.45	50.55	-	Al$_{12}$Mg$_{17}$
9	65.65	34.35	-	Al$_{12}$Mg$_{17}$
10	77.70	22.30	-	δ-Mg
11	38.74	61.26	-	Al$_3$Mg$_2$
12	61.87	22.09	16.03	Mg$_2$Si
13	49.28	50.72	-	Al$_{12}$Mg$_{17}$
14	63.43	36.57	-	Al$_{12}$Mg$_{17}$
15	85.78	14.22	-	δ-Mg

Figure 4(a2–c2) shows the microstructures of the Mg–Al interfaces in layer II, which is composed of the black Mg$_2$Si phase and the Al$_{12}$Mg$_{17}$ substrate. These results indicate that the vibration also affects the distribution and size of the Mg$_2$Si phase in the Al$_{12}$Mg$_{17}$ substrate. Without applying the vibration, the aggregation of the Mg$_2$Si phase is observed in Figure 5b. As shown in Figure 5b,c, fewer large Mg$_2$Si bulks are observed in the Al$_{12}$Mg$_{17}$ substrate when the vibration is applied. It indicates that the Mg$_2$Si phases in layer II are dispersed and refined after the mechanical vibration is brought to the manufacturing process. As to the microstructures in layer III, there is no significant difference in the δ-Mg and Al$_{12}$Mg$_{17}$ eutectic structure, after applying the vibration.

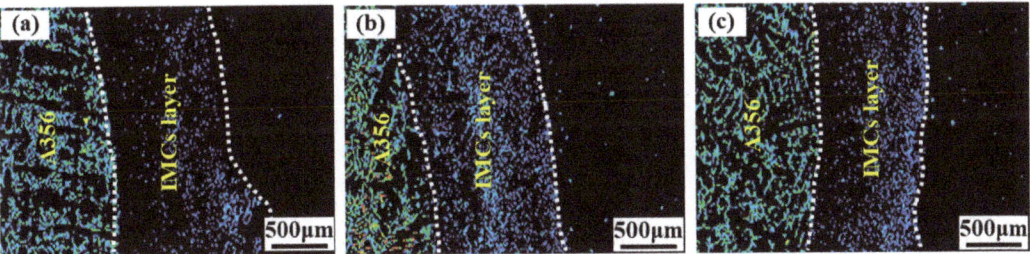

Figure 5. The distributions of the Si element in the Mg–Al bimetals prepared with different conditions: (**a**) without vibration; (**b**) with the vibration acceleration of 0.3 g; (**c**) applying the vibration acceleration of 0.9 g.

Figure 5 shows the distributions of the Si element in the Mg–Al bimetals prepared under different conditions. According to the microstructures observed in Figure 4, the Si elements distributed in the Mg–Al interface may mainly exist in the precipitates (the Mg$_2$Si phase) located in the IMCs layer. After the vibration, the distribution of the Si element in the IMCs layer becomes more uniform, and the composition of the Si element in the eutectic layer increases slightly. It indicates that the vibration can improve the uniformity of the Mg$_2$Si phase in the Mg–Al interface.

The above experimental results show that the Mg–Al interface can be divided into two parts, the IMCs layer (composed of layers I and II) and the eutectic layer (layer III). There are a lot of Mg$_2$Si precipitates in the IMCs layer. In the eutectic layer, eutectic structures and some primary dendrites can be observed. These two parts can be distinguished and measured according to their differences in microstructure and contrast. Figure 6 summarizes the measurements of the thickness of the different parts of the Mg–Al interface. Compared with the Mg–Al bimetal without vibration, the thickness of region I decreases by 29.6%, from 914 µm to 643 µm, after applying the vibration of 0.9 g. However, the change of the thickness of the eutectic layer presents a different phenomenon. It decreases when the vibration is applied. Then, it increases when the acceleration of the vibration increases to about 0.9 g.

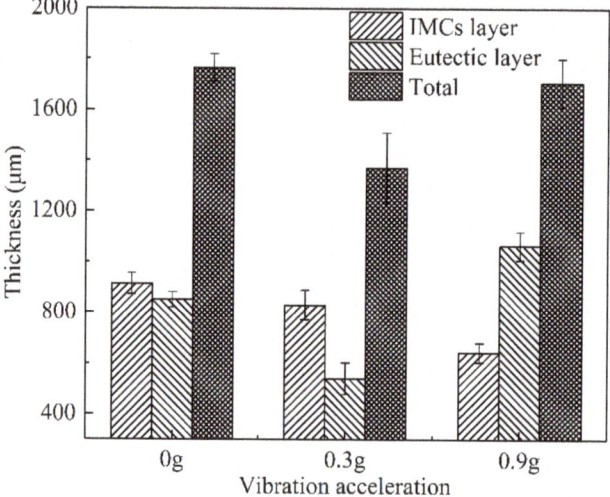

Figure 6. The thicknesses of the Mg–Al interfaces obtained with different conditions.

The microstructures shown in Figure 4 indicate that the Mg$_2$Si phase mainly exists in layers I and II in the Mg–Al interface. To quantify the influence of vibration acceleration on the Mg$_2$Si phase in the Mg–Al interface, we observed and measured the size of the Mg$_2$Si phase in layers I and II, respectively, according to the locations shown in Figure 4a–c. Figure 7 shows the sizes of the Mg$_2$Si phase in the Mg–Al interfaces. The results indicates that the size of the Mg$_2$Si phase in layer II is larger than that in layer I. The size of the Mg$_2$Si phase in the IMCs layer decreases with the enhancement of the acceleration of the vibration. Compared with the Mg–Al bimetal without vibration, the size of the Mg$_2$Si phase decreases from 4.3 µm to 1.8 µm in layer I, and drops from 4.7 µm to 3.3 µm in layer II, after applying the vibration of 0.9 g. Figure 4(a3–c3) shows the microstructures of layer III in the Mg–Al interfaces.

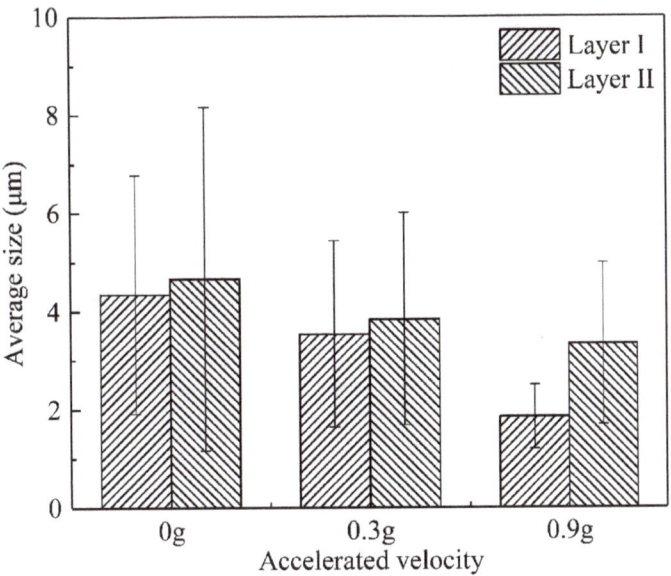

Figure 7. The sizes of the Mg$_2$Si phase in the IMCs layers of the Mg–Al interfaces.

To further confirm the phase composition of the Mg–Al interface, the Mg–Al interface was observed by TEM. Figure 8a shows the bright-field image in layer I. There are many large granular phases distributed in the substrate. Figure 8c,d shows the analysis result of the diffraction spots. The results confirm that layer I in the Mg–Al interface is composed of the Al$_3$Mg$_2$ substrate and the Mg$_2$Si precipitates. Research has shown the appearance of the bend contours is related to the strain field led by the residual stresses [38,39]. Since the linear expansion coefficient of Mg$_2$Si (13 × 10^{-6} K^{-1}) [40] is lower than that of the substrate (Al$_3$Mg$_2$, 22 × 10^{-6} K^{-1}) [41], it will generate the compressive stress in the precipitates and the substrate. The stress may lead to the deformation of the substrate in the area near the Mg$_2$Si phase. It may result in the occurrence of the bend contours, as shown in Figure 8a. On the other hand, the compressive stress in the Mg$_2$Si phase may lead to a large number of dislocations in the particles, as shown in Figure 8b. These results indicate that the presence of Mg$_2$Si in the interface may lead to the generation of residual stress in the interface. Therefore, the oversized Mg$_2$Si phase may lead to the increase in residual stress and adversely affect the interface properties.

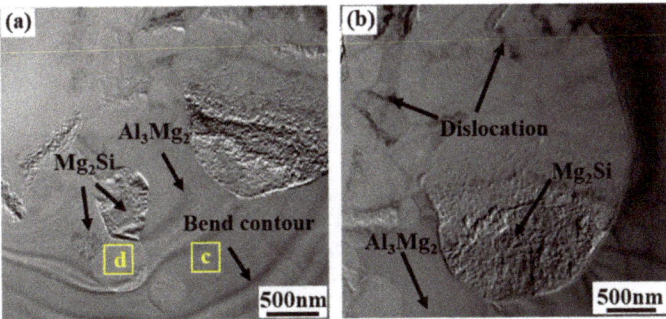

Figure 8. *Cont.*

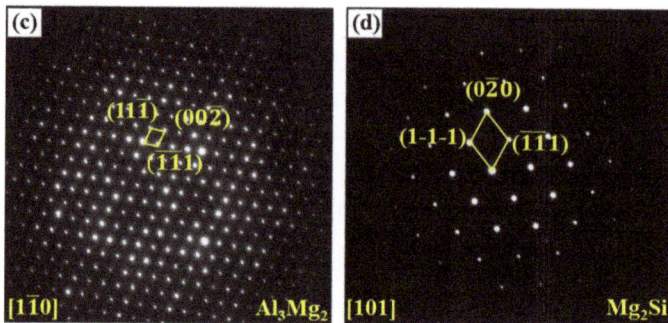

Figure 8. TEM analysis results of layer I in the Mg–Al interface: (**a,b**) TEM bright-field images; (**c**) SAED pattern of the Al$_3$Mg$_2$ phase; (**d**) SAED pattern of the Mg$_2$Si phase.

3.2. Effect of the Vibration Acceleration on Bonding Strength of the Mg–Al Bimetal

Figure 9a shows stress-displacement curves of the Mg–Al bimetals with different conditions, and the average shear strength of the Mg–Al bimetal prepared with different conditions is presented in Figure 9b. With the increase in the vibration acceleration, the shear strength of the Mg–Al bimetal gradually increases from 32.2 MPa to 41.5 MPa and 45.1 MPa. Compared with the Mg–Al bimetal obtained without vibration, it increases by 30% and 40%, respectively, after applying the vibration with the accelerations of 0.3 g and 0.9 g.

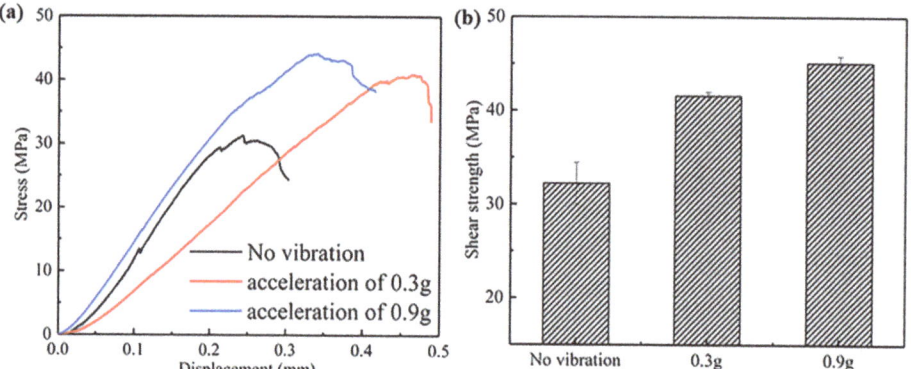

Figure 9. (**a**) Stress-displacement curves of the Mg–Al bimetals with different conditions; (**b**) Shear strengths of Mg–Al bimetals with different conditions.

Figure 10 presents the SEM fractographies of the Mg–Al bimetals with different conditions, and the compositions of the phases observed on the fracture surface were analyzed. The cleavage planes and river patterns are observed in the SEM fractographies, as shown in Figure 10b,e,h, demonstrating that the Mg–Al bimetal fractures by a brittle fracture. Figure 10a shows the macroscopic morphology of the fracture surface of the Mg–Al bimetal without vibration. It indicates a noticeable slope on the surface of the fracture. Region b and c, shown in Figure 10a, were selected for observation. The results show that the Mg–Al bimetal fractures in different Mg–Al interface areas. In region b, shown in Figure 10b, the composition of the flat region is the Al$_3$Mg$_2$ phase, and the composition of the granular structure and pit area is the Mg$_2$Si phase, indicating that region b belongs to the IMCs layer. The Mg$_2$Si phase on the fracture surface aggregates and distributes in a reticular form when the vibration is not applied.

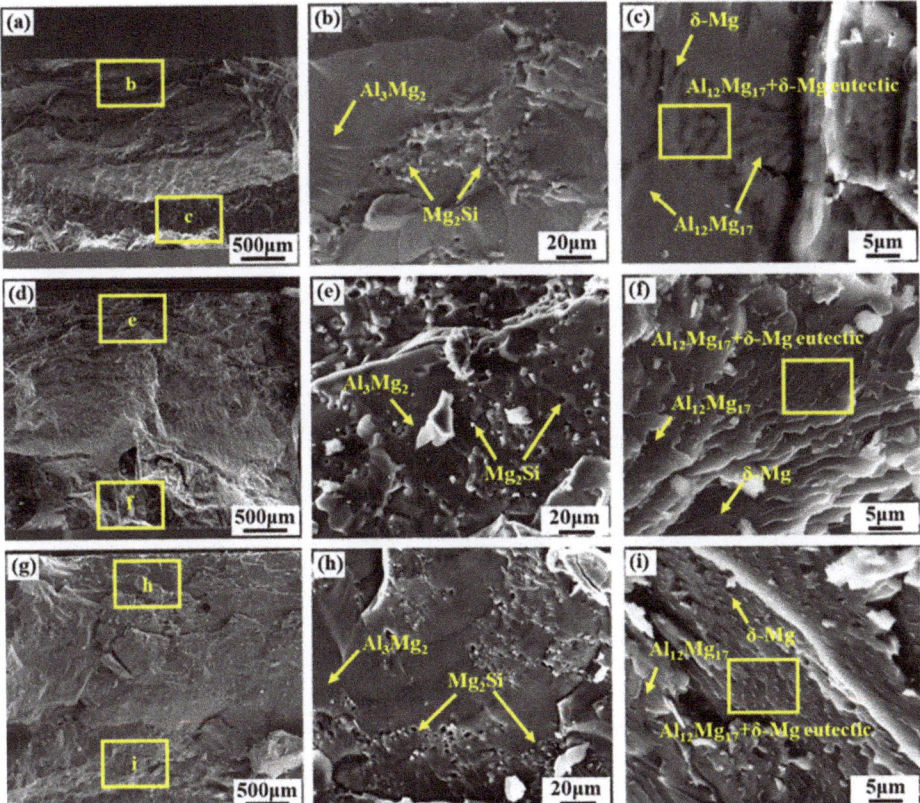

Figure 10. SEM morphologies of the fracture surfaces of the Mg–Al bimetals at the Mg substrate side: (**a**) macroscopic morphology of the fracture surface of the Mg–Al bimetal without vibration; (**b,c**) enlarged images of region b and c in (**a**), respectively; (**d**) macroscopic morphology of the fracture surface of the Mg–Al bimetal with the vibration acceleration of 0.3 g; (**e,f**) enlarged images of region e and f in (**d**), respectively; (**g**) macroscopic morphology of the fracture surface of the Mg–Al bimetal with the vibration acceleration of 0.9 g; and (**h,i**) enlarged images of region h and i in (**g**), respectively.

In region c of Figure 10a, the surface of the fracture is relatively straight and flat, and the $Al_{12}Mg_{17}$ + δ-Mg eutectic structure can be observed, indicating that it may belong to the eutectic layer. However, it is noteworthy that no plastic deformation is observed in the eutectic structure, as shown in Figure 10c. The fracture morphology of the Mg–Al bimetal with vibration demonstrates a similar fracture pattern. The Al_3Mg_2 phase, Mg_2Si phase, and the eutectic structure are also found on the fracture surface. Compared with the fracture morphology of the Mg–Al bimetal without vibration, the aggregated Mg_2Si phase is dispersed after applying the vibration, as shown in Figure 10e,h. Moreover, the δ-Mg in the eutectic structure is elongated, which is beneficial to improve the ductility of the Mg–Al interface.

After the shear testing, the fragments of broken interfacial structures were used for XRD testing. The fragments were broken and screened again before testing. The XRD testing result, shown in Figure 11, also confirms the presence of the Al_3Mg_2, $Al_{12}Mg_{17}$, and Mg_2Si phases in the IMCs layers. Because the IMCs is more brittle, it is more likely to break and fall off during shear fracture. Due to the presence of a large number of δ-Mg phases in the eutectic structure, its plasticity is relatively better, and it is not easy for it to break and fall off, so it may not be included in the tested samples. This may explain the result that δ-Mg was not detected in the Figure 11.

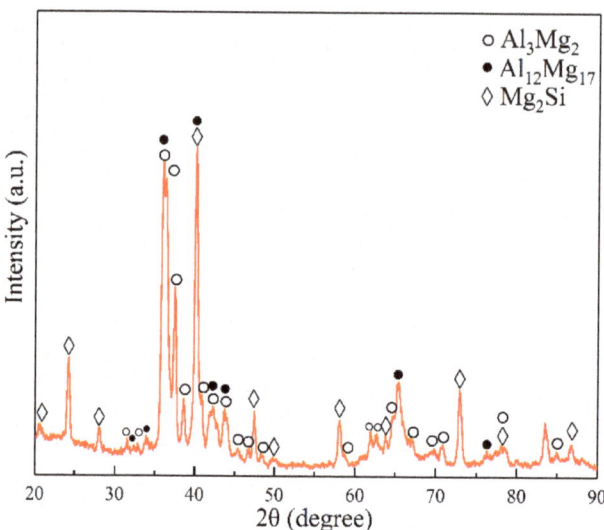

Figure 11. XRD diagram of the IMCs layer of the Mg–Al interface.

4. Discussion

4.1. The Effect of Vibration Acceleration on the Microstructure of Mg–Al Bimetallic Interface

Figure 6 indicates that the thickness of the IMCs layer in the Mg–Al interface gradually decreases with the increase in the vibration acceleration. To investigate the influence mechanism of vibration acceleration on the thickness of the IMCs layer, the solidification curve of the interfacial region was measured by the thermocouple, located on the surface of the aluminum insert, and the results are shown in Figure 12. Previous research has shown that under the vibration, the disturbance and convection of the molten metal promoted the heat exchange at the solid–liquid interface [42]. As a result, it can increase the cooling rate during the solidification process. It can be observed that the cooling rate of the solidification process significantly increased after the application of the vibration, as shown in Figure 12a. According to the research of Haq et al. [43], the increase in the derivate curve indicates the generation of a new phase. Combining the Al–Mg binary phase diagram with the solidification curve shown in Figure 12, the reaction duration of the IMCs (both the Al_3Mg_2 and $Al_{12}Mg_{17}$) phases and the cooling rates during the formation of the IMCs were measured. Then, they are used to estimate the time taken to form the IMCs layer. The cooling rate during the formation of the IMCs was about 0.27 K/s, under the condition of no vibration. After applying the vibration with the accelerations of 0.3 g and 0.9 g, the cooling rates were 0.38 K/s and 0.40 K/s, respectively, during the formation of the IMCs. It increased significantly when the vibration was applied. Without vibration, the reaction duration of the IMCs layer (t_{IMCs}) was about 47.7 s. After applying the vibration with the accelerations of 0.3 g and 0.9 g, it decreased to 34.5 s and 28.76 s, respectively. Compared with the Mg–Al bimetal obtained without vibration, it was reduced by 27.7% and 39.7%, respectively, after applying the vibration with the accelerations of 0.3 g and 0.9 g. The decrease in the t_{IMCs} may be the reason why the thickness of the IMCs layer in the Mg–Al bimetallic interface decreased, with the increasing of the vibration acceleration.

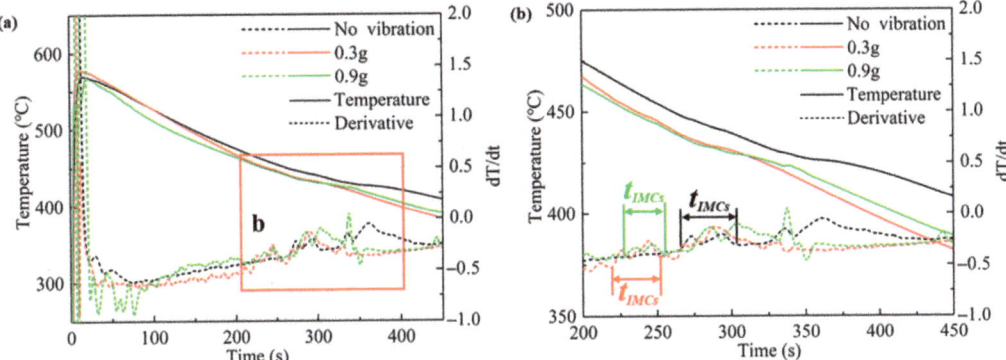

Figure 12. (**a**) Results of thermocouple measurement; (**b**) Temperature curve zoom in area b in (**a**).

Moreover, the disturbance and convection led by the vibration may also affect the size and distribution of the Mg_2Si phase. Figure 13 shows the influence mechanism of the vibration on the size and distribution of the Mg_2Si phase. The Si element in the Mg_2Si phase comes from the Si phase in the A356 insert. Figure 13a shows the initial state of the Si element in the manufacturing process of the Mg–Al bimetal. In the beginning, the Si existed in the needle-like or slate-like Si phase on the aluminum substrate. After pouring the molten AZ91D alloys, the molten metal filled the position of the foam mold and came into contact with the A356 insert.

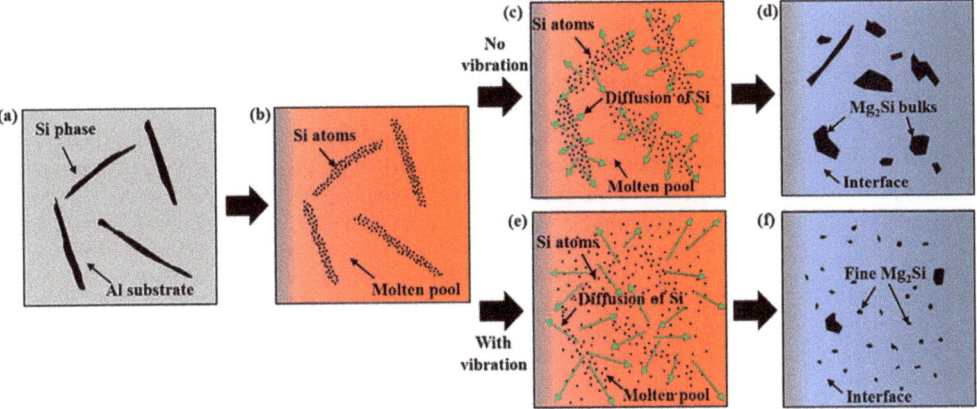

Figure 13. The influence mechanism of the vibration on the size and distribution of the Mg_2Si phase: (**a**) The initial state of the Si element; (**b**) The insert melted to form a molten pool on the surface of the solid insert; (**c**) The diffusion of the Si in the molten pool under the condition of no vibration; (**d**) The interface formed under the condition of no vibration; (**e**) The diffusion of the Si in the molten pool after applying the vibration; (**f**) The interface formed after applying the vibration.

During the casting process, the highest temperature measured in the insert surface region was about 571 °C, as shown in Figure 12, close to the Al–Si eutectic reaction temperature of 577 °C [35]. The actual temperature of the insert surface area should be higher than the measured temperature, so when the AZ91D melt was in contact with the A356 matrix, the surface area of the insert might be melted. In addition, under the high-temperature condition, the mutual diffusion of the Al and Mg elements occurred between the solid aluminum insert and the molten magnesium alloy. It changed the composition of the insert surface, decreased the melting point of the insert surface [44], and promoted the melting of the surface region of the solid insert. So, the insert melted to form a molten pool on the surface of the solid insert, as shown in Figure 13b. In the molten pool, there were some Si-rich regions where the eutectic Si phase initially existed. Subsequently, the Si element in the molten pool gradually diffused due to the concentration gradient, as shown in Figure 13c. Without vibration, the diffusion distance of the Si element was relatively short due to the brief solidification time of the interface region during the manufacturing process. Finally, as shown in Figure 13c, the Si element aggregated in a small area and precipitated the large Mg_2Si bulks from the molten metal, as shown in Figure 13d.

After applying the vibration, the disturbance and convection led by the vibration promoted the diffusion of the Si element. The diffusion distance of the Si element increased, and the Si element was dispersed into a larger region, as shown in Figure 13e. Finally, the size of the Mg_2Si phase was refined, and its distribution became more uniform, as shown in Figure 13f. Therefore, it can be seen from the above experiment results that the size of the Mg_2Si phase in the IMCs layer is gradually refined as the vibration acceleration increases to 0.9 g.

However, the improvement of the distribution of the Mg_2Si phase in layer I is mainly observed after applying the vibration. Since the A356 insert melts from the outside to the inside during the preparation process, the Si element in layer II has more diffusion time. As a result, the distribution of the Mg_2Si phase in that region is relatively uniform. Therefore, the effect of the vibration on the distribution of the Mg_2Si phase in layer II is not as significant as that of layer I.

4.2. Strengthening Mechanism of the Mg–Al Bimetal

The IMCs layer in the Mg–Al interface is mainly composed of the IMCs substrate and the Mg_2Si phase distributed on the substrate. Compared with A356 and AZ91D matrixes, the IMCs layer has significantly higher hardness and lower plastic deformation ability [45]. Therefore, the thickness of the IMCs layer has an important influence on the number of brittle phases and bonding properties of the Mg–Al bimetal. In the preparation process of the Mg–Al bimetal, reducing the content of these brittle phases is the key to improving the bonding property of the Mg–Al bimetal [46,47]. After applying the vibration, the thinning of the IMCs layer means decreasing the mass of the brittle phases, thereby improving the mechanical properties of the Mg–Al bimetal. Studies have shown that the size and morphology of the Mg_2Si phase have an important effect on the properties of the material [48,49]. The refined Mg_2Si phase may also contribute to enhancing the shear strength of the Mg–Al bimetal. Without vibration, there are many Mg_2Si bulks in the IMCs layer. Since the linear expansion coefficient of the Mg_2Si is lower than that of the Al_3Mg_2 substrate, it will generate compressive stress in the Mg_2Si precipitates. Existing research has found that for the large-sized Mg_2Si phase, its ability to carry external loads is weaker because the larger size reinforcement particles are more likely to fracture under the action of the residual stress [50]. Therefore, under the action of the stress, the Mg_2Si bulks in the IMCs layer are more likely to form cracks when it is under load, leading to the formation of crack propagation channels, which will adversely affect the ability of the Mg_2Si phase to withstand and disperse loads [51]. When the vibration is applied, the thickness of the IMCs layer decreases, and the Mg_2Si phase is refined. Therefore, the shear strength of the Mg–Al bimetal is improved from 32.2 MPa to 41.5 MPa, after applying the vibration with the acceleration of 0.3 g. As the acceleration increases to 0.9 g, the shear strength of the Mg–Al bimetal continues to rise to 45.1 MPa, due to the further reduction in the thickness of the IMCs layer and the further refinement of the Mg_2Si phase.

5. Conclusions

In the present work, the Mg–Al bimetal was fabricated by the LFCC process, and the effect of the vibration acceleration on the interfacial microstructure and mechanical properties was studied. The main conclusions are presented in the following:

- The interface of the Mg–Al bimetal fabricated by the LFCC process was divided into three areas, named layer I (Al_3Mg_2 and Mg_2Si phases), layer II ($Al_{12}Mg_{17}$, and Mg_2Si phases), and layer III ($Al_{12}Mg_{17}$ + δ-Mg eutectic structure). With the increase in the vibration acceleration, the cooling rate of the Mg–Al bimetal increased, and the reaction duration of the IMCs layer (including layers I and II) decreased. In addition, the thickness of the IMCs reduced.
- The vibration promoted the refinement and dispersion of the Mg_2Si phase. After applying the vibration, the distribution of the Mg_2Si in the IMCs layer became more uniform, and the size of the Mg_2Si phase decreased with the increase in the vibration acceleration.
- The shear strength of the Mg–Al bimetal increased with the increase in the vibration acceleration. As the acceleration grew to 0.9 g, the shear strength of the Mg–Al bimetal continued to rise to 45.1 MPa, which was 40% higher than that of the Mg–Al bimetal without vibration. The significant improvement of the shear strength of the Mg–Al bimetal might be attributed to the decrease in the IMC's thickness, as well as the refinement and uniform distribution of the Mg_2Si phase.

Supplementary Materials: The following supporting information can be downloaded at: https://www.mdpi.com/article/10.3390/met12050766/s1, Figure S1: The original SEM image., Figure S2: The standard for the measurement., Figure S3: The processed image used for measurement., Figure S4: Space calibration, Figure S5: Select the area to be measured., Figure S6: Using the "perform segmentation" function to convert images to black and white image., Figure S7: The black and white image, Figure S8: Click on the "Count and measure objects" option, Figure S9: Click on the "Select Measurements" option., Figure S10: Setup the measurement options., Figure S11: Measurement results.

Author Contributions: Conceptualization, J.W. and G.L.; methodology, Z.Z.; investigation, F.G. and J.W.; writing—original draft preparation, F.G.; writing—review and editing, W.J. and S.F.; visualization, S.F.; Supervision, W.J. All authors have read and agreed to the published version of the manuscript.

Funding: This research was funded by National Natural Science Foundation of China [grant number 52075198], National Key Research and Development Program of China [grant number 2020YFB2008304] and State Key Lab of Advanced Metals and Materials [grant number 2021-ZD07].

Data Availability Statement: Data presented in this article are available at request from the corresponding author.

Acknowledgments: The authors gratefully acknowledge the support provided by the Analytical and Testing Center, HUST.

Conflicts of Interest: The authors declare no conflict of interest.

References

1. Yu, H.M.; Li, W.; Tan, Y.; Tan, Y.B. The Effect of Annealing on the Microstructure and Properties of Ultralow-Temperature Rolled Mg–2Y–0.6Nd–0.6Zr Alloy. *Metals* **2021**, *11*, 315. [CrossRef]
2. Yang, X.K.; Xiong, B.Q.; Li, X.W.; Yan, L.Z.; Li, Z.H.; Zhang, Y.A.; Li, Y.N.; Wen, K.; Liu, H.W. Effect of Li addition on mechanical properties and ageing precipitation behavior of extruded Al−3.0Mg−0.5Si alloy. *J. Cent. South Univ.* **2021**, *28*, 2636–2646. [CrossRef]
3. Li, Z.C.; Deng, Y.L.; Yuan, M.F.; Zhang, J.; Guo, X.B. Effect of isothermal compression and subsequent heat treatment on grain structures evolution of Al-Mg-Si alloy. *J. Cent. South Univ.* **2021**, *28*, 2670–2686. [CrossRef]
4. Zhang, D.D.; Liu, C.M.; Wan, Y.C.; Jiang, S.N.; Zeng, G. Microstructure and anisotropy of mechanical properties in ring rolled AZ80-Ag alloy. *J. Cent. South Univ.* **2021**, *28*, 1316–1323. [CrossRef]
5. Suhuddin, U.F.H.; Fischer, V.; dos Santos, J.F. The Thermal Cycle during the Dissimilar Friction Spot Welding of Aluminum and Magnesium Alloy. *Scr. Mater.* **2013**, *68*, 87–90. [CrossRef]
6. Li, G.; Jiang, W.; Guan, F.; Zhu, J.; Zhang, Z.; Fan, Z. Microstructure, Mechanical Properties and Corrosion Resistance of A356 Aluminum/AZ91D Magnesium Bimetal Prepared by a Compound Casting Combined with a Novel Ni-Cu Composite Interlayer. *J. Mater. Process. Technol.* **2021**, *288*, 116874. [CrossRef]
7. Xu, Z.; Li, Z.; Li, J.; Ma, Z.; Yan, J. Control Al/Mg Intermetallic Compound Formation during Ultrasonic-Assisted Soldering Mg to Al. *Ultrason. Sonochem.* **2018**, *46*, 79–88. [CrossRef]

8. Feng, B.; Xin, Y.; Guo, F.; Yu, H.; Wu, Y.; Liu, Q. Compressive Mechanical Behavior of Al/Mg Composite Rods with Different Types of Al Sleeve. *Acta Mater.* **2016**, *120*, 379–390. [CrossRef]
9. Zhu, Z.; Shi, R.; Klarner, A.D.; Luo, A.A.; Chen, Y. Predicting and Controlling Interfacial Microstructure of Magnesium/Aluminum Bimetallic Structures for Improved Interfacial Bonding. *J. Magnes. Alloys* **2020**, *8*, 578–586. [CrossRef]
10. He, K.; Zhao, J.; Li, P.; He, J.; Tang, Q. Investigation on Microstructures and Properties of Arc-Sprayed-Al/AZ91D Bimetallic Material by Solid–Liquid Compound Casting. *Mater. Des.* **2016**, *112*, 553–564. [CrossRef]
11. Hajjari, E.; Divandari, M.; Razavi, S.H.; Emami, S.M.; Homma, T.; Kamado, S. Dissimilar Joining of Al/Mg Light Metals by Compound Casting Process. *J. Mater. Sci.* **2011**, *46*, 6491–6499. [CrossRef]
12. Emami, S.M.; Divandari, M.; Hajjari, E.; Arabi, H. Comparison between Conventional and Lost Foam Compound Casting of Al/Mg Light Metals. *Int. J. Cast Met. Res.* **2013**, *26*, 43–50. [CrossRef]
13. Jiang, W.; Fan, Z.; Li, G.; Yang, L.; Liu, X. Effects of Melt-to-Solid Insert Volume Ratio on the Microstructures and Mechanical Properties of Al/Mg Bimetallic Castings Produced by Lost Foam Casting. *Metall. Mater. Trans. A* **2016**, *47*, 6487–6497. [CrossRef]
14. Jiang, W.; Li, G.; Fan, Z.; Wang, L.; Liu, X. Investigation on the Interface Characteristics of Al/Mg Bimetallic Castings Processed by Lost Foam Casting. *Metall. Mater. Trans. A* **2016**, *47*, 2462–2470. [CrossRef]
15. Fan, S.; Jiang, W.; Li, G.; Mo, J.; Fan, Z. Fabrication and Microstructure Evolution of Al/Mg Bimetal Using a near-Net Forming Process. *Mater. Manuf. Processes* **2017**, *32*, 1391–1397. [CrossRef]
16. Li, G.; Yang, W.; Jiang, W.; Guan, F.; Jiang, H.; Wu, Y.; Fan, Z. The Role of Vacuum Degree in the Bonding of Al/Mg Bimetal Prepared by a Compound Casting Process. *J. Mater. Process. Technol.* **2019**, *265*, 112–121. [CrossRef]
17. Abdollahzadeh, A.; Shokuhfar, A.; Cabrera, J.M.; Zhilyaev, A.P.; Omidvar, H. The Effect of Changing Chemical Composition on Dissimilar Mg/Al Friction Stir Welded Butt Joints Using Zinc Interlayer. *J. Manuf. Processes* **2018**, *34*, 18–30. [CrossRef]
18. Zhang, J.; Luo, G.; Wang, Y.; Xiao, Y.; Shen, Q.; Zhang, L. Effect of Al Thin Film and Ni Foil Interlayer on Diffusion Bonded Mg–Al Dissimilar Joints. *J. Alloys Compd.* **2013**, *556*, 139–142. [CrossRef]
19. Wang, Y.; Prangnell, P.B. Evaluation of Zn-Rich Coatings for IMC Reaction Control in Aluminum-Magnesium Dissimilar Welds. *Mater. Charact.* **2018**, *139*, 100–110. [CrossRef]
20. Xu, Z.; Li, Z.; Zhao, D.; Liu, X.; Yan, J. Effects of Zn on Intermetallic Compounds and Strength of Al/Mg Joints Ultrasonically Soldered in Air. *J. Mater. Process. Technol.* **2019**, *271*, 384–393. [CrossRef]
21. Mola, R.; Bucki, T.; Gwoździk, M. The Effect of a Zinc Interlayer on the Microstructure and Mechanical Properties of a Magnesium Alloy (AZ31)–Aluminum Alloy (6060) Joint Produced by Liquid–Solid Compound Casting. *JOM* **2019**, *71*, 2078–2086. [CrossRef]
22. Liu, N.; Liu, C.; Liang, C.; Zhang, Y. Influence of Ni Interlayer on Microstructure and Mechanical Properties of Mg/Al Bimetallic Castings. *Met. Mater. Trans. A* **2018**, *49*, 3556–3564. [CrossRef]
23. Xu, G.; Chen, Y.; Liu, L.; Luo, A.; Ma, L. Effect of La on Structures and Properties of the Liquid-Solid Diffusion Bonding Interface of Magnesium/Aluminum. *Chin. J. Nonferrous Met.* **2014**, *24*, 2743–2748.
24. Wu, L.; Kang, H.; Chen, Z.; Liu, N.; Wang, T. Horizontal Continuous Casting Process under Electromagnetic Field for Preparing AA3003/AA4045 Clad Composite Hollow Billets. *Trans. Nonferrous Met. Soc. China* **2015**, *25*, 2675–2685. [CrossRef]
25. Tayal, R.K.; Kumar, S.; Singh, V. Experimental Investigation and Optimization of Process Parameters for Shear Strength of Compound Cast Bimetallic Joints. *Trans. Indian Inst. Met.* **2018**, *71*, 2173–2183. [CrossRef]
26. Tayal, R.K.; Kumar, S.; Singh, V.; Gupta, A.; Ujjawal, D. Experimental Investigation and Evaluation of Joint Strength of A356/Mg Bimetallic Fabricated Using Compound Casting Process. *Int. Metalcast.* **2019**, *13*, 686–699. [CrossRef]
27. Babaee, M.H.; Maleki, A.; Niroumand, B. A Novel Method to Improve Interfacial Bonding of Compound Squeeze Cast Al/Al−Cu Macrocomposite Bimetals: Simulation and Experimental Studies. *Trans. Nonferrous Met. Soc. China* **2019**, *29*, 1184–1199. [CrossRef]
28. Wang, C.; Cui, J. Application and Research of Vibration Technology in Metal Casting Process. *Foundry Technol.* **2018**, *39*, 2240–2243.
29. Yuanyang, L.I.; Chen, W.; Zheng, K.; Wang, J.; Zheng, Z. Effect of Mechanical Vibration on Bimetallic Compound Casting Behavior. *Foundry Technol.* **2017**, *38*, 377–381.
30. Zhao, J.; Shang, Z.; Wang, L.; Hou, Z.; Wang, Z. Effect of Mechanical Vibration on Heat Transfer of Casting-Mold Interface in Filling and Solidification Process of AZ91 Magnesium Alloy. *Rare Met. Mater. Eng.* **2015**, *44*, 3141–3146.
31. Promakhov, V.; Khmeleva, M.; Zhukov, I.; Platov, V.; Khrustalyov, A.; Vorozhtsov, A. Influence of Vibration Treatment and Modification of A356 Aluminum Alloy on Its Structure and Mechanical Properties. *Metals* **2019**, *9*, 87. [CrossRef]
32. Meng, X.; Jin, Y.; Ji, S.; Yan, D. Improving Friction Stir Weldability of Al/Mg Alloys via Ultrasonically Diminishing Pin Adhesion. *J. Mater. Sci. Technol.* **2018**, *34*, 1817–1822. [CrossRef]
33. Li, G.; Jiang, W.; Guan, F.; Zhu, J.; Yu, Y.; Fan, Z. Improving mechanical properties of AZ91D magnesium/A356 aluminum bimetal prepared by compound casting via a high velocity oxygen fuel sprayed Ni coating. *J. Magnes. Alloys* **2021**, in press. [CrossRef]
34. Wen, F.; Zhao, J.; Yuan, M.; Wang, J.; Zheng, D.; Zhang, J.; He, K.; Shangguan, J.; Guo, Y. Influence of Ni interlayer on interfacial microstructure and mechanical properties of Ti-6Al-4V/AZ91D bimetals fabricated by a solid–liquid compound casting process. *J. Magnes. Alloys* **2021**, *9*, 1382–1395. [CrossRef]
35. Feng, B.; Feng, X.; Yan, C.; Xin, Y.; Wang, H.; Wang, J.; Zheng, K. On the rule of mixtures for bimetal composites without bonding. *J. Magnes. Alloys* **2020**, *8*, 1253–1261. [CrossRef]
36. Jia, B.R.; Liu, L.B.; Yi, D.Q.; Jin, Z.P.; Nie, J.F. Thermodynamic Assessment of the Al–Mg–Sm System. *J. Alloys Compd.* **2008**, *459*, 267–273. [CrossRef]
37. Yan, X.-Y.; Chang, Y.A.; Zhang, F. A Thermodynamic Analysis of the Mg-Si System. *J. Phase Equilibria* **2000**, *21*, 379. [CrossRef]

38. Zhao, X.; Bhushan, B. Material Removal Mechanisms of Single-Crystal Silicon on Nanoscale and at Ultralow Loads. *Wear* **1998**, *223*, 66–78. [CrossRef]
39. Rodríguez-González, B.; Pastoriza-Santos, I.; Liz-Marzán, L.M. Bending Contours in Silver Nanoprisms. *J. Phys. Chem. B* **2006**, *110*, 11796–11799. [CrossRef]
40. Imai, M.; Isoda, Y.; Udono, H. Thermal Expansion of Semiconducting Silicides β-FeSi2 and Mg2Si. *Intermetallics* **2015**, *67*, 75–80. [CrossRef]
41. Trebin, H.-R. *Quasicrystals: Structure and Physical Properties*; John Wiley & Sons: Hoboken, NJ, USA, 2003; ISBN 3-527-40399-X.
42. Lyubimova, T.P.; Parshakova, Y.N. Effect of Rotational Vibrations on Directional Solidification of High-Temperature Binary SiGe Alloys. *Int. J. Heat Mass Transf.* **2018**, *120*, 714–723. [CrossRef]
43. Ihsan-ul-haq; Shin, J.-S.; Lee, Z.-H. Computer-Aided Cooling Curve Analysis of A356 Aluminum Alloy. *Met. Mater. Int.* **2004**, *10*, 89–96. [CrossRef]
44. Xiao, Y.; Ji, H.; Li, M.; Kim, J. Ultrasound-Induced Equiaxial Flower-like CuZn5/Al Composite Microstructure Formation in Al/Zn–Al/Cu Joint. *Mater. Sci. Eng. A* **2014**, *594*, 135–139. [CrossRef]
45. Li, G.; Jiang, W.; Guan, F.; Zhu, J.; Jiang, H.; Fan, Z. Effect of Insert Materials on Microstructure and Mechanical Properties of Al/Mg Bimetal Produced by a Novel Solid-Liquid Compound Process. *J. Manuf. Process.* **2019**, *47*, 62–73. [CrossRef]
46. Chen, Y.; Zhang, H.; Zhu, Z.; Xu, G.; Luo, A.A.; Liu, L. Inhibiting Brittle Intermetallic Layer in Magnesium/Aluminum Bimetallic Castings via In Situ Formation of Mg2Si Phase. *Met. Mater. Trans. B* **2019**, *50*, 1547–1552. [CrossRef]
47. Zamani, R.; Mirzadeh, H.; Emamy, M. Mechanical Properties of a Hot Deformed Al-Mg2Si in-Situ Composite. *Mater. Sci. Eng. A* **2018**, *726*, 10–17. [CrossRef]
48. Maleki, M.; Mirzadeh, H.; Emamy, M. Improvement of Mechanical Properties of in Situ Mg-Si Composites via Cu Addition and Hot Working. *J. Alloy Compd.* **2022**, *905*, 164176. [CrossRef]
49. Guo, Y.; Quan, G.; Ren, L.; Liu, B.; Al-Ezzi, S.; Pan, H. Effect of Zn Interlayer Thickness on the Microstructure and Mechanical Properties of Two-Step Diffusion Bonded Joint of ZK60Mg and 5083Al. *Vacuum* **2019**, *161*, 353–360. [CrossRef]
50. Lloyd, D.J. Aspects of Fracture in Particulate Reinforced Metal Matrix Composites. *Acta Met. Mater.* **1991**, *39*, 59–71. [CrossRef]
51. Wang, Z.; Song, M.; Sun, C.; He, Y. Effects of Particle Size and Distribution on the Mechanical Properties of SiC Reinforced Al–Cu Alloy Composites. *Mater. Sci. Eng. A* **2011**, *528*, 1131–1137. [CrossRef]

Article

Study on Friction and Wear Properties and Mechanism at Different Temperatures of Friction Stir Lap Welding Joint of SiCp/ZL101 and ZL101

Bei Yuan [1], Dunming Liao [1,*], Wenming Jiang [1,*], Han Deng [2] and Guangyu Li [1]

[1] State Key Laboratory of Materials Processing and Die & Mould Technology, School of Materials Science and Engineering, Huazhong University of Science and Technology, Wuhan 430074, China
[2] CRRC Qishuyan Locomotive & Rolling Stock Technology Research Institute Co., Ltd., Changzhou 213011, China
* Correspondence: liaodunming@hust.edu.cn (D.L.); wmjiang@hust.edu.cn (W.J.); Tel./Fax: +86-(27)-87558134 (D.L. & W.J.)

Abstract: In order to achieve the goal of lightening the braking system of urban rail trains, SiCp/ZL101 and ZL101 plates were welded by friction stir lap welding (FSLW) to prepare a new type of brake disc material. The friction and wear properties of the friction-stir-processed composite material were studied at different temperatures (30 °C, 100 °C, 150 °C, 200 °C, 250 °C, 300 °C) to provide a theoretical basis for the evaluation of braking performance. The experimental results showed that the sliding friction processes at each temperature were relatively stable, the friction coefficients did not vary much and the average friction coefficients changed slightly, stabilizing at about 0.4. The wear extent and the depth of wear scars increased with the increase in the temperature, reaching the highest at 150 °C and then began to decrease. At room temperature, the wear forms were mainly oxidative wear and abrasive wear; as the temperature rose, under the cyclic shearing action of the grinding ball, the abrasive debris fell off under the expansion of fatigue cracks and fatigue wear was the main form at this stage. When the temperature reached 200 °C, it began to show the characteristics of adhesive wear; after 250 °C, due to the gradual formation of a mechanical mixed layer containing more SiC particles and oxides on the wear surface, it exhibited high-temperature lubrication characteristics, and the wear extent was equivalent to 35% of the wear extent at normal temperature, indicating that the composite material had good high-temperature friction and wear properties.

Keywords: FSLW; SiCp/ZL101 and ZL101 joint; temperature; friction and wear properties; wear mechanism

1. Introduction

Aluminum matrix composites (AMCs) have the advantages of light weight, high tensile strength, high specific stiffness and specific strength, better fatigue strength, corrosion resistance and a low thermal expansion coefficient, etc. [1–6]. They have broad application prospects in electronics, new energy vehicles, aerospace and other fields [7–9]. Traditional cast iron and cast steel brake discs are not conducive to improving the braking performance of vehicles due to defects such as heavy weight, poor thermal conductivity and poor fatigue performance [4,10]. Aluminum-based composite materials are ideal substitutes for traditional brake disc materials [4,11]. The application in the brake disk has been widely considered [12]. For example, Venkatachalam et al. [13] prepared Al6082 composite material by stir casting, and verified through performance research that the composite material had a small friction coefficient and wear rate, and could be used to prepare automobile brake discs. Firouz et al. [14] prepared Al-9Si-SiC composite automobile brake discs with 10% and 20% SiC volume fractions by stir casting, and conducted thermal fatigue research. The results showed that the thermal fatigue performance of composite brake discs was better than that of cast iron brake discs. Sadagopan et al. [15] prepared the Al 6061 metal matrix composite brake rotor with 20% SiC volume fraction by stir casting, and verified through experiments that the composite brake disc had better efficiency in terms of braking distance and heat dissipation than the cast iron disc. Daoud et al. [16] prepared the A359 particle composite brake rotor with 20% SiC volume fraction by sand casting, and verified that the AMC brake disc had the advantages of wear resistance and higher thermal conductivity compared with the cast iron brake disc, lighter weight and a more uniform coefficient of friction; these characteristics reduced braking distance and braking noise. In addition, there are many studies on the friction and wear behavior of AMCs. For example, Jin et al. [5] conducted a high-temperature friction and wear test on as-cast SiCp/A356 composites. The results showed that the wear rate of the as-cast SiCp/A356 was very sensitive to temperature changes, and the friction stability decreased sharply with the increase in temperature. Hekner et al. [17] used molecular dynamic simulations to study the nanoscale wear behavior of SiC particle-reinforced AMCs (SiC/Al NCs), and the results showed that the wear mechanism was changed during high temperature.

Particle-reinforced composite castings prepared by traditional casting technology have defects such as uneven particle distribution and large porosity [18–20], and do not have high strength and ductility compared with the base material [21], resulting in the inability to fully exert the performance of composite materials. To overcome this problem, Friction Stir Processing (FSP) is widely used in the preparation of composite materials [22–24]; through the continuous stirring motion of the stirring tool, the reinforced particles are evenly distributed throughout the matrix, which reduces the porosity and improves the friction and wear properties of the composite [25]. Based on the above research results, this study proposed a new method for the preparation of composite brake discs, which used friction stir welding to lap the AMC sheet on the aluminum alloy substrate, and at the same time used FSP to modify the AMC, to prepare a functionally graded brake disc material with both wear resistance and toughness. At present, there is much research on FSW for AMCs. For example, Avettand-Fènoël et al. [26] reviewed the microstructure, friction stir welding performance and other indicators of the FSW joints of various AMC materials, and proposed ways to improve them. Zuo et al. [27] reviewed the weldability, macrostructure and microstructure of joints, mechanical properties of joints, tool wear and monitoring of SiCp/Al composites, and looked forward to the future development direction. In addition, the research on the use of FSP to prepare AMCs is also a hot topic that has attracted much attention. For example, Vijayavel et al. [28] used FSP to process the surface of the lm25 composite material with a volume fraction of 5% SiC. The experimental results showed that when the shaft-to-shoulder ratio of the stirring pin was 3.0, the obtained equiaxed grains were finer [29] and the microstructures processed at a tool traverse speed of 40 mm/min showed excellent wear resistance. Mohamadigangaraj et al. [30] evaluated the effects of friction stir processing parameters on the properties of A390-10 wt% SiC composite using response surface methodology, and the results showed that the speed of rotation had a higher impact on hardness than other parameters. Kumar et al. [31] investigated the mechanism for improving the tensile properties, wear properties and corrosion resistance of stir-cast Al7075–2 wt.% SiC composites by friction stir processing (FSP); the results showed that nanoparticle-reinforced composites after FSP exhibited better wear resistance than microparticle-reinforced composites. Kurtyka et al. [32] studied the effect of the plastic deformation generated in the FSP process on the concentration and distribution of SiC particles in the cast composite A339/SiCp, and the study showed that the FSP process significantly improved the distribution of reinforced particles in the composite. Butola et al. [33] conducted pin-on-disk friction and wear tests on AA7075–2 wt.% SiC composites prepared by FSP, and the results showed that FSP can produce surface composites with no defects and the uniform distribution of reinforcement materials, which helped to improve the wear resistance of composites. Aruri et al. [34] studied the effect of tool speed on the wear properties of aluminum-based surface hybrid composites manufactured by FSP, and the results showed that reducing the tool speed appropriately could reduce the wear rate of Al-SiC/Al2O3 surface composites. Devaraju et al. [35] studied the effects of rotational speed and reinforced particles such as SiC and Al2O3 on the wear and mechanical properties of aluminum alloy-based surface hybrid composites prepared by FSP, and found that the size of the reinforced particles was reduced, and the wear resistance was greatly improved after FSP. Rana et al. [36] used FSP to prepare Al 7075-T651-B4C surface composite material, and found that the wear resistance of the composite material increased by 100% compared with the base material. The FSW process research on AMCs mentioned in the above research rarely involves the FSLW process. The existing research on the preparation of AMCs by FSP mainly focuses on the optimization of process parameters, and the volume fraction of the reinforcing phase is not high ($\leq$10%). The research on the friction and wear properties of the prepared surface composites is also mainly concentrated on the normal temperature in the environment. Since the sliding friction will generate a lot of heat during the braking process, the temperature of the brake disc will change sharply, which will affect the braking effect (thermal stability, the vibration of the braking system, braking noise, braking safety and so on [37,38]). The above research results rarely involve the influence of temperature on the friction and wear properties of FSPed composite material. The friction and wear properties of this brake disc material at different temperatures are of great significance for exploring the wear mechanism of the brake materials at different temperatures and evaluating their braking performance.

ZL101 has excellent casting performance and good weldability, and is widely used in the preparation and welding of AMCs [39–44]. In this study, the 20% volume of the stir-casted SiCp/ZL101 composite sheet and the ZL101 sheet was used for the preparation of the composite material using the FSLW. The friction and wear performances and mechanisms of the SiCp/ZL101 and ZL101 composite material at 30 °C, 100 °C, 150 °C, 200 °C, 250 °C and 300 °C were studied, providing a theoretical basis for the evaluation of the braking performance of the brake disc material.

2. Experimental Procedure

2.1. Experimental Materials

In this experiment, the SiCp/ZL101 composite material with 20% SiCp volume fraction was prepared by stir casting method, and the size of the SiC particles was 1000 mesh. The material composition of ZL101 aluminum alloy is shown in Table 1. The prepared composite materials were cut into sheets of size 6 mm × 180 mm × 90 mm by wire cut, ZL101 sheets were cut into sheets of size 9 mm × 180 mm × 90 mm by the wire cut, and the surface roughness of the two sheets was processed to Ra 0.8 by machining technology.

Table 1. Chemical composition of the ZL101 aluminum alloy (wt.%).

Aluminum Alloy	C	Si	Mn	Mo	Cr	Ni	Mg	Al	Fe
ZL101	-	6.5~7.5	$\leq$0.35	-	-	-	0.25~0.45	Bal	-

The experimental equipment in this study was a FSW equipment modified from a X35K milling machine from Zhengling (Liuzhou, China), and the FSLW process was adopted to weld SiCp/ZL101 composite plate (upper plate) and ZL101 plate (lower plate) to prepare multiple sets of welding samples. Figure 1a is a schematic diagram of the FSLW process. The overlapping rate of adjacent welding passes was 50%. The detailed welding parameters are shown in Table 2. After the welding samples were prepared, natural aging was carried out for 7 days.

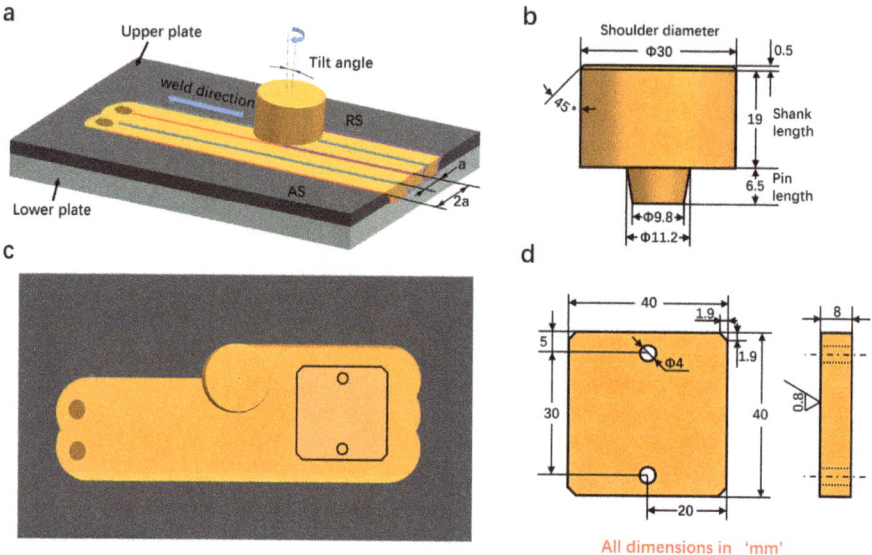

Figure 1. (a) Schematic diagram of FSLW; (b) Schematic diagram of the size of the stirring pin; (c) Schematic diagram of the sampling location of the friction and wear specimen; (d) Schematic diagram of the sizes of the friction and wear specimen.

Table 2. The parameters of the FSLW process.

No.	Speed (rpm)	Welding Speed (mm/min)	Pressing Amount (mm)	Welding Pass	Tool Tilt Angle (°)
1	375	35.5	0.15	1	3.5

Six friction and wear test specimens were cut from the weld joint of the sample by wire cut. The sampling location is shown in Figure 1c, and the sample size is shown in Figure 1d. The surface roughness of the specimens was treated to Ra0.8 by the machining technology.

2.2. Friction and Wear Test

The equipment used in this experiment was MMQ-02G ball-on-disk high-temperature friction and wear testing machine from Yihua (Jinan, China), and six specimens were subjected to friction and wear tests in air atmosphere at different temperatures. Figure 2 is a schematic diagram of the experimental device. As can be seen, the specimen was fixed on the rotating disk, and the counter-grinding ball was in contact with the surface of the specimen under the specified load. The distance between the counter-grinding ball and the center of the rotating disc is the friction radius. After the device was heated to the specified temperature in the incubator, the rotating disk drove the friction specimen to perform relative frictional motion with the grinding ball at a specified speed. Si_4N_3 balls with a diameter of 6 mm were used as the counter-grinding balls, and the test time was 60 min. The test parameters of the six specimens are shown in Table 3. During the experiment, the test parameters such as test force, rotational speed, friction coefficient, temperature, time, etc., were collected and calculated by the computer in real time, and the wear debris was collected after each test.

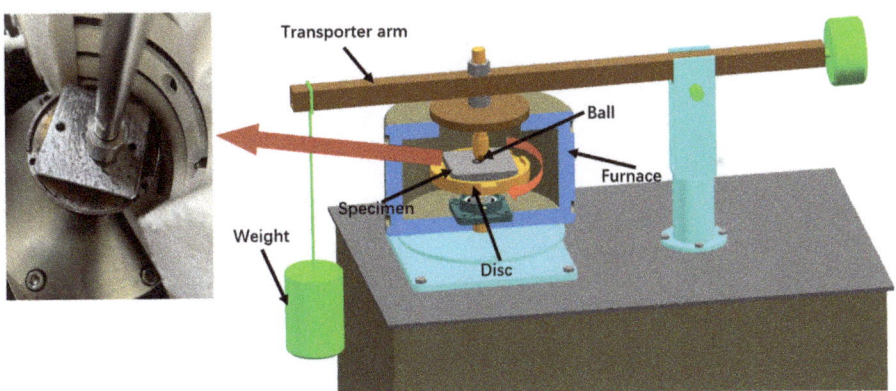

Figure 2. Schematic diagram of the friction and wear test device.

Table 3. The parameters of the friction and wear test.

No.	Speed (r/min)	Temperature (°C)	Load (N)	Radius of Friction (mm)
1	150	30	6	5
2	150	100	6	5
3	150	150	6	5
4	150	200	6	5
5	150	250	6	5
6	150	300	6	5

2.3. Wear Detection and Structural Characterization

After cleaning and drying the friction and wear specimens with alcohol, the MS-M9000 multifunctional friction tester from Huahui (Lanzhou, China) was used to measure the wear extent and wear scar depth of the friction and wear specimens by the surface profile method. Each specimen was measured at four symmetrical parts of the friction ring, and the arithmetic mean value of the four groups of data was taken.

In order to study the microstructure of the wear surface and wear debris, a Quanta 400 scanning electron microscope (SEM) equipped with an energy-dispersive X-ray spectrometer (EDS) (FEI, Eindhoven, The Netherlands) was used to observe the microstructure of the wear surface and wear debris, and the elemental compositions of the wear debris were analyzed by EDS.

3. Results

3.1. Macrostructure Morphology of Wear Surface

Figure 3 is a schematic diagram of the macrostructure of the wear surface at 30 °C, 100 °C, 150 °C, 200 °C, 250 °C and 300 °C. As shown in the figure, the wear scar of the 30 °C specimen (Figure 3a) is relatively wide, and the wear surface is rough and uneven, showing a silvery white luster. The wear surface of the specimens from 100 °C to 300 °C is black; the wavy folds caused by extrusion deformation can be clearly observed on the wear surface of the specimen at 100 °C (Figure 3b), and the distribution is relatively dense. The distribution of wavy folds on the wear surface of the 150 °C specimen (Figure 3c) is relatively sparse. The wear surface of the 200 °C specimen (Figure 3d) is relatively flat, with only one obvious wavy fold observed. The wear surfaces of the 250 °C and 300 °C specimens (Figure 3e,f) are relatively flat.

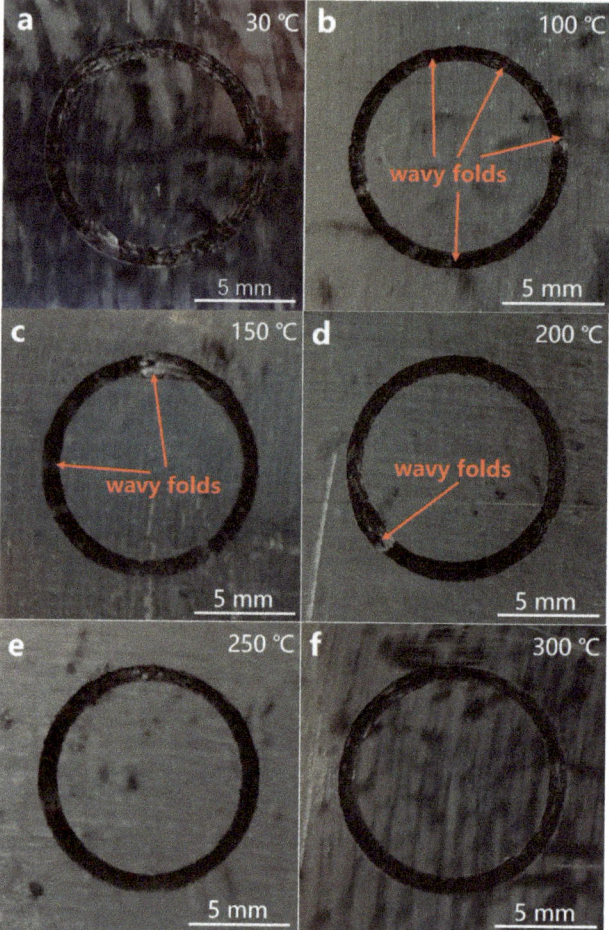

Figure 3. (a–f) Macrostructures of the wear surfaces of 30 °C, 100 °C, 150 °C, 200 °C, 250 °C and 300 °C.

3.2. Microstructure Morphology of Wear Surface

Figure 4 is a schematic diagram of the microstructure morphology of the wear surface at 30 °C, 100 °C, 150 °C, 200 °C, 250 °C and 300 °C. As shown in Figure 4a, there are a large number of pits on the wear surface at 30 °C due to the shedding of wear debris, and these pits are connected to each other, resulting in uneven wear surfaces. The high-magnification image of the SEM (Figure 4a1) shows that there are fine furrows on the surface of the wear scar. The wear surface at 100 °C (Figure 4b,b1) shows obvious traces of plastic deformation due to extrusion, and the amount of wear debris falling off is reduced compared with that at 30 °C, and it begins to show obvious peeling marks. At 150 °C, the amount and area of the wear debris shedding on the wear surface (Figure 4c) increase significantly, and the SEM high-magnification image (Figure 4c1) shows obvious peeling marks. The wear surface above 200 °C shows significant plastic deformation traces. Due to the extrusion of the grinding ball, the metal softened at high temperature overflows at the edge of the wear scar to form an obvious flash-like structure (Figure 4d–f). The high-magnification image of the SEM (Figure 4f1) shows that fatigue cracks are generated on the wear scar surface under the action of cyclic stress.

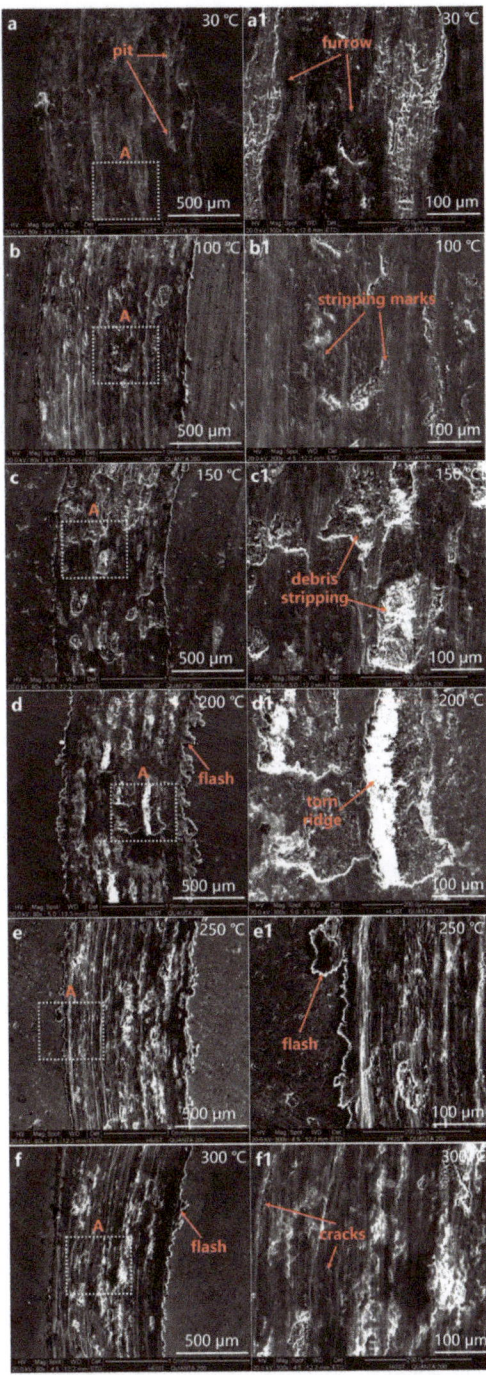

Figure 4. (**a–f**) Low magnification SEM micrographs of the microstructure of the wear surface at 30 °C, 100 °C, 150 °C, 200 °C, 250 °C and 300 °C; (**a1–f1**) High magnification SEM micrographs of the structure of area A in a–f.

Figure 5 is a graph of the wear extent curve, wear scar depth curve and average friction coefficient curve at each experimental temperature. As shown in Figure 5a,b, as the temperature increases, the wear extent and wear scar depth first decrease and then increase, and then show a downward trend after reaching the highest level at 150 °C. The wear extent above 200 °C is equivalent to about 35% of the wear extent at room temperature; the wear extent and wear scar depth show the smallest dispersion at 30 °C, and the largest dispersion at 150 °C. As shown in Figure 5c, the average friction coefficient does not show significant differences with the change of temperature, and is stable at around 0.4; the average friction coefficient has the largest dispersion at 300 °C.

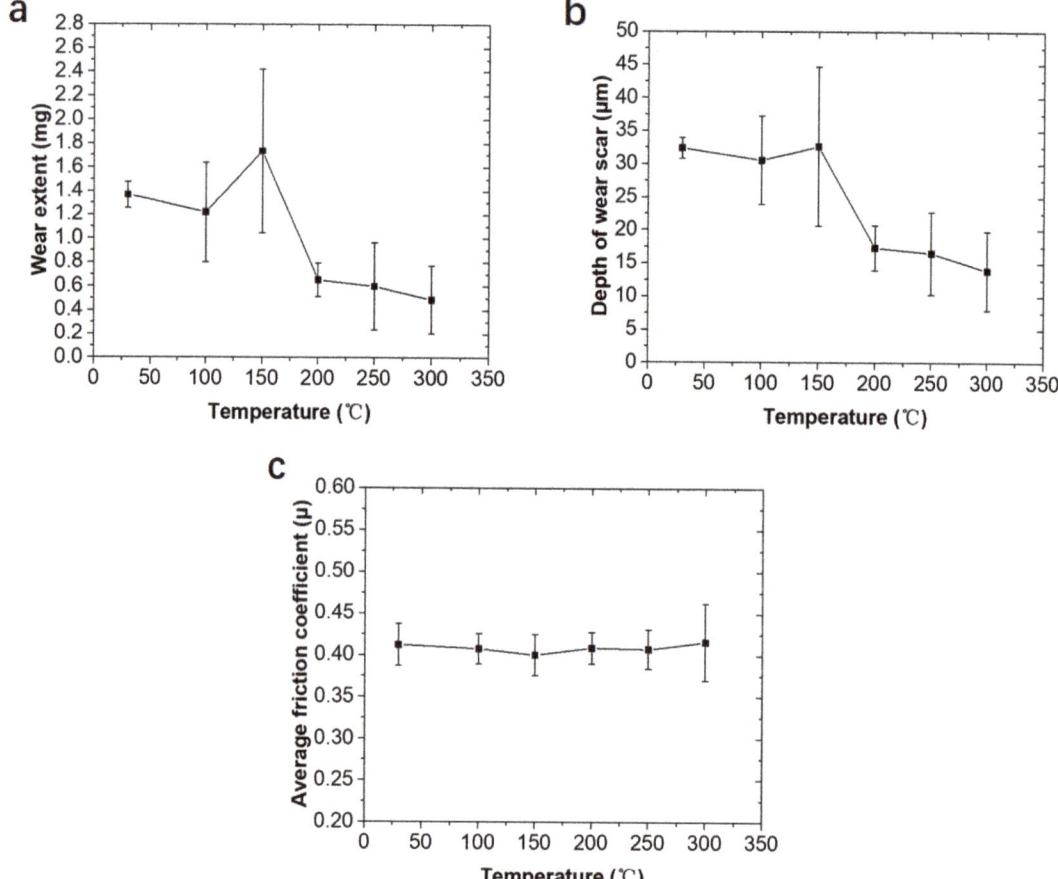

Figure 5. (a) Wear curves at 30 °C, 100 °C, 150 °C, 200 °C, 250 °C and 300 °C; (b) Wear scar depth curves at 30 °C, 100 °C, 150 °C, 200 °C, 250 °C and 300 °C; (c) Curves of the average friction coefficient at 30 °C, 100 °C, 150 °C, 200 °C, 250 °C and 300 °C.

Figure 6 is a graph of the friction coefficient changing with time at various experimental temperatures. As shown in the figure, at each experimental temperature, the friction coefficient does not show large fluctuations with time, reflecting a stable friction performance.

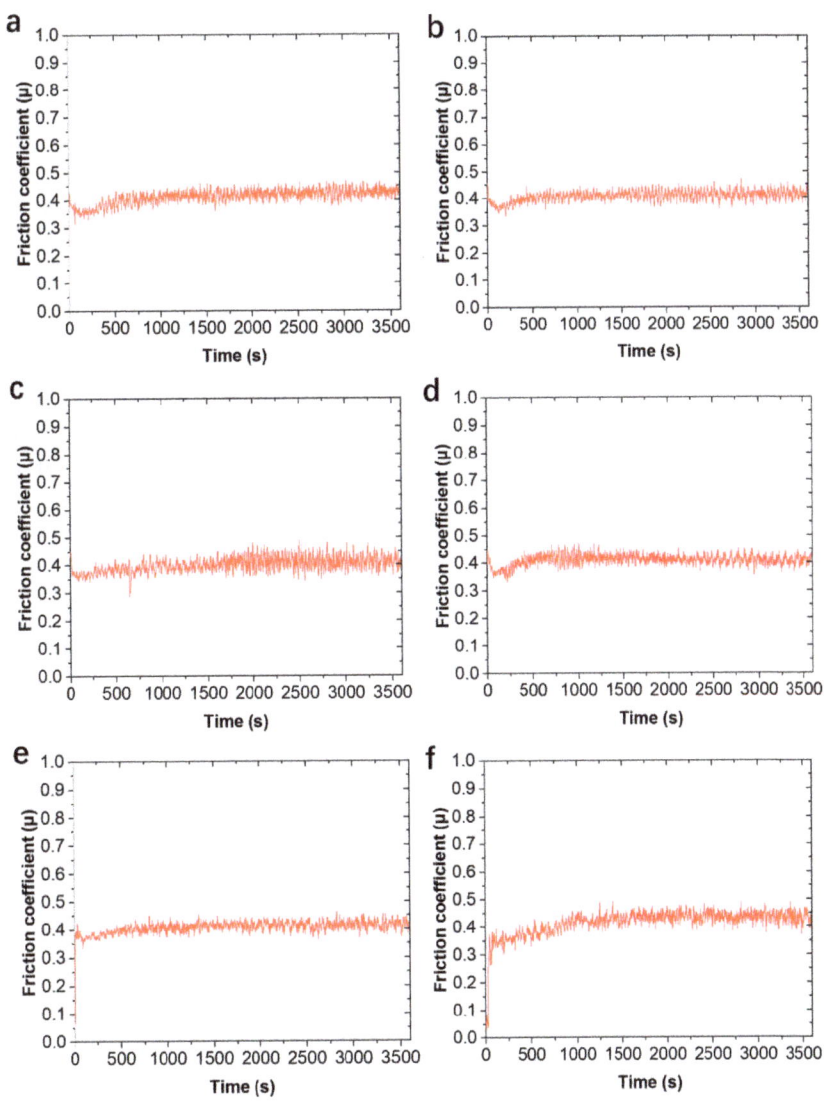

Figure 6. (**a–f**) Curves of friction coefficient versus time at 30 °C, 100 °C, 150 °C, 200 °C, 250 °C and 300 °C.

3.3. Morphology of Wear Debris

Figure 7 shows SEM micrographs of wear debris at 30 °C, 100 °C, 150 °C, 200 °C, 250 °C and 300 °C. As shown in Figure 7a, the abrasive debris consists of a small amount of large-sized massive abrasive debris and a large amount of powdery abrasive debris at 30 °C. As the temperature increases, the powdery abrasive debris decreases greatly (as shown in Figure 7b). The size of the massive abrasive debris reaches the maximum at 150 °C. As the temperature rises further, the powdery abrasive debris begins to increase, and the size of the debris tends to be consistent (as shown in Figure 7e,f); obvious cracks appear on the surface of the massive wear debris at 150 °C and 200 °C.

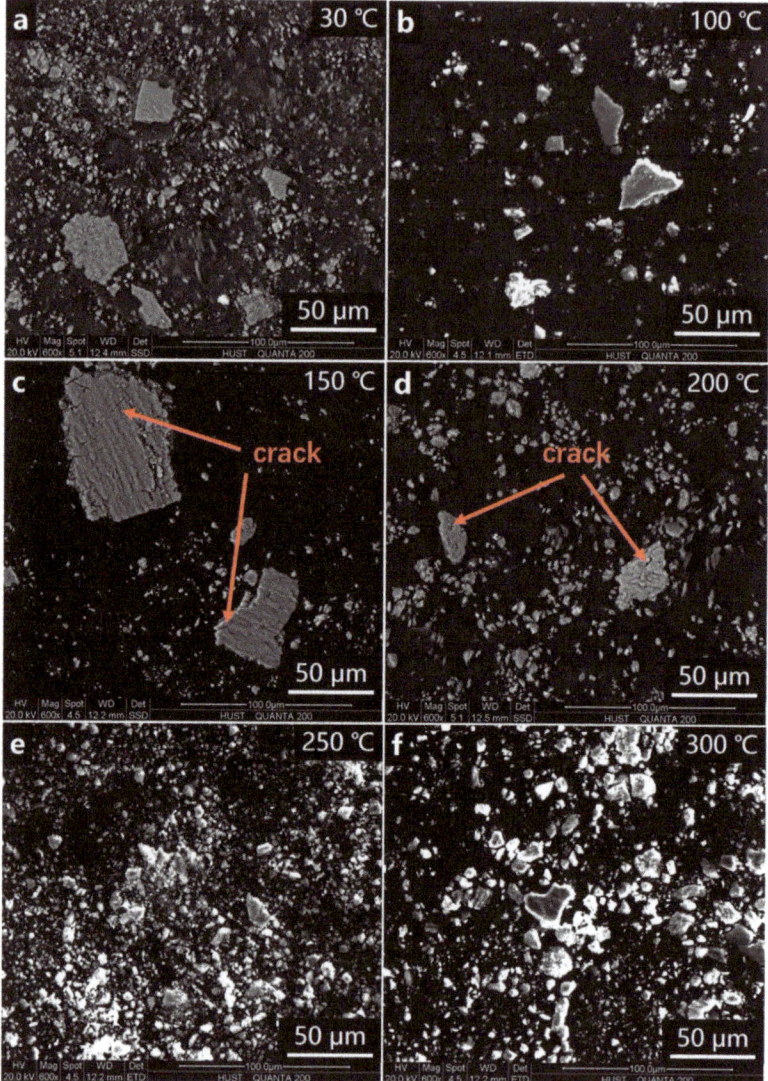

Figure 7. (**a**–**f**) SEM micrographs of wear debris of 30 °C, 100 °C, 150 °C, 200 °C, 250 °C and 300 °C.

An energy-dispersive X-ray spectrometer was used to analyze the composition of the wear debris of 30 °C and 150 °C. As shown in Figure 8, besides the Al and Si elements contained in the matrix material, there are more O elements in the wear debris. Since the grinding balls are Si_4N_3 ceramic balls with high hardness and stable chemical properties, it shows that the O element in the grinding debris comes from the oxidation of the composite material during the wear process.

An energy-dispersive X-ray spectrometer was used to analyze the composition of the wear debris at 300 °C. As shown in Figure 9, the EDS component analysis shows the presence of Al, Si, O and C. Since the friction and wear experiments were carried out in air atmosphere, the surface of the exfoliated SiC particles was easily stained with the oxide produced by the Al matrix. Therefore, according to the EDS analysis results in Figure 9b, it is judged that the particles in Figure 9a are SiC particles.

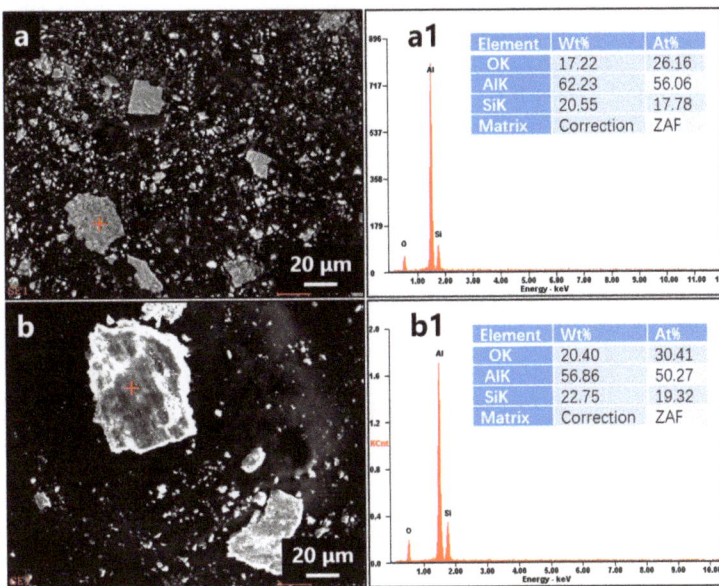

Figure 8. (**a**,**a1**) The EDS composition analysis of wear debris of 30 °C, the red plus sign in (**a**) is the EDS sampling point; (**b**,**b1**) EDS composition analysis of wear debris of 150 °C, the red plus sign in (**b**) is the EDS sampling point.

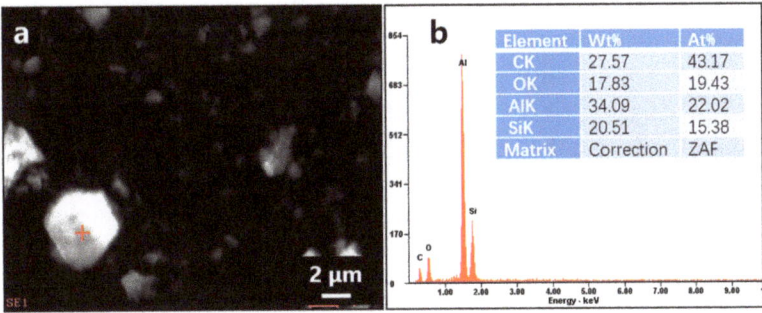

Figure 9. (**a**,**b**) The EDS analysis result of wear debris at 300 °C, the red plus sign in (**a**) is the EDS sampling point.

4. Discussion

4.1. Effect of Temperature on the Morphology of Wear Scars

The roughness of sliding friction surfaces plays a crucial role in the wear process [45]. When the grinding balls first come into contact with the composite surface, they only make contact at a few rough points where these micro-protrusions cover only a small portion of the surface area. As a result, very high stresses are generated in these small surface areas and wear occurs at these points [46]. Figure 10 is a schematic diagram of the wear surface morphology change from low temperature to high temperature. As shown in Figure 10a, when the temperature is low, the plastic deformation of the contact surface is small, and the uneven contact point cracks and breaks under the cyclic shearing action of the grinding ball; the wear process at 30 °C fits this type (Figure 3a). In addition, due to the higher fracture energy [46], the plastic deformation of the metal caused by wear at 30 °C is small; a silver-white rough wear surface is finally formed. As the temperature increases, the plasticity of the friction surface of the composite material improves; the uneven contact points on the friction surface are fractured due to the large plastic deformation under the sliding shear of the grinding ball and accumulate to the advancing side of the grinding ball. With the increase in accumulation, due to insufficient shear force, the grinding ball will move on over the accumulated metal. Wavy folds are formed on the wear surface under this cyclic friction (as shown at point A in Figure 10b). Since the plasticity of the composite material is further enhanced with the increase in temperature, more metal needs to be accumulated on the advancing side of the grinding ball to generate sufficient shear resistance (Figure 10c). Therefore, the spacing of the wavy folds on the wear surface gradually increases (as shown in Figure 3b–d).

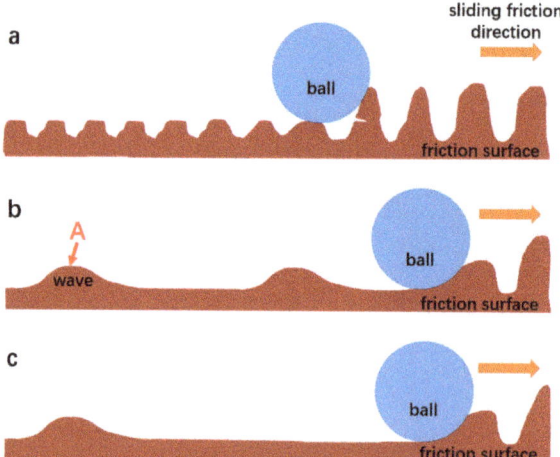

Figure 10. (**a**) Schematic diagram of the change of the wear surface morphology during the friction and wear process at room temperature; (**b**) Schematic diagram of the formation of wavy folds on the wear surface when the temperature of the friction and wear test increased; (**c**) The temperature of the friction and wear test was further increased; the spacing of the wavy folds on the wear surface increases.

When the temperature reaches 200 °C, the increased plasticity of the composite leads to the improvement of the fluidity, leading to flashing at the edge of the wear scar due to composite spillage (as shown in Figure 4d–f). When the temperature exceeds 250 °C, the wear surface exhibits extensive plastic deformation. Therefore, the wear groove track formed parallel to the sliding direction is relatively clear (as shown in Figure 4e,f).

4.2. Effect of Temperature on Wear Mechanisms

Friction Stir Processing improves the defects in the as-cast composites [25] and promotes the performance improvement of the composites [47]. Figure 11a is the photo of the cross-section of the FSLW joint; Figure 11b is the SEM photograph of the as-cast composite material area at point A not affected by FSP in Figure 11a; Figure 11c is the SEM photo of the nugget area at point B in Figure 11a. As shown in Figure 11b, the SiC particles are agglomerated and unevenly distributed, and there are casting shrinkage defects in the as-cast SiCp/ZL101. The microstructure of the as-cast SiCp/ZL101 is significantly improved after FSP (Figure 11c). The SiC particles are refined and evenly distributed, and casting shrinkage cavities are eliminated. The composite material after stirring is denser, and the wear surface maintains better compactness and continuity [46,48]. Therefore, the friction process at each temperature is relatively stable, and the fluctuation range of the friction coefficient is small. A stable friction coefficient can lead to better braking effects, such as better thermal stability, the reduction of the vibration of the braking system, low noise, controllable braking safety and so on [37,38].

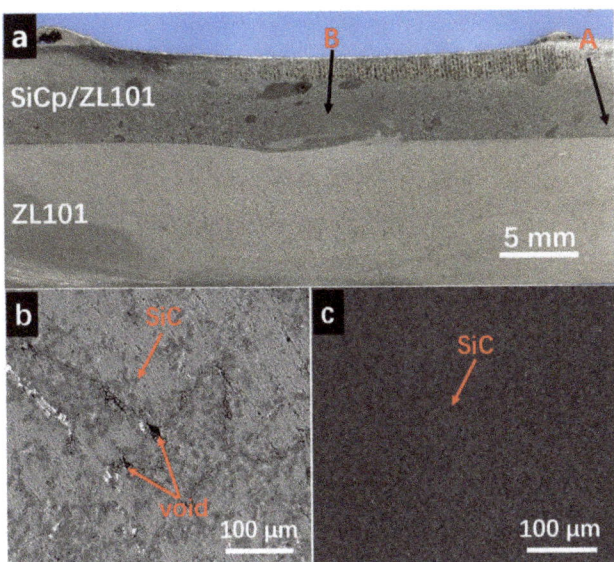

Figure 11. (**a**) Schematic diagram of the cross-section of FSLW joint; (**b**) SEM photograph of the as-cast composite material area at point A not affected by FSP in Figure 11a; (**c**) SEM photo of the nugget area at point B in Figure 11a.

When the friction and wear test was carried out at 30 °C, the micro-protrusions on the surface of the composite material were plastically deformed under the cutting action of the grinding ball and the rotating torque; a large amount of heat was generated locally, resulting in the partial oxidation of the material and it falling off the surface to form wear debris, leading to pitting on the wear surface. This phenomenon exhibited characteristics of oxidation wear. The fine scratches on the wear surface along the sliding direction were mainly caused by the reinforcement particles shed from the matrix [49], indicating that abrasive wear also occurred during the sliding friction process [50]. It is worth noting that the stirred composite materials show very shallow grooves after sliding wear, which may be attributed to the following two factors. First, the SiC particle becomes smaller and has a good interfacial bond with the Al matrix after the stirring treatment. Good interfacial bonding persists at 30 °C and allows load transfer across the particle/matrix interface. Therefore, only a small number of SiC particles fracture and detach from the Al matrix. Secondly, since there is no obvious accumulation of reinforced particles after the stirring treatment [51], the evenly distributed small-sized and high-hardness SiC particles improve the hardness, strength and wear resistance of the composite material [52], which can act as a hard barrier against scratching and plowing by abrasive particles [53]. These two factors lead to the formation of shallow grooves in the wear surface.

The plasticity of the wear surface is improved with the increase in temperature. Figure 12 is a schematic diagram of fatigue crack growth on the wear surface. As shown in the figure, due to the repeated extrusion of the grinding ball, microcracks nucleate at the stress concentration of the friction surface, gradually break through the wear surface, and then grow and connect with each other during the sliding process. Eventually, the surface metal will fall off and become wear debris. This statement is confirmed by the cracks on the wear debris in Figure 7c,d and the pits with irregular edges formed by the shedding of wear debris in Figure 4c1,d1. The wear surface exhibits fatigue wear behavior at this stage. The initiation and growth of the fatigue cracks are further accelerated with the increase in temperature; the size of the wear debris is larger; the wear extent and wear scar depth reach the maximum at 150 °C. The fatigue wear leads to the shedding of large pieces of wear debris at 150 °C (Figure 7c), and the wear surface is rough and uneven. Therefore, the fluctuation range of the friction coefficient becomes significantly larger after 1500s (Figure 6c), which shows significant differences with the friction coefficient curves at other temperatures. As shown in Figure 3c, there were wavy folds caused by plastic deformation on the wear surface at 150 °C. When the surface profiling method was used to detect the wear extent and wear scar depth, if the sampling part was just near the wavy folds, it would cause a large deviation between different measurements. Therefore, the large dispersion of wear extent and wear scar depth at 150 °C can be attributed to the combination of the high roughness of the wear surface and the morphology of the measurement sampling part.

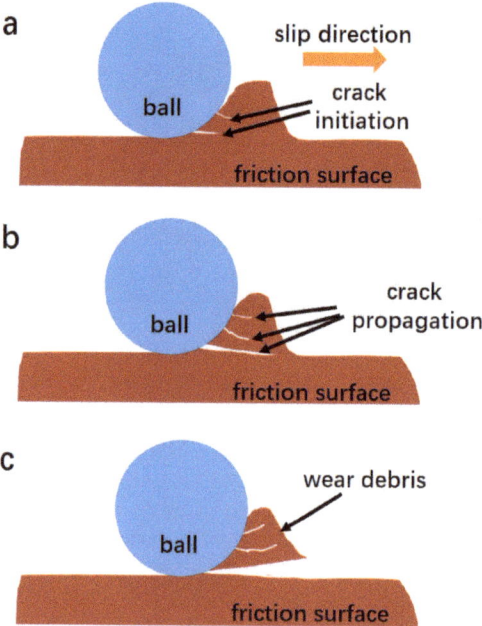

Figure 12. (**a**) Fatigue crack initiation on the wear surface; (**b**) Fatigue crack growth on the wear surface; (**c**) Fatigue cracks penetrated through the wear surface, and wear debris fell off.

In Figure 4d1, the irregular tear ridges are shown after the wear debris breaks away at 200 °C, and the wear scar depth is significantly smaller, showing the characteristics of adhesive wear [53], indicating that the further improvement of the plasticity of the composite material at high temperature leads to the transition of the wear behavior to adhesive wear. As the temperature further increases, the softening of the Al matrix leads to the cracking or loosening of the SiC particles. As shown in Figure 9, the EDS component analysis of the abrasive debris at 300 °C shows the presence of small-sized SiC particles and oxide particles. Figure 13 is a schematic diagram of the sliding friction process at high temperature. As shown in Figure 13b, these small-sized SiC particles and oxides aggregate between the wear surface and the grinding ball, and gradually form a mechanically mixed layer on the wear surface under the action of cyclic load [54]. Elastic deformation occurs on the wear surface at high temperature [55], and this mechanically mixed layer is more likely to cause shear instability on the sliding friction surface [56,57]. Meanwhile, the wear surface maintains better compactness and continuity due to the denser composite material after stirring [46,48], and is not easily damaged by the shear force of the mechanical mixing layer, which further reduces the severity of the microplowing action caused by interactions between abrasive particles. The combined effect of appealing factors leads to a reduced wear rate at elevated temperatures. On the other hand, the plastic deformation resistance of the composite decreases due to the increase in temperature. The wear debris adhered to the grinding ball forms a clear furrow on the wear surface; the wear does not progress further, due to the lubrication of the high-hardness mechanical mixed layer [58]. Therefore, there were no pits on the wear surface caused by the shedding of large-sized wear debris at 250 °C and 300 °C, and the depth of the wear scars did not continue to increase.

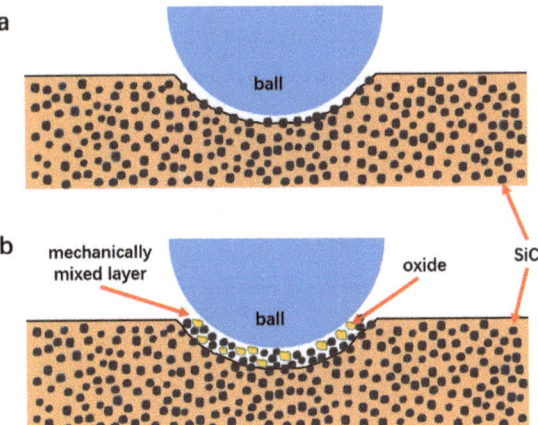

Figure 13. (**a**) The SiC particles started to loosen and detach from the matrix during sliding friction at high temperature; (**b**) SiC particles and oxides aggregated on the wear surface to form a mechanically mixed layer.

5. Conclusions

(1) The wear scars of the specimen at 30 °C were relatively wide, and the wear surface was rough and uneven. The wear scars of the specimens at 100 °C to 300 °C were black, and obvious wavy folds could be observed on the wear surface. The distribution of wavy folds was denser and decreased with increasing temperature. The wear surfaces showed significant signs of plastic deformation at temperatures above 200 °C. The size of the wear debris reached its maximum at 150 °C, and obvious cracks appeared on the surface of the wear debris. Then, the powdery abrasive debris began to increase, and the size of the wear debris tended to be consistent. The O element was detected in the wear debris at each temperature.

(2) The friction process at each temperature was relatively stable; the friction coefficient did not change much. The average friction coefficient changed slightly and was stable at around 0.4. The wear extent and the depth of wear scars increased with increasing temperature, reaching the highest at 150 °C, and then began to decrease. The wear extent above 200 °C was equivalent to about 35% of the wear extent at room temperature.

(3) The wear mechanisms were mainly oxidation wear and abrasive wear at 30 °C. As the temperature increased, the wear debris fell off under the propagation of the fatigue cracks caused by the action of the cyclic shearing of the grinding ball; fatigue wear was the main form at this stage. When the temperature reached 200 °C, it began to show the characteristics of adhesive wear. Due to the gradual formation of a mechanical mixed layer containing SiC particles and oxides on the wear surface at high temperature, it exhibited high-temperature lubrication characteristics and better high-temperature friction and wear performance.

Author Contributions: Conceptualization, B.Y. and G.L.; methodology, H.D.; investigation, B.Y. and G.L.; writing—original draft preparation, B.Y.; writing—review and editing, W.J. and D.L.; visualization, W.J.; Supervision, D.L. All authors have read and agreed to the published version of the manuscript.

Funding: This research was funded by National Key Research and Development Program of China [No. 2020YFB2008300, 2020YFB2008304], the National Natural Science Foundation of China [Nos. 52075198 and 52271102].

Institutional Review Board Statement: Not applicable.

Informed Consent Statement: Not applicable.

Data Availability Statement: Data presented in this article are available at request from the corresponding author.

Acknowledgments: The authors gratefully acknowledge the support provided by the Analytical and Testing Center, HUST.

Conflicts of Interest: The authors declare no conflict of interest.

References

1. Jiang, W.; Zhu, J.; Li, G.; Guan, F.; Yu, Y.; Fan, Z. Enhanced mechanical properties of 6082 aluminum alloy via SiC addition combined with squeeze casting. *J. Mater. Sci. Technol.* **2021**, *88*, 119–131. [CrossRef]
2. Zhu, J.; Jiang, W.; Li, G.; Guan, F.; Yu, Y.; Fan, Z. Microstructure and mechanical properties of SiCnp/Al6082 aluminum matrix composites prepared by squeeze casting combined with stir casting. *J. Mater. Process. Technol.* **2020**, *283*, 116699. [CrossRef]
3. Geng, R.; Qiu, F.; Jiang, Q.C. Reinforcement in Al Matrix Composites: A Review of Strengthening Behavior of Nano-Sized Particles. *Adv. Eng. Mater.* **2018**, *20*, 13. [CrossRef]
4. Wasekar, M.K.; Khond, M.P. A Composite Material an alternative for manufacturing of Automotive Disc Brake: A Review. *IOP Conf. Ser. Mater. Sci. Eng.* **2021**, *1126*, 012067. [CrossRef]
5. Jin, Y.X.; Wang, X.Y.; Tong, Q.Q.; Chen, H.M.; Lee, J.M. High-Temperature Dry Sliding Friction and Wear Characteristics of As-Cast SiCp/A356 Composite. *Adv. Mater. Res.* **2013**, *690–693*, 318–322. [CrossRef]
6. Dasgupta, R. Aluminium alloy-based metal matrix composites: A potential material for wear resistant applications. *Int. Sch. Res. Not.* **2012**, *2012*, 594573. [CrossRef]
7. Moustafa, E.B. Dynamic characteristics study for surface composite of AMMNCs matrix fabricated by friction stir process. *Materials* **2018**, *11*, 1240. [CrossRef]
8. Khalil, A.M.; Loginova, I.S.; Solonin, A.N.; Mosleh, A.O. Controlling liquation behavior and solidification cracks by continuous laser melting process of AA-7075 aluminum alloy. *Mater. Lett.* **2020**, *277*, 128364. [CrossRef]
9. Khalil, A.M.; Loginova, I.S.; Pozdniakov, A.V.; Mosleh, A.O.; Solonin, A.N. Evaluation of the Microstructure and Mechanical Properties of a New Modified Cast and Laser-Melted AA7075 Alloy. *Materials* **2019**, *12*, 3430. [CrossRef]
10. Ahmed, F.; Srivastava, S.; Agarwal, A.B. Synthesis & Characterization of Al-Ti-Cr MMC as friction material for disc brakes application. *Mater. Today Proc.* **2017**, *4*, 405–414. [CrossRef]
11. Mhaske, M.S.; Shirsat, D.U.M. Investigations of Tribological Behaviour of Al-SiC MMC for Automobile Brake Pad. In Proceedings of the TRIBOINDIA-2018: An International Conference on Tribology, Mumbai, India, 13–15 December 2018. [CrossRef]
12. Kolli, M.; Devaraju, A.; Saikumar, G.; Kosaraju, S. Electrical Discharge Machining of SiC Reinforced 6061-T6 Aluminum Alloy Surface Composite Fabricated by Friction Stir Processing. *E3S Web Conf.* **2021**, *309*, 01044. [CrossRef]
13. Venkatachalam, G.; Arumugam, K. Mechanical Behaviour of Aluminium Alloy Reinforced with Sic/Fly Ash/Basalt Composite for Brake Rotor. *Polym. Polym. Compos.* **2017**, *25*, 203–208. [CrossRef]
14. Firouz, F.M.; Mohamed, E.; Lotfy, A.; Daoud, A.; Abou El-Khair, M.T. Thermal expansion and fatigue properties of automotive brake rotor made of AlSi–SiC composites. *Mater. Res. Express* **2020**, *6*, 1265d1262. [CrossRef]
15. P, S.; Natarajan, H.K.; Kumar J., P. Study of silicon carbide-reinforced aluminum matrix composite brake rotor for motorcycle application. *Int. J. Adv. Manuf. Technol.* **2018**, *94*, 1461–1475. [CrossRef]
16. Daoud, A.; Abou El-khair, M.T. Wear and friction behavior of sand cast brake rotor made of A359-20vol% SiC particle composites sliding against automobile friction material. *Tribol. Int.* **2010**, *43*, 544–553. [CrossRef]
17. Hekner, B.; Myalski, J.; Valle, N.; Botor-Probierz, A.; Sopicka-Lizer, M.; Wieczorek, J. Friction and wear behavior of Al-SiC(n) hybrid composites with carbon addition. *Compos. Part B Eng.* **2017**, *108*, 291–300. [CrossRef]
18. Sika, R.; Rogalewicz, M.; Popielarski, P.; Czarnecka-Komorowska, D.; Przestacki, D.; Gawdzińska, K.; Szymański, P. Decision Support System in the Field of Defects Assessment in the Metal Matrix Composites Castings. *Materials* **2020**, *13*, 3552. [CrossRef]
19. Sharma, V.; Singla, Y.; Gupta, Y.; Raghuwanshi, J. Post-processing of metal matrix composites by friction stir processing. *AIP Conf. Proc.* **2018**, *1953*, 090062. [CrossRef]
20. Fadavi Boostani, A.; Tahamtan, S.; Jiang, Z.Y.; Wei, D.; Yazdani, S.; Azari Khosroshahi, R.; Taherzadeh Mousavian, R.; Xu, J.; Zhang, X.; Gong, D. Enhanced tensile properties of aluminium matrix composites reinforced with graphene encapsulated SiC nanoparticles. *Compos. Part A Appl. Sci. Manuf.* **2015**, *68*, 155–163. [CrossRef]
21. Zhou, D.; Qiu, F.; Jiang, Q. The nano-sized TiC particle reinforced Al-Cu matrix composite with superior tensile ductility. *Mater. Sci. Eng. A* **2015**, *622*, 189–193. [CrossRef]

22. Liu, Q.; Ke, L.; Liu, F.; Huang, C.; Xing, L. Microstructure and mechanical property of multi-walled carbon nanotubes reinforced aluminum matrix composites fabricated by friction stir processing. *Mater. Des.* **2013**, *45*, 343–348. [CrossRef]
23. Velavan, K.; Palanikumar, K. Effect of Silicon Carbide (SiC) on Stir Cast Aluminium Metal Matrix Hybrid Composites—A Review. *Appl. Mech. Mater.* **2015**, *766–767*, 293–300. [CrossRef]
24. Eskandari, H.; Taheri, R. A Novel Technique for Development of Aluminum Alloy Matrix/TiB2/Al2O3 Hybrid Surface Nanocomposite by Friction Stir Processing. *Procedia Mater. Sci.* **2015**, *11*, 503–508. [CrossRef]
25. Moustafa, E.B.; Mikhaylovskaya, A.V.; Taha, M.A.; Mosleh, A.O. Improvement of the microstructure and mechanical properties by hybridizing the surface of AA7075 by hexagonal boron nitride with carbide particles using the FSP process. *J. Mater. Res. Technol.* **2022**, *17*, 1986–1999. [CrossRef]
26. Avettand-Fènoël, M.-N.; Simar, A. A review about Friction Stir Welding of metal matrix composites. *Mater. Charact.* **2016**, *120*, 1–17. [CrossRef]
27. Zuo, L.; Zhao, X.; Li, Z.; Zuo, D.; Wang, H. A review of friction stir joining of SiCp/Al composites. *Chin. J. Aeronaut.* **2020**, *33*, 792–804. [CrossRef]
28. Vijayavel, P.; Rajkumar, I.; Sundararajan, T. Surface characteristics modification of lm25 aluminum alloy–5% sic particulate metal matrix composites by friction stir processing. *Met. Powder Rep.* **2021**, *76*, 140–151. [CrossRef]
29. Vijayavel, P.; Sundararajan, T.; Rajkumar, I.; Ananthakumar, K. Effect of tool diameter ratio of tapered cylindrical profile pin on wear characteristics of friction stir processing of Al-Si alloy reinforced with SiC ceramic particles. *Met. Powder Rep.* **2021**, *76*, 75–89. [CrossRef]
30. Mohamadigangaraj, J.; Nourouzi, S.; Jamshidi Aval, H. Statistical modelling and optimization of friction stir processing of A390-10 wt% SiC compo-cast composites. *Measurement* **2020**, *165*, 108166. [CrossRef]
31. Kumar, A.; Pal, K.; Mula, S. Simultaneous improvement of mechanical strength, ductility and corrosion resistance of stir cast Al7075-2% SiC micro- and nanocomposites by friction stir processing. *J. Manuf. Process.* **2017**, *30*, 1–13. [CrossRef]
32. Kurtyka, P.; Rylko, N.; Tokarski, T.; Wójcicka, A.; Pietras, A. Cast aluminium matrix composites modified with using FSP process – Changing of the structure and mechanical properties. *Compos. Struct.* **2015**, *133*, 959–967. [CrossRef]
33. Butola, R.; Tyagi, L.; Singari, R.M.; Murtaza, Q.; Kumar, H.; Nayak, D. Mechanical and wear performance of Al/SiC surface composite prepared through friction stir processing. *Mater. Res. Express* **2021**, *8*, 016520. [CrossRef]
34. Aruri, D.; Adepu, K.; Adepu, K.; Bazavada, K. Wear and mechanical properties of 6061-T6 aluminum alloy surface hybrid composites [(SiC+Gr) and (SiC+Al2O3)] fabricated by friction stir processing. *J. Mater. Res. Technol.* **2013**, *2*, 362–369. [CrossRef]
35. Devaraju, A.; Kumar, A.; Kumaraswamy, A.; Kotiveerachari, B. Influence of reinforcements (SiC and Al2O3) and rotational speed on wear and mechanical properties of aluminum alloy 6061-T6 based surface hybrid composites produced via friction stir processing. *Mater. Des.* **2013**, *51*, 331–341. [CrossRef]
36. Rana, H.G.; Badheka, V.J.; Kumar, A. Fabrication of Al7075/B4C surface composite by novel friction stir processing (FSP) and investigation on wear properties. *Procedia Technol.* **2016**, *23*, 519–528. [CrossRef]
37. Shin, K.; Brennan, M.J.; Oh, J.E.; Harris, C.J. Analysis of disc brake noise using a two-degree-of-freedom model. *J. Sound Vib.* **2002**, *254*, 837–848. [CrossRef]
38. Brake vibration and noise: Reviews, comments, and proposals. *Int. J. Mater. Prod. Technol.* **1997**, *12*, 496–513. [CrossRef]
39. Xia, P.; Wu, Y.; Yin, T.; Xie, K.; Tan, Y. Formation mechanism of TiC–Al2O3 ceramic reinforcements and the influence on the property of ZL101 composites. *Ceram. Int.* **2022**, *48*, 2577–2584. [CrossRef]
40. Luo, X.; Han, Y.; Li, Q.; Hu, X.; Xue, L. Microstructure and Properties of ZL101 Alloy Affected by Substrate Movement Speed of a Novel Semisolid Continuous Micro Fused-Casting for Metal Process. *J. Wuhan Univ. Technol. -Mater. Sci. Ed.* **2018**, *33*, 715–719. [CrossRef]
41. Cheng, S.-j.; Zhao, Y.-h.; Hou, H.; Jin, Y.-c.; Guo, X.-x. Preparation of semi-solid ZL101 aluminum alloy slurry by serpentine channel. *Trans. Nonferrous Met. Soc. China* **2016**, *26*, 1820–1825. [CrossRef]
42. Cui, H.-c.; Lu, F.-g.; Peng, K.; Tang, X.-h.; Yao, S. Research on electron beam welding of in situ TiB2p/ZL101 composite. *J. Shanghai Jiaotong Univ.* **2010**, *15*, 479–483. [CrossRef]
43. Wan, J.; Yan, H.; Xu, D. Rheological study of semi-solid TiAl3/ZL101 composites prepared by ultrasonic vibration. *Int. J. Mater. Res.* **2015**, *106*, 1244–1249. [CrossRef]
44. Chen, L.; Zhao, Y.; Yan, F.; Hou, H. Statistical investigations of serpentine channel pouring process parameters on semi-solid ZL101 aluminum alloy slurry using response surface methodology. *J. Alloys Compd.* **2017**, *725*, 673–683. [CrossRef]
45. Rabinowicz, E.; Tanner, R.I. Friction and Wear of Materials. *J. Appl. Mech.* **1966**, *33*, 479. [CrossRef]
46. Saadatmand, M.; Mohandesi, J.A. Comparison Between Wear Resistance of Functionally Graded and Homogenous Al-SiC Nanocomposite Produced by Friction Stir Processing (FSP). *J. Mater. Eng. Perform.* **2014**, *23*, 736–742. [CrossRef]
47. Manickam, A.; Kuppusamy, R.; Jayaprakasham, S.; Santhanam, S.K.V. Multi Response Optimization of Friction Stir Process Parameters on AA2024/SiC Composite Fabricated Using Friction Stir Processing. In Proceedings of the ASME 2021 International Mechanical Engineering Congress and Exposition, Online, 1–5 November 2021.
48. Cui, G.; Qian, Y.; Bian, C.; Gao, G.; Hassani, M.; Liu, Y.; Kou, Z. CoCrNi matrix high-temperature wear resistant composites with micro- and nano-Al2O3 reinforcement. *Compos. Commun.* **2020**, *22*, 100461. [CrossRef]
49. Roy, P.; Singh, S.; Pal, K. Enhancement of mechanical and tribological properties of SiC- and CB-reinforced aluminium 7075 hybrid composites through friction stir processing. *Adv. Compos. Mater.* **2019**, *28*, 1–18. [CrossRef]

50. Kumar, A.; Mahapatra, M.M.; Jha, P.K. Modeling the abrasive wear characteristics of in-situ synthesized Al–4.5%Cu/TiC composites. *Wear* **2013**, *306*, 170–178. [CrossRef]
51. Gangil, N.; Maheshwari, S.; Siddiquee, A.N. Novel Use of Distribution Facilitators and Time–Temperature Range for Strengthening in Surface Composites on AA7050-T7451. *Metallogr. Microstruct. Anal.* **2018**, *7*, 561–577. [CrossRef]
52. Ostovan, F.; Amanollah, S.; Toozandehjani, M.; Shafiei, E. Fabrication of Al5083 surface hybrid nanocomposite reinforced by CNTs and Al2O3 nanoparticles using friction stir processing. *J. Compos. Mater.* **2019**, *54*, 1107–1117. [CrossRef]
53. Huang, G.; Hou, W.; Li, J.; Shen, Y. Development of surface composite based on Al-Cu system by friction stir processing: Evaluation of microstructure, formation mechanism and wear behavior. *Surf. Coat. Technol.* **2018**, *344*, 30–42. [CrossRef]
54. Muratoğlu, M.; Aksoy, M. Abrasive wear of 2124Al–SiC composites in the temperature range 20–200°C. *J. Mater. Process. Technol.* **2006**, *174*, 272–276. [CrossRef]
55. Kumar, S.; Panwar, R.S.; Pandey, O.P. Effect of dual reinforced ceramic particles on high temperature tribological properties of aluminum composites. *Ceram. Int.* **2013**, *39*, 6333–6342. [CrossRef]
56. Rosenfield, A.R. A shear instability model of sliding wear. *Wear* **1987**, *116*, 319–328. [CrossRef]
57. Alexeyev, N.M. On the motion of material in the border layer in solid state friction. *Wear* **1990**, *139*, 33–48. [CrossRef]
58. Tyagi, L.; Butola, R.; Jha, A.K. Mechanical and tribological properties of AA7075-T6 metal matrix composite reinforced with ceramic particles and aloevera ash via Friction stir processing. *Mater. Res. Express* **2020**, *7*, 066526. [CrossRef]

Disclaimer/Publisher's Note: The statements, opinions and data contained in all publications are solely those of the individual author(s) and contributor(s) and not of MDPI and/or the editor(s). MDPI and/or the editor(s) disclaim responsibility for any injury to people or property resulting from any ideas, methods, instructions or products referred to in the content.

Article

Effect of Rubidium on Solidification Parameters, Structure and Operational Characteristics of Eutectic Al-Si Alloy

Igor A. Petrov [1], Anastasiya D. Shlyaptseva [1], Alexandr P. Ryakhovsky [1], Elena V. Medvedeva [2] and Victor V. Tcherdyntsev [3,*]

1. Moscow Aviation Institute, Orshanskaya 3, 125993 Moscow, Russia; petrovia2@mai.ru (I.A.P.); shlyaptsevaad@mai.ru (A.D.S.); ryahovskiyap@mai.ru (A.P.R.)
2. Institute of Electrophysics, Ural Branch, Russian Academy of Science, Amudsena Str., 106, 620016 Yekaterinburg, Russia; lena@iep.uran.ru
3. Laboratory of Functional Polymer Materials, National University of Science and Technology "MISIS", Leninskii prosp, 4, 119049 Moscow, Russia
* Correspondence: vvch@misis.ru; Tel.: +7-9104002369

Abstract: Modification of the eutectic silicon in Al–Si alloys causes a structural transformation of the silicon phase from a needle-like to a fine fibrous morphology and is carried out extensively in the industry to improve mechanical properties of the alloys. The theories and mechanisms explaining the eutectic modification in Al–Si alloys are considered. We discuss the mechanism of eutectic rubidium modification in the light of experimental data obtained via quantitative X-ray spectral microanalysis and thermal analysis. X-ray mapping revealed that rubidium, which theoretically satisfies the adsorption mechanisms of silicon modification, had an effect on the silicon growth during solidification. Rubidium was distributed relatively homogeneously in the silicon phase. Microstructural studies have shown that rubidium effectively refines eutectic silicon, changing its morphology. Modification with rubidium extends the solidification range due to a decrease in the solidus temperature. The highest level of mechanical properties of the alloy under study was obtained with rubidium content in the range of 0.007–0.01%. We concluded that rubidium may be used as a modifier in Al–Si eutectic and pre-eutectic alloys. The duration of the modifying effect of rubidium in the Al-12wt%Si alloy melt and porosity in the alloy modified with rubidium were evaluated.

Keywords: cast aluminum alloys; modification; rubidium; microstructure; eutectic silicon; solidification process; porosity

1. Introduction

Cast aluminum alloys are used in the automotive and aerospace industries and occupy a special position among structural materials. This is because of the possibility of achieving an optimal combination of basic operational properties (strength, ductility, corrosion resistance, density, etc.) with technological properties, including excellent casting characteristics.

The phase composition, the structural components and the nature of solidification of any cast aluminum alloy are the most important criteria that determine either the operational or technological properties [1].

During the solidification of an alloy, its internal structure is formed, which is one of the determining factors of its operational properties [2]. Solidification is a complex physical and chemical process that can proceed at different rates and under the influence of various external factors. The course of solidification is affected by the physical characteristics of an alloy and the cooling conditions of the casting. Technological factors, such as the temperature of casting the alloy and various types of melt treatment (including modification), significantly affect the solidification process.

In most cases, modification can be considered as the adding of modifying additives into an alloy—elements are added in an amount from 0.001 to 0.3%. Generally, a large number of elements that have a modifying effect on the structure of Al-Si alloys are known.

Modifying elements change the characteristics of the origin, growth and shape of eutectic silicon crystals. Such elements primarily include some alkaline (Na [3–5] and K [6–8]) and alkaline earth metals (Sr [9–11], Ba [12–14] and Ca [12,15,16]). To date, the modifying effect of a number of REMs (Y [12,17], La [18], Ce [19], Sm [20], Eu [21], Ho [22], Er [23,24] and Yb [12,24,25]) on eutectic silicon crystals is known. Such elements are introduced into aluminum alloys with Si content from 6 to 13%, where the eutectic is the main structural component of the alloy. To explain the modification of Al-Si alloys, the theory of supercooling by Edwards and Archer [26] as well as adsorption theories considering the adsorption mechanisms of silicon modification are used. Thus, according to the "Twin plane re-entrant edge" (TPRE) "poisoning" mechanism proposed by Day and Hellawell [27], the modifying element is adsorbed at active growth points in the re-entrant edges of twins, retarding the growth of silicon crystals and changing their growth directions. This deactivates the TPRE mechanism, causing silicon crystals to grow in a more isotropic manner. The basic idea of the "Impurity-induced twinning" (IIT) mechanism proposed by Lu and Hellawell [28] is that the atoms of the modifying element are absorbed on the steps of a growing silicon crystal at the interface between the solid and liquid phases. This leads to the formation of new twins and consequently ensures their growth in other directions. According to the IIT mechanism, only elements showing the "ideal" ratio of atomic radii $r_i/r \sim 1.646$ (where r_i is the atomic radius of the element and r is the radius of silicon) are modifiers.

The interest of researchers in the processes of structure formation during modification of Al-Si alloys has led to the advancement of modification technologies and the production of high-quality castings as well as the development and confirmation of various modification theories.

Of great interest is the use of Rb for modifying Al-Si alloys. There is insufficient literature data on the use of rubidium as a modifier in aluminum–silicon alloys. Rubidium is a chemically active alkali metal with a melting point of 39.3 °C [29]. It is close in its physical and chemical properties to sodium; therefore, it is assumed that the modifying effect of rubidium on a eutectic (α + Si) is similar to the action of sodium. Based on the adsorption theory of modification, the most effective modifiers that refine a eutectic (α + Si) are elements with a low surface tension. Rubidium has a surface tension equal to 83 mN/m [30]. It is significantly lower than the surface tension of aluminum (914 mN/m) and silicon (865 mN/m) [30].

Rubidium has not received practical application in the foundry of eutectic Al-Si alloys. This is due to the complexity of adding this metal into the melt, and the lack of information regarding its effect on the properties of Al-Si alloys. Data on its modifying effect on Al-Si alloys are contradictory [6,28,31]. Rubidium in its pure form is difficult to add to the melt; therefore, it was decided to use salt to add it.

The effect of rubidium on the structure, mechanical properties and solidification of a eutectic Al-Si alloy was examined in our study. The duration of the modifying effect of rubidium in the Al-Si alloy melt was determined, and the porosity of an alloy modified with rubidium was evaluated.

2. Materials and Methods

2.1. Materials and Modification Technology

The object of the study was a eutectic Al-12wt%Si alloy. The chemical compositions of the experimental alloys are presented in Table 1.

Table 1. Chemical compositions of experimental alloys.

Al-12wt%Si Alloy	Estimated Quantities Rb %	Mass Fraction, % (Al-Base)						
		Si	Cu	Mn	Ti	Zn	Fe	Rb
Unmodified	—	11.42	0.0020	0.002	0.007	0.0097	0.22	—
Alloy 1	0.1	11.39	0.0017	0.0023	0.004	0.0090	0.22	0.002
Alloy 2	0.3	11.51	0.0013	0.002	0.006	0.01	0.26	0.0052
Alloy 3	0.5	11.36	0.0021	0.0026	0.006	0.0092	0.20	0.0075
Alloy 4	1.0	11.48	0.0018	0.0042	0.009	0.01	0.25	0.01

Salt ($RbNO_3$) was chosen as the compound to add rubidium into the melt. The estimated amounts of rubidium added to the alloy were chosen to be equal to 0.1%, 0.3%, 0.5%, and 1.0% from the weight of the melt. The salt was dried at 150–200 °C for 2 h before melting.

Experimental melting was carried out in an electric resistance furnace. The weight of one melt was 1 kg. The melt was previously degassed with argon. Rubidium salt ($RbNO_3$) was added into the melt at the bottom of the crucible at a temperature of $750 \pm 5°$ with the help of a "bell". The treatment of the melt with the salt was accompanied by bubbling. Then, the melt stood for 10 min, slag was removed from its surface, and samples of the experimental alloys were poured into a sandy-clay mold at a temperature equal to 710 ± 5 °C. Samples were cast separately to determine porosity.

The weight of the melt to determine the duration of the modification effect was equal to 5 kg. The melt was kept at a constant temperature equal to $730 \pm 5°C$ for up to 240 min. In the process of holding the melt, samples were poured into a sandy-clay mold at the time intervals of 15, 30, 45, 60, 120, 180 and 240 min.

2.2. Methods of Studying the Microstructure and Mechanical Properties

The mechanical properties of the experimental alloys (ultimate strength and relative elongation) were determined in accordance with ASTM B557M-15 on an Instron 5982 testing machine (Instron, Norwood, MA, USA). In each experiment, 4 samples were tested. Each experiment was repeated twice.

Microstructural studies were carried out on a Carl Zeiss-brand Imager.Z2m AXIO universal research motorized microscope (Carl Zeiss, Microscopy GmbH, Göttingen, Germany).

Quantitative analysis of the microstructure (average length, area of eutectic silicon and α-Al dendrites) was carried out using the specialized program ImageExpert Pro 3.7, version 3.7.5.0, NEXSYS, (Moscow, Russia) over three images per each sample. The photographs were processed through the use of the following operations: changing the size of the photograph, selecting the scale, binarization, determining the object of study by color, etc. For image processing, the ImageExpert Pro 3.7 program uses built-in algorithms (the methods correspond to the ASTM E112-10 international standard).

To measure the size of the α-Al dendrite, the secondary dendrite arm spacing (SDAS) was determined. The SDAS was evaluated (as reported in [32]) by measuring thirty dendrites for each sample using three images at $50\times$ magnification.

The microstructure and elemental composition and distribution of modifying elements in the structure were studied using a Phenom XL scanning electron microscope (SEM) with an integrated energy-dispersive spectrometry (EDS) system (Phenom-World BV, Eindhoven, Netherlands). To determine the elemental composition at a point and according to area, the method of quantitative X-ray spectral microanalysis was used using special Phenom Element Identification v. 3.8.0.0 software. To carry out X-ray mapping, the Elemental Mapping software was used.

2.3. Methods of Chemical and Thermal Analysis

The chemical (elemental) composition of the samples was studied using a CCD-based Q4 TASMAN-170 spark optical emission spectrometer. The Q4 TASMAN-170 spectrometer was controlled from a desktop computer using special QMatrix software version 3.8.1 (Bruker Quantron GmbH, Kalkar, Germany).

The actual rubidium content was obtained using an ICAP 6300 inductively coupled plasma atomic emission spectrometer (ICP-AES) (Thermo Fisher Corporation, Cambridge, UK).

Thermal analysis of the alloys' solidification was carried out using a Netzsch DSC404 F3 Pegasus differential scanning calorimeter (Netzsch-Geratebau GmbH, Bavaria, Germany). The samples were subjected to exposure at room temperature for 24 h before the test. The analyzed sample and the platinum standard were placed with thermocouples (Pt–Rh) in platinum crucibles and put into the heating chamber of the installation. The tests were carried out in an inert argon atmosphere. NETZSCH Proteus® v. 7.1 software was used to process the data obtained (Netzsch-Geratebau GmbH, Bavaria, Germany).

2.4. Porosity Investigation Method

The porosity of the alloys was evaluated on a five-level standard scale in macrosections of samples cut from castings in sandy-clay molds in accordance with ISO 10049:2019(E) [33]. To disclose the pores, the sample was sanded and etched in a 20% NaOH solution in accordance with ISO 10049:2019(E).

3. Results

3.1. Thermodynamic Analysis of the Interaction of Rubidium Nitrate with Melt

The study of an alloy modification with rubidium using $RbNO_3$ salt is impossible without a preliminary thermodynamic analysis. In this regard, the thermodynamic probability of the salt-decomposition process and its interaction with the melt up to a temperature of 1200 K (927 °C) was determined. Calculations of the free reaction energy were carried out while taking into account phase transformations based on the literature data [34,35].

It is known [34] that rubidium nitrate $RbNO_3$ is a low-melting salt (melting point t = 313 °C) and that when heated above the melting point, it decomposes to form a nitrite and oxygen according to Reaction (1):

$$RbNO_3 = RbNO_2 + 0.5O_2 \tag{1}$$

The resulting $RbNO_2$ reacts with aluminum and is reduced to rubidium according to Reaction (2):

$$RbNO_2 + \frac{4}{3}Al = \frac{2}{3}Al_2O_3 + \frac{1}{2}N_2 + Rb \tag{2}$$

At the same time, direct interaction of the nitrate with aluminum according to Reaction (3) with the formation of rubidium is also possible:

$$RbNO_3 + 2Al = Al_2O_3 + \frac{1}{2}N_2 + Rb \tag{3}$$

According to the calculations obtained, the free reaction energy for Reactions (2) and (3) at the temperatures of 600–1200 K was negative (Figure 1). Consequently, at the salt input temperature of 1023 K (750 °C), the formation of rubidium in the aluminum melt was possible according to Reactions (2) and (3), for which the energy of the system reached values of −432 and −367 kJ/mol, respectively.

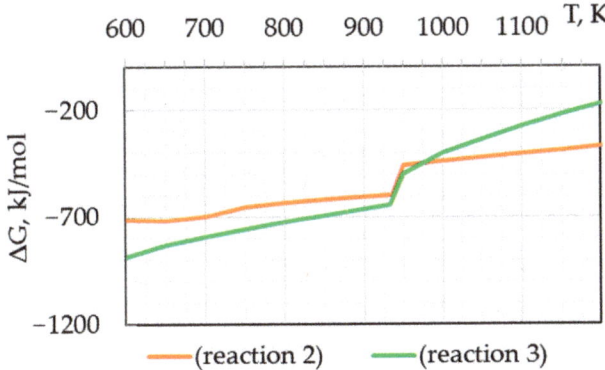

Figure 1. Temperature dependences of the free reaction energy of Reactions (2) and (3) on temperature.

3.2. Mechanical Properties

The results of mechanical tests of the Al-12wt%Si alloy after modification with rubidium are presented in Table 2. It was experimentally established that adding rubidium caused an increase in the mechanical properties (ultimate strength σ_B and relative elongation δ).

Table 2. Mechanical properties of Al-12wt%Si alloy and average dimensions of eutectic Si and SDAS.

Al-12wt%Si Alloy	Features			
	σ_v, MPa	δ, %	$l_{Si\ eut}$, µm	SDAS, µm
Unmodified	145 ± 2	2.6 ± 0.2	14 − 49.4	42 ± 10
Alloy 1	149 ± 2	4.1 ± 0.1	5.42 − 28.58	40 ± 9
Alloy 2	166 ± 3	5.5 ± 0.4	$\frac{1.88 - 7.88}{0.7 - 22}$	37 ± 10
Alloy 3	169 ± 1	9.3 ± 0.5	$\frac{0.36 - 2.29}{0.2 - 7.6}$	37 ± 9
Alloy 4	171 ± 2	8.6 ± 0.5	$\frac{0.72 - 3.97}{0.29 - 11.7}$	39 ± 8

Note: the numerator shows the average size of the structural components and the denominator shows the minimum and maximum size.

The largest increase in the mechanical properties of the alloy 3 is explained by the modified microstructure of the alloy. An increase in the Rb content in alloy 4 did not lead to a further increase in mechanical properties, which remained at the level of alloy 3.

3.3. Microstructural Studies

The results of the optical microscopy of the structures of the unmodified Al-12wt%Si alloy and those modified with different rubidium content are shown in Figure 2. For alloy 3, the results of the X-ray spectral microanalysis and EDS elementary mapping are presented in Figure 3.

In the Al-12wt%Si alloy under study, the silicon concentration was 11.4%. The microstructure of the unmodified alloy was characterized as coarse eutectic ($l_{Si\ eut}$ = 14–49.4 µm) with large α-Al dendrites (SDAS 42 µm). Due to the presence of impurities and low cooling rates (casting into a sandy-clay mold), primary silicon crystals could be observed in the structure of the alloy under study. The iron concentration in the alloy was about 0.3%, which can lead to the β-FeSiAl$_5$ phase formation.

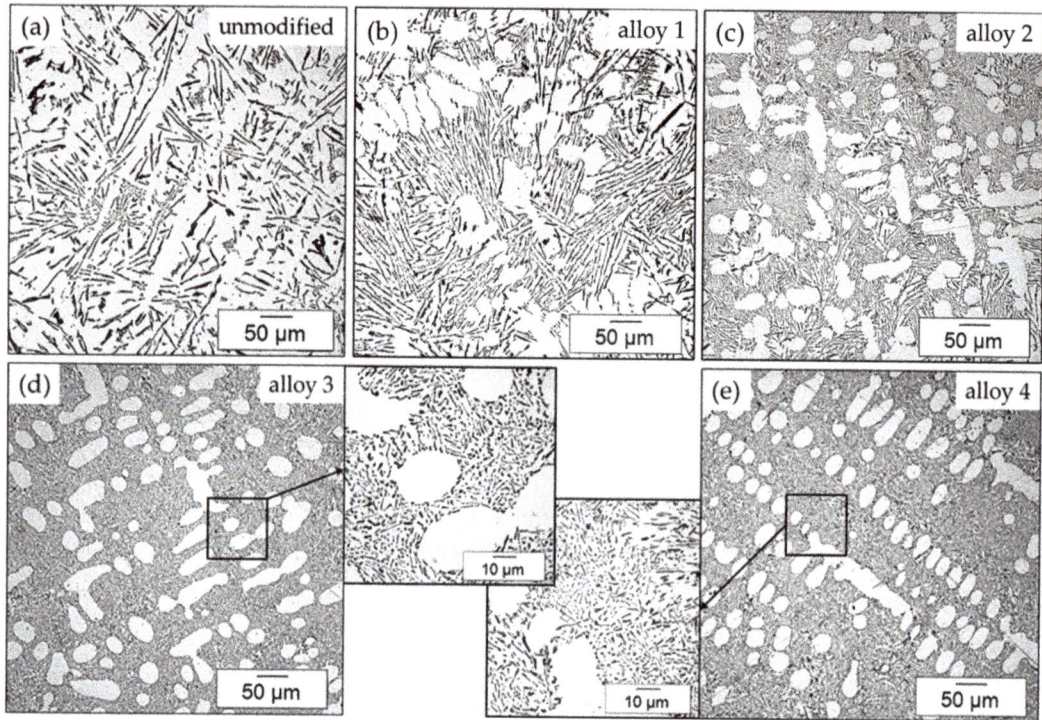

Figure 2. Microstructure of Al-12wt%Si alloys: (**a**) unmodified; (**b**) alloy 1; (**c**) alloy 2; (**d**) alloy 3; (**e**) alloy 4.

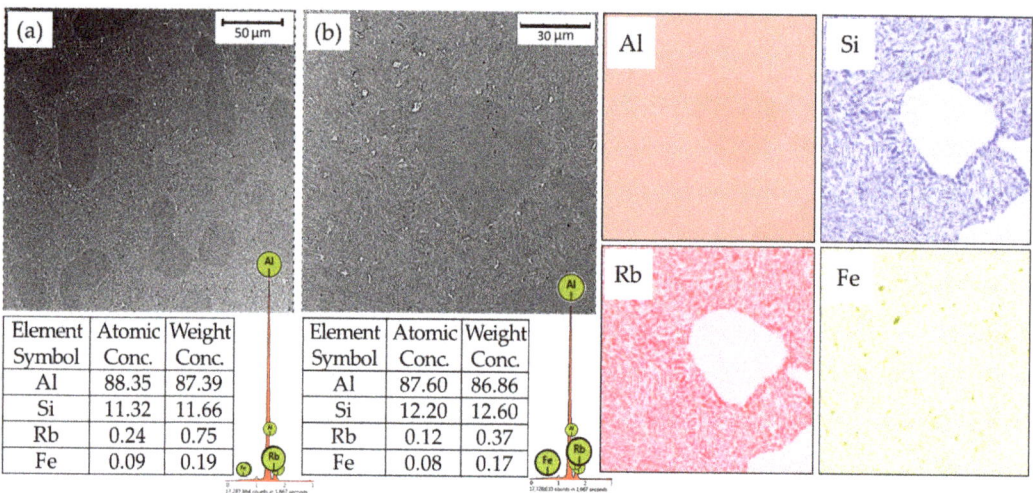

Figure 3. *Cont.*

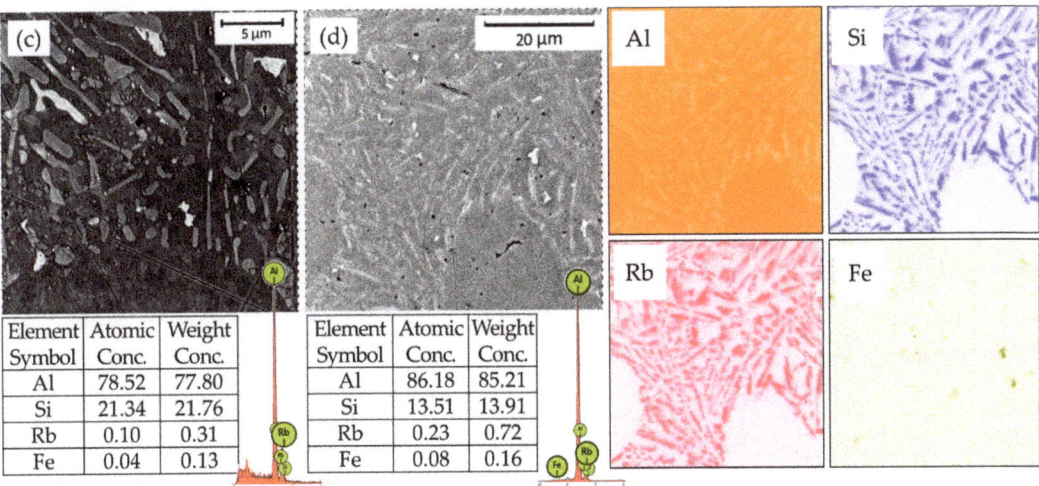

Figure 3. Energy-dispersive X-ray spectral microanalysis of a sample of alloy 3 and elementary mapping of the studied area of a sample modified with rubidium. (**a**) Magnification at 1100× (composition according to area); (**b**) magnification at 2150× (composition according to area and mapping according to elements); (**c**) magnification at 11,500× (composition according to area); (**d**) magnification at 4300× (composition according to area and mapping according to elements).

Studies of the microstructure of alloy 1 showed that modification with a minimal amount of rubidium led to a change in morphology, a decrease in the linear dimensions of the eutectic silicon length to 5.42–22.58 µm, and a decrease in α-Al dendrites by 5%. With an increase in the rubidium content in alloys 2 and 3, the respective linear dimensions of the eutectic silicon decreased to 1.33–5.02 and 0.36–2.29 µm (Table 2). As a result, the Si particles acquired a finer fibrous morphology, as shown in Figure 2d,e. An increase in the rubidium content in alloy 4 did not lead to greater refinement of eutectic silicon (Figure 2d). Different contents of rubidium weakly affected the size of the α-Al dendrites, and the SDAS in the alloy decreased by an average of 11–12%. Therefore, in order to obtain a modified eutectic in the alloy under study, it was necessary to ensure the actual rubidium content in the range of 0.007–0.01 wt %. Therefore, further studies were carried out with an estimated rubidium content of 0.5% (corresponding to 0.0075 wt %).

Based on the results of studying the microstructure of the sample of alloy 3, the presence of rubidium in the studied samples was studied using X-ray spectral microanalysis over the area of the image. The structure, X-ray spectrum and composition of the elements are shown in Figure 3a,c.

To confirm the elemental composition, elementary mapping of a sample of alloy 3 modified with rubidium was carried out. The elementary maps of the rubidium-modified sample in the eutectic area (containing aluminum, silicon and the Fe-containing phase) obtained by scanning at a low resolution are shown in Figure 3b, and a higher-resolution map is shown in Figure 3d. The mapping results showed that rubidium atoms were present in the eutectic silicon and had a low concentration.

3.4. Thermal Analysis

The solidification parameters of the alloys under study (depending on the melt treatment) are presented in Table 3.

Table 3. Solidification parameters of the studied alloys.

Al-12wt%Si,	Solidification Parameters, Temperature (°C)			
	Liquidus (t_{liq})	Solidus (t_{sol})	Solidification Range (Δt)	Beginning of Solidification of Eutectic ($t_{eu.sol.st}$)
Unmodified	582.0	557.5	24.2	576.0
Alloy 1	581.0	551.0	30.0	571.9
Alloy 2	581.5	549.1	32.2	570.6
Alloy 3	581.8	539.3	42.5	568.4
Alloy 4	581.7	538.2	43.5	567.9

A DSC curves analysis (Figure 4) showed two-stage solidification of the Al-12wt%Si alloy samples from the liquid state up to complete solidification. In the first stage, the nucleation of aluminum solid solution crystals and their growth occurred, which was recorded in the DSC thermogram as the first thermal effect. The second thermal effect (second stage) was caused by the solidification of the binary eutectic according to the reaction L→α + Si. Since the alloy contained an iron impurity, during solidification of this alloy, not only the formation of binary eutectic (α + Si) was possible but also a more complex iron-containing eutectic L→α + Si +Al$_5$FeSi could form. The solidification temperatures of these eutectics were close [1].

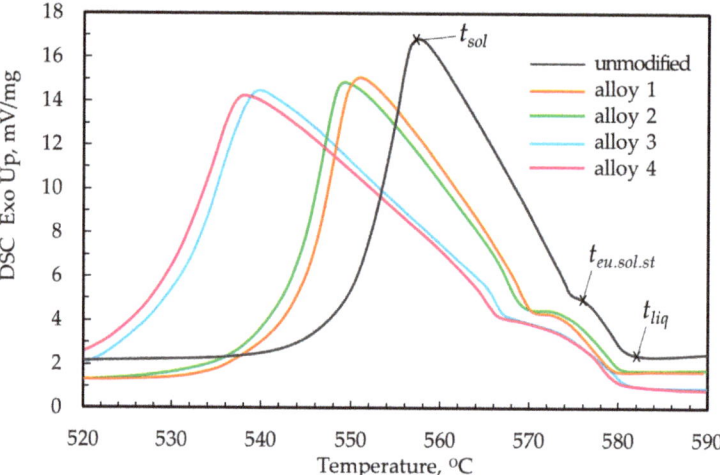

Figure 4. Thermograms of experimental alloys.

In the alloys studied, the solidification parameters changed with an increase in the rubidium concentration. Alloy modification with rubidium decreased the solidus temperatures and the temperatures at which the solidification of the eutectic began compared to the unmodified alloy. As a result, the solidification range of the Al-12wt%Si alloys under study expanded (Table 3). According to the data obtained (Figure 4), the most pronounced change in solidification parameters occurred in alloys 3 and 4.

An analysis of the alloy 3 sample's solidification parameters showed that rubidium did not change the liquidus temperature, which remained at the level of the unmodified alloy, while at the same time the solidus temperature decreased by 18.2 °C. The solidification range of the alloy with rubidium under study expanded by 18.3 °C compared to the unmodified alloy. It should be noted that the temperature of the beginning of the second thermal effect on the alloys was considered as the temperature of the beginning of solidification of the eutectic (Table 3). Rubidium contributed to a significant decrease in the temperature of the beginning of the solidification of the eutectic (α + Si) compared to the unmodified alloy (equal to 7.6 °C).

3.5. Study of the Effect of Rubidium on the Porosity of the Alloy

A qualitative assessment of the effect of rubidium on the formation of gas porosity in the experimental alloys at different concentrations of rubidium was carried out (Figure 5).

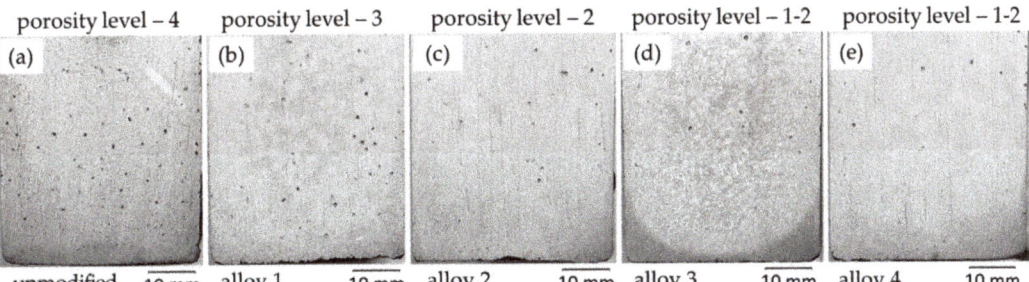

Figure 5. Porosity distribution in Al-12wt%Si alloy castings when modified with rubidium: (**a**) unmodified; (**b**) alloy 1; (**c**) alloy 2; (**d**) alloy 3; (**e**) alloy 4.

The studies showed that the samples of the Al-12wt%Si alloys treated with rubidium had a decrease in gas porosity compared to an unmodified alloy, the porosity of which corresponded to four levels on the standard ISO 10049:2019(E) porosity scale. In alloy 1, the porosity corresponds to level 3, and with an increase in the rubidium content in alloy 4, the porosity decreased to level 1 according to the porosity scale (Figure 5).

Studies were also conducted on the effect of modification of the Al-12wt%Si alloy with rubidium on the manifestation of volumetric shrinkage of the experimental alloys. The formation of shrinkage defects in a massive element, which was the thermal center of the casting, was studied. Since the unmodified Al-12wt%Si alloy had a narrow crystallization interval of 24.2 degrees, it crystallized frontally (successively) (Figure 6a). In this case, a shrinkage cavity was formed on the upper surface of the sample, and the shrinkage porosity was formed on the opposite side. Alloy 3 modified with rubidium had a larger crystallization interval that equaled 42.5 degrees. Therefore, it was more prone to bulk crystallization. At the same time, a shrinkage cavity was formed at the top, and distributed porosity was formed below it (Figure 6b).

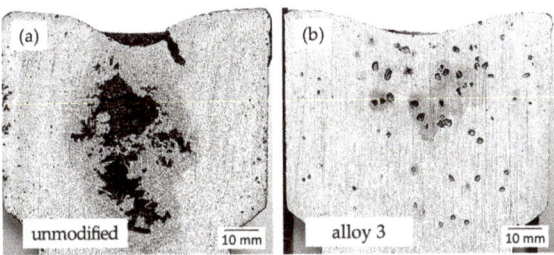

Figure 6. The manifestation of volumetric shrinkage in the Al-12wt%Si alloy: (**a**) unmodified; (**b**) modified with Rb.

3.6. Evaluation of the Duration of the Modifying Effect of Rubidium in the Melt

Based on the studies shown above, alloy 3 was selected to determine the duration of the modifying effect. Treatment of the melt with rubidium led to an increase in the duration of the effect of modifying the Al-Si alloy (Figure 7a,b) compared to sodium modifiers [7].

During the first 30 min after the melt treatment, the σ value of the Al-12wt%Si alloy modified with rubidium was 176 MPa. However, with a longer holding time of up to 60 and 120 min, the ultimate strength of the rubidium-modified alloy decreased slightly to 173 and 170 MPa, respectively. With further holding of the melt up to 240 min, the ultimate strength of the rubidium-modified alloy decreased to 155 MPa.

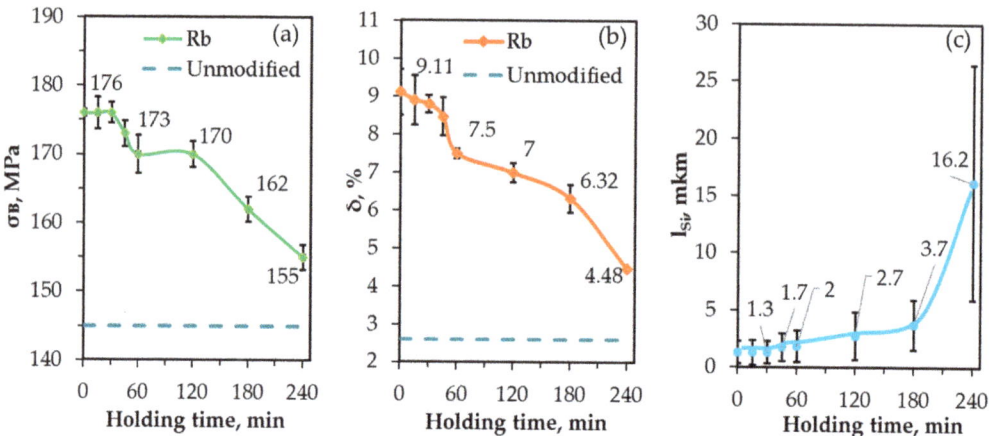

Figure 7. Effect of melt holding time after rubidium modification on: (**a**) ultimate strength; (**b**) relative elongation; (**c**) the average length of eutectic silicon particles.

The values of the relative elongation of the Al-12wt%Si alloy modified with Rb remained at approximately the same level (from 9.1 to 8.46%) during the first 45 min. However, with a longer holding time from 60 to 180 min, the relative elongation of the Al-12wt%Si alloy modified with rubidium decreased from 7.5% to 6.32%.

When holding the melt up to 240 min, the relative elongation of the rubidium-modified alloy decreased significantly to 4.48%.

When modifying the Al-12wt%Si alloy with rubidium, the refined eutectic silicon phase endured for 45 min. A thin fibrous morphology of eutectic silicon was observed (Figure 8a–d). With the increase in the holding time from 60 to 120 min, a smooth size enlargement and coarsening of the eutectic (α + Si) was observed in the microstructure of the alloy (Figure 8e,f). With further holding of the melt up to 180 min, changes to the thin-plate shape in the morphology of the eutectic silicon were observed (Figure 8g). With the proceeding increase in the holding time to 240 min, an abrupt enlargement in size and coarsening of the eutectic (α + Si) was observed in the microstructure of the alloy. The morphology of eutectic silicon changed to needle-like (Figure 8h). However, even with such a long melt holding time, the resulting structure could be characterized as partially modified.

Quantitative analysis of the microstructure of the Al-12wt%Si alloy modified with rubidium showed that an increase in the melt holding time led to an increase in the linear size of the eutectic silicon (Figure 7c). After holding the melt with rubidium modification for 60 min, the average length of eutectic silicon particles increased from 1.3 to 2 microns, after 120 min—to 2.7 microns, and after 180 min—to 3.7 microns. With further exposure of the melt for another 60 min, a sharp increase in the linear sizes of the silicon particles to 16.2 microns occurred.

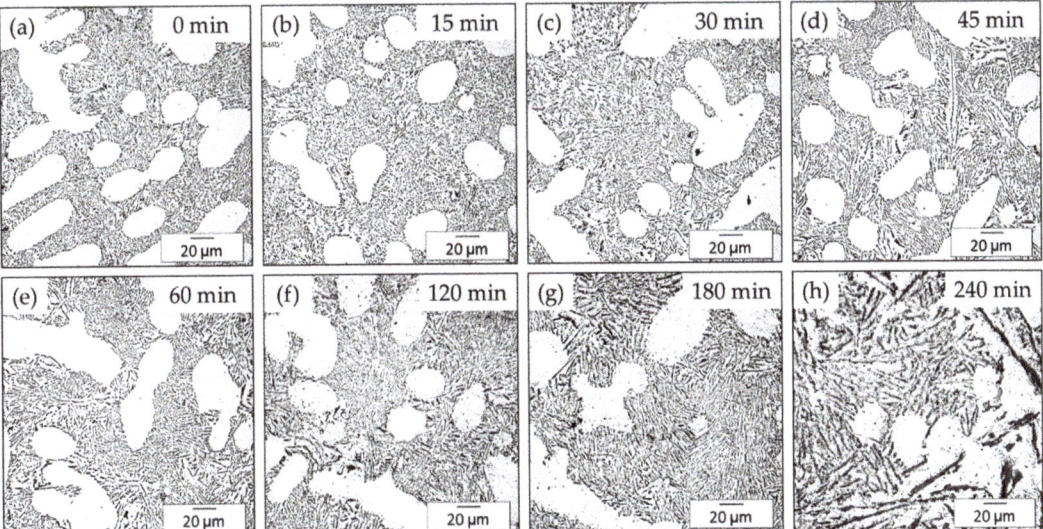

Figure 8. Effect of the melt holding time after modification of the microstructure of alloy 3. (**a**) 0 min.; (**b**) 15 min.; (**c**) 30 min.; (**d**) 45 min.; (**e**) 60 min.; (**f**) 120 min.; (**g**) 180 min.; (**h**) 240 min.

4. Discussion

To determine the possibility of modifying the Al-12wt%Si alloy under study with RbNO$_3$ salt, a thermodynamic analysis of the probable chemical Reactions (2) and (3) was carried out. The formation of rubidium and nitrogen in the melt was thermodynamically possible and confirmed experimentally. During the experimental melting, the release of gaseous compounds during the adding of the salt into the melt was noted.

Based on the obtained data from the atomic emission spectral analysis and energy-dispersive X-ray spectral microanalysis of the samples modified with Rb, it can be concluded that rubidium passed into the melt when added with its nitrate.

The assimilated amount of the modifying element in the range of 0.007–0.01% was sufficient to modify the structure and increase the mechanical properties of the alloy under study.

When modified with rubidium, a thin fibrous eutectic structure was obtained, which is typical of a structure with standard Al-Si alloy modifiers such as sodium and strontium [4,5,36].

The modification mechanism of eutectic silicon by rubidium can be explained using the theory of supercooling and adsorption theory; in particular, the adsorption mechanism of the modification of eutectic silicon.

Firstly, the basis of the theory of supercooling is the data of thermal analysis. According to the theory of supercooling [26], modification of Al-Si alloys lowers the temperature of the beginning of the solidification of the eutectic. An analysis of the DSC thermograms confirmed that rubidium contributed to a significant decrease in the temperature of the beginning of the solidification of the eutectic (α + Si), which led to supercooling of the melt contributed to the appearance of more embryos, and consequently led to the modification of the eutectic.

Secondly, since rubidium is close in its physical and chemical properties to sodium and is a surface-active element, it can be assumed that it will work via a "poisoning" (TPRE) mechanism [27].

On the other hand, according to the IIT mechanism, only elements showing the "ideal" ratio of atomic radii $r_i/r \sim 1.646$ (where r_i is the atomic radius of the element and r is the radius of silicon) are modifiers.

However, rubidium has a ratio of its atomic radii of $r_i/r = 2.08$ [28] and does not meet this criterion, as do some other modifying elements (for example, Sr, Ba and Na). This indicates that the ratio of the atomic radii of a potential modifying element by itself is not capable of predicting the spheroidization of eutectic silicon.

Thus, besides the atomic radius, other factors that are important for the modification of eutectic silicon (the vapor pressure, the formation of secondary compounds or oxides as well as phase equilibria with aluminum and silicon phases, and the cooling rate) probably exist.

The concept of the "poisoning" mechanism (TPRE) and IIT suggests the formation of a high density of twinning in modified silicon crystals [26,28,37,38]. In order for the impurity rubidium atoms to cause such a high twinning density, it is necessary that they are relatively evenly distributed (at least at the nucleation stage) throughout the silicon crystal, as was observed in the results of the energy-dispersive X-ray spectral microanalysis and elementary mapping in our studies. Thus, these results confirmed the operation of the "poisoning" mechanism (TPRE) and IIT in the rubidium-modified 12wt%Si alloy.

It should be noted that the mechanism of rubidium modification has not yet been fully determined and requires further research.

The observed decrease in gas porosity when modifying the studied alloys with rubidium can be caused by the degassing effect of nitrogen formed during the decomposition of the $RbNO_3$ salt according to Reactions (2) and (3). The hydrogen dissolved in the melt diffused into the floating nitrogen bubbles and was removed from the liquid metal. An increase in the amount of salt added to the melt for its modification led to an increase in the degree of degassing of the 12wt%Si alloy.

The formation of shrinkage defects in a casting is affected by the nature of the solidification of the alloy. Under the same cooling conditions, the main effect on the nature of the solidification of alloys is exerted by the temperature range of solidification. Narrow-interval alloys, which include the 12wt%Si alloy, are characterized by sequential or frontal solidification with slight supercooling and the formation of a concentrated shrink shell [39–41].

With an increase in the solidification range of an alloy, the tendency the alloy to lose porosity increases. It is known that when modifying Al-Si alloys with surfactants, which include rubidium, the degree of supercooling of such alloys increases [40,41]. When modifying the 12wt%Si alloy with rubidium, the liquidus temperature of the alloy practically did not change compared to the unmodified alloy (Table 3). At the same time, as a result of the adsorption of rubidium atoms onto the solidification centers of the eutectic (Figure 3), their deactivation occurred. The solidus temperature of the experimental alloy decreased, and its degree of hypothermia increased. According to the authors of [42,43], due to the neutralization of solidification, eutectic colonies grow in the form of spheroids, forming isolated pores when closing. As a result, volumetric solidification and the formation of shrinkage porosity occurred in the thermal center of the casting (Figure 6b).

The results of the studies conducted to determine the duration of the modifying effect showed that when treated with rubidium, the modifying effect in the melt lasted up to 180 min. This was indirectly confirmed by the mechanical properties and quantitative analysis of the microstructure at different melt holding times. Apparently, the duration of the modifying effect depended on the kinetics of oxidation of rubidium in the melt, which requires additional research. Compared to commonly used eutectic silicon modifiers [44,45], the retention time of the modifying effect of rubidium is higher than that of Na but lower than that of Sr.

5. Conclusions

1. The results of the studies conducted showed that the use of rubidium as a modifier is an interesting direction for improving the structure of cast aluminum alloys and improving their mechanical properties. The most stable and highest level of mechanical properties was obtained via modification with rubidium using an actual content in the range of 0.007–0.01%.
2. Microstructural studies showed that rubidium effectively refined eutectic silicon and changed its morphology but had little effect on α-Al (SDAS) dendrites. Energy-

3. Thermal analysis showed that modification with rubidium changed the solidification parameters of the 12wt%Si alloy, causing an extension of the solidification range. The solidus temperature decreased by 18.2 °C. Rubidium contributed to a significant decrease in the temperature of the beginning of the solidification of the eutectic by 7.6 °C.
4. When the 12wt%Si alloy was treated with $RbNO_3$ salt, the melt was degassed and the gas porosity decreased. Modification with rubidium allowed the duration of the modifying effect to be maintained in the melt for up to 180 min.

Author Contributions: Supervision, A.P.R.; writing—original draft preparation, A.D.S. and I.A.P.; writing—review and editing, A.D.S., I.A.P. and V.V.T.; investigations, A.D.S., I.A.P., E.V.M. and A.P.R. All authors have read and agreed to the published version of the manuscript.

Funding: This work was carried out within the framework of the state task (Theme No. 122011200363-9).

Data Availability Statement: Data presented in this article are available at request from the corresponding author.

Conflicts of Interest: The authors declare no conflict of interest.

References

1. Belov, N.A. *Phase Composition of Industrial and Promising Aluminium Alloys: Monograph*; MISiS Publishing House: Moscow, Russia, 2010; 511p.
2. Petrov, I.A.; Shlyaptseva, A.D. Effect of REE on the solidification of eutectic silumin. *Russ. Metall. Met.* **2022**, *3*, 204–210. [CrossRef]
3. Fredriksson, H.; Hillert, M.; Lange, N. The modification of aluminium-silicon alloys by sodium. *J. Inst. Met.* **1973**, *101*, 285–299.
4. Flood, S.C.; Hunt, J.D. Modification of Al-Si eutectic alloys with Na. *Met. Sci.* **1981**, *15*, 287–294. [CrossRef]
5. Liu, Q.Y.; Li, Q.C.; Liu, Q.F. Modification of aluminum-silicon alloys with sodium. *Acta Metall. Mater.* **1991**, *39*, 2497–2502.
6. Hegde, S.; Prabhu, K.N. Modification of eutectic silicon in Al–Si alloys. *J. Mater. Sci.* **2008**, *43*, 3009–3027. [CrossRef]
7. Moniri, S.; Shahani, A.J. Chemical modification of degenerate eutectics: A review of recent advances and current issues. *J. Mater. Res.* **2019**, *34*, 20–34. [CrossRef]
8. Ashtari, P.; Tezuka, H.; Sato, T. Modification of Fe-containing intermetallic compounds by K addition to Fe-rich AA319 aluminum alloys. *Scr. Mater.* **2005**, *53*, 937–942. [CrossRef]
9. Jenkinson, D.C.; Hogan, L.M. The modification of aluminium-silicon alloys with strontium. *J. Cryst. Growth* **1975**, *28*, 171–187. [CrossRef]
10. Dahle, A.K.; Nogita, K.; McDonald, S.D.; Dinnis, C.; Lu, L. Eutectic modification and microstructure development in Al-Si Alloys. *Mater. Sci. Eng. A* **2005**, *413–414*, 243–248. [CrossRef]
11. Fracchia, E.; Gobber, F.S.; Rosso, M. Effect of Alloying Elements on the Sr Modification of Al-Si Cast Alloys. *Metals* **2021**, *11*, 342. [CrossRef]
12. Knuutinen, A.; Nogita, K.; McDonald, S.D.; Dahle, A.K. Modification of Al-Si alloys with Ba, Ca, Y and Yb. *J. Light Met.* **2001**, *1*, 229–240. [CrossRef]
13. Rao, J.; Zhang, J.; Liu, R.; Zheng, J.; Yin, D. Modification of eutectic Si and the microstructure in an Al-7Si alloy with barium addition. *Mater. Sci. Eng. A* **2018**, *728*, 72–79. [CrossRef]
14. Shlyaptseva, A.D.; Petrov, I.A.; Ryakhovsky, A.P.; Medvedeva, E.V.; Tcherdyntsev, V.V. Complex structure modification and improvement of properties of aluminium casting alloys with various silicon content. *Metals* **2021**, *11*, 1946. [CrossRef]
15. Sreeja Kumari, S.S.; Pillai, R.M.; Pai, B.C. Structure and properties of calcium and strontium treated Al–7Si–0.3Mg alloy: A comparison. *J. Alloys Compd.* **2008**, *460*, 472–477.
16. Abdollahi, A.; Gruzleski, J.E. An evaluation of calcium as a eutectic modifier in A357 alloy. *Int. J. Cast Met. Res.* **1998**, *11*, 145–155. [CrossRef]
17. Li, B.; Wang, H.; Jie, J.; Wei, Z. Effects of yttrium and heat treatment on the microstructure and tensile properties of Al–7.5Si–0.5Mg alloy. *Mater. Des.* **2011**, *32*, 1617–1622. [CrossRef]
18. Tsai, Y.C.; Chou, C.Y.; Lee, S.L.; Lin, C.K.; Lin, J.C.; Lim, S.W. Effect of trace La addition on the microstructures and mechanical properties of A356 (Al-7Si-0.35Mg) aluminum alloys. *J. Alloys Compd.* **2009**, *487*, 157–162. [CrossRef]
19. Tsai, Y.C.; Lee, S.L.; Lin, C.K. Effect of trace Ce addition on the microstructures and mechanical properties of A356 (AL-7SI-0.35Mg) aluminum alloys. *J. Chin. Inst. Eng.* **2011**, *34*, 609–616. [CrossRef]
20. Qiu, H.; Yan, H.; Hu, Z. Effect of samarium (Sm) addition on the microstructures and mechanical properties of Al–7Si–0.7Mg alloys. *J. Alloys Compd.* **2013**, *567*, 77–81. [CrossRef]
21. Li, J.H.; Wang, X.D.; Ludwig, T.H.; Tsunekawa, Y.; Arnberg, L.; Jiang, J.Z.; Schumacher, P. Modification of eutectic Si in Al–Si alloys with Eu addition. *Acta Mater.* **2015**, *84*, 153–163. [CrossRef]

22. Wang, Q.; Shi, Z.; Li, H.; Lin, Y.; Li, N.; Gong, T.; Zhang, R.; Liu, H. Effects of Holmium Additions on Microstructure and Properties of A356 Aluminum Alloys. *Metals* **2018**, *8*, 849. [CrossRef]
23. Shi, Z.M.; Wang, Q.; Zhao, G.; Zhang, R.Y. Effects of erbium modification on the microstructure and mechanical properties of A356 aluminum alloys. *Mater. Sci. Eng. A* **2015**, *626*, 102–107. [CrossRef]
24. Nogita, K.; McDonald, S.D.; Dahle, A.K. Eutectic modification of Al-Si alloys with rare earth metals. *Mater. Trans.* **2004**, *45*, 323–326. [CrossRef]
25. Li, J.H.; Suetsugu, S.; Tsunekawa, Y.; Schumacher, P. Refinement of eutectic Si phase in Al-5Si alloys with Yb additions. *Metall. Mater. Trans. A* **2012**, *44*, 669–681. [CrossRef]
26. Edwards, J.D.; Archer, R.S. The new aluminum-silicon alloys—An important process of "modification" and the remarkable improvement in properties it brings about. *Chem. Metall. Eng.* **1924**, *31*, 504–508.
27. Day, M.G.; Hellawell, A. The microstructure and crystallography of aluminium silicon eutectic alloys. *Proc. Royal Soc. A Math. Phys. Eng. Sci.* **1968**, *305*, 473–491.
28. Lu, S.; Hellawell, A. The mechanism of silicon modification in aluminum-silicon alloys: Impurity induced twinning. *Metall. Trans. A.* **1987**, *18*, 1721–1733.
29. Haynes, W.M.; Lide, D.R. (Eds.) *CRC Handbook of Chemistry and Physics*, 95th ed.; CRC Press: Boca Raton, FL, USA, 2014.
30. Gale, W.F.; Totemeier, T.C. (Eds.) *Smithells Metals Reference Book*, 8th ed.; Totemeier Imprint: Oxford, UK, 2003; p. 2080.
31. Altman, M.B.; Stromskaya, N.P.; Guskova, N.V. Method of Simultaneous Refining and Modification of Silumins. Patent SU 423867 A1, 15 April 1974.
32. Merkus, H.G. *Particle Size Measurements: Fundamentals, Practice, Quality*; Springer: Berlin/Heidelberg, Germany, 2009; 534p.
33. ISO 10049:2019(E); Aluminium alloy castings—Visual method for assessing porosity. International Organization for Standardization: Geneva, Switzerland, 2019.
34. Stern, K.H. high temperature properties and decomposition of inorganic salts part 3, nitrates and nitrites. *J. Phys. Chem. Ref. Data* **1972**, *1*, 747–772. [CrossRef]
35. Gurevich, V.L.; Veyts, I.V. *Thermodynamic Properties of Individual Substances: A Handbook*; Nauka: Moscow, Russia, 1978; Volume 4.
36. Shin, S.S.; Kim, E.S.; Yeom, G.Y.; Lee, J.C. Modification effect of Sr on the microstructures and mechanical properties of Al–10.5Si–2.0Cu recycled alloy for die casting. *Mater. Sci. Eng. A* **2012**, *532*, 151. [CrossRef]
37. Hogan, L.M.; Song, H. Interparticle spacings and undercoolings in Al-Si eutectic microstructures. *Met. Trans. A* **1987**, *18*, 707–713. [CrossRef]
38. Nogita, K.; Yasuda, H.; Yoshiya, M.; McDonald, S.D.; Uesugi, K.; Takeuchi, A.; Suzuki, Y. The role of trace element segregation in the eutectic modification of hypoeutectic Al–Si alloys. *J. Alloys. Compd.* **2010**, *489*, 415–420. [CrossRef]
39. Sigworth, G.K. Modification of Aluminum-Silicon Alloys, Casting. In *ASM Handbook*; Viswanathan, S., Apelian, D., Donahue, R.J., DasGupta, B., Gywn, M., Jorstad, J.L., Monroe, R.W., Sahoo, M., Prucha, T.E., Twarog, D., Eds.; ASM International: Detroit, MI, USA, 2008; pp. 240–254. [CrossRef]
40. Ardo, D. Porosity in aluminum foundry alloys—The effect of modification. In Proceedings of the International Symposium on Reduction and Casting of Aluminum, Montreal, QC, Canada, 28–31 August 1988; Ardo, D., Gruzleski, J.E., Eds.; Pergamon Press: Oxford, UK, 1988; pp. 263–282.
41. Andrushevich, A.A.; Sadokha, M.A. Shrinkage phenomena in silumins when treated with long-acting modifiers. *Foundry Prod. Metall.* **2022**, *3*, 30–35. [CrossRef]
42. Nogita, K.; McDonald, S.D.; Zindel, J.W.; Dahle, A.K. Eutectic solidification mode in sodium modified Al-7 mass%Si-3.5 mass%Cu-0.2 mass%Mg casting alloys. *Mater. Trans.* **2001**, *42*, 1981–1986. [CrossRef]
43. Knuutinen, A.; Nogita, K.; McDonald, S.S.; Dahle, A.K. Porosity formation in aluminium alloy A356 modified with Ba, Ca, Y and Yb. *J. Light Met.* **2001**, *1*, 241–249. [CrossRef]
44. Huang, C.; Liu, Z.; Li, J. Influence of Alloying Element Mg on Na and Sr Modifying Al-7Si Hypoeutectic Alloy. *Materials* **2022**, *15*, 1537. [CrossRef]
45. Ganesh, M.S.; Reghunath, N.; Levin, M.J.; Prasad, A.; Doondi, S.; Shankar, K.V. Strontium in Al–Si–Mg Alloy: A Review. *Met. Mater. Int.* **2022**, *28*, 1–40. [CrossRef]

Disclaimer/Publisher's Note: The statements, opinions and data contained in all publications are solely those of the individual author(s) and contributor(s) and not of MDPI and/or the editor(s). MDPI and/or the editor(s) disclaim responsibility for any injury to people or property resulting from any ideas, methods, instructions or products referred to in the content.

Article

Microstructure and Compressive Properties of Porous 2024Al-Al$_3$Zr Composites

Wenchang Zhang [1,2,3], Kun Xu [1,2,3], Wei Long [1,2,3,*] and Xiaoping Zhou [1,2,3]

1. School of Materials and Chemical Engineering, Hubei University of Technology, Wuhan 430068, China
2. Key Laboratory of Green Materials for Light Industry of Hubei Provincial, Wuhan 430068, China
3. Hubei Engineering Laboratory of Automotive Lightweight Materials and Processing, Wuhan 430068, China
* Correspondence: maillong1982@126.com

Abstract: Porous 2024Al-Al$_3$Zr composites were prepared by in situ and spatial scaffolding methods. As the Al$_3$Zr content increased from 5 wt.% to 30 wt.%, the binding of the powder in the pore wall increased and the defects in the composites decreased. The yield strength of the composites reached 28.11 MPa and the energy absorption capacity was 11.68 MJ/m^3 at a Zr content of 20 wt.%, when the composites had the best compression and energy absorption performance. As the space scaffold content increased from 50% to 70%, the porosity of the composites then increased from 53.51% to 70.70%, but the apparent density gradually decreased from 1.46 g/cm^3 to 0.92 g/cm^3, leading to a gradual decrease in their compressive properties. In addition, by analysing the compression fracture morphology, the increase of Al$_3$Zr will reduce the stress concentration and hinder the crack growth, while too much Al$_3$Zr will lead to brittleness and reduce the performance.

Keywords: porous composite; space holder method; compression property; energy absorption

Citation: Zhang, W.; Xu, K.; Long, W.; Zhou, X. Microstructure and Compressive Properties of Porous 2024Al-Al$_3$Zr Composites. *Metals* **2022**, *12*, 2017. https://doi.org/10.3390/met12122017

Academic Editors: Noé Cheung and Matej Vesenjak

Received: 13 September 2022
Accepted: 22 November 2022
Published: 24 November 2022

Publisher's Note: MDPI stays neutral with regard to jurisdictional claims in published maps and institutional affiliations.

Copyright: © 2022 by the authors. Licensee MDPI, Basel, Switzerland. This article is an open access article distributed under the terms and conditions of the Creative Commons Attribution (CC BY) license (https://creativecommons.org/licenses/by/4.0/).

1. Introduction

Porous aluminium, as a widely used structural and functional material [1], has become a highly practical topic due to its low density, high specific surface area, high specific strength and light weight [2–5]. Its properties such as vibration damping, energy absorption, flame retardancy, sound absorption and heat dissipation have been widely studied [6–8]. The preparation methods of porous aluminium generally include the gas injection method [9], foaming method [10], mould casting method [11], deposition method [12] and space holder method [13]. However, the mechanical strength of pure aluminium and aluminium alloys prepared by the above methods is poor, which limits the application of porous materials. Therefore, to enhance the mechanical strength of the material, some researchers added the second phase to the porous aluminium to prepare composites. The in situ synthesis method is widely used because of its uniform distribution and tight bonding. Inoguchi [14] prepared in situ Al$_3$Ti/Al porous composites and the compression properties of porous materials improved by the combination of Al phase and Al$_3$Ti. Atturan [15] prepared an A357-TiB$_2$ porous composite and the mechanical strength is much higher than Al porous materials. In situ synthesis of reinforced particles in porous aluminium enhances the mechanical properties of the composites.

In recent years, in situ generation of Al$_3$X, such as Al$_3$Ni, Al$_3$Fe, Al$_3$Ti, and Al$_3$Zr, has been used in different composites due to their attractive properties. But among these, Al$_3$Zr has attracted much attention due to its low density (4.11 g/cm^3), high melting point (1580 °C), high specific strength, high specific stiffness, excellent corrosion and wear resistance [16,17]. The lattice matching values of Al$_3$Zr (tetragonal structure) and α-Al (FCC structure) in the a and c/2 directions are 93% and 99.2%. The relatively high wettability contributes to a better dispersion of particles in the matrix, better grain refinement and better properties of composite [18]. Gupta and Danie [19] added K$_2$ZrF$_6$ salt into the aluminium melt to produce Al$_3$Zr particles, the Al$_3$Zr particles dispersed uniformly and refined microstructure during the melting process, which significantly improved the strength of the material. Pourkhorshid [20] used mechanical alloying (MA) and a hot extrusion process to prepare Al/Al$_3$Zr composites, the Al$_3$Zr began with the nucleation of the metastable phase and then transformed into a stable tetragonal Al$_3$Zr structure. The tensile yield stress of obtained Al-10 wt.%Al$_3$Zr composites is 103 MPa, which is about twice that of pure aluminium (53 MPa). It can be seen that the in situ Al$_3$Zr has a good effect on the mechanical properties of composites. Nevertheless, the studies on the application of Al$_3$Zr to improve the mechanical properties of porous aluminium have not been fully investigated. In the present study, the effects of Zr and NaCl space holder content on the microstructure of porous Al$_3$Zr/2024Al composites with space holder and in situ synthesis method were investigated. Changes in the compressive and energy absorption characteristics were also studied.

2. Materials and Methods

Materials: Zr powder (China, MACKLIN, purity > 99%, d50: 48 µm), 2024Al alloy powder (China, Sinopharm Group Chemical Reagent Co., average particle size: 48 µm), the alloy composition is shown in Table 1 and NaCl space-holder powders (China, Shanghai Wokai Biotechnology Co., purity: A.R., average size: 100~500 µm) were used, the morphology of powders is shown in Figure 1a–c.

Table 1. Chemical composition of the 2024Al alloy (mass fraction, %).

Cu	Mg	Mn	Si	Fe	Zn	Others	Al
3.949	1.283	0.452	0.087	0.123	0.137	0.050	Bal.

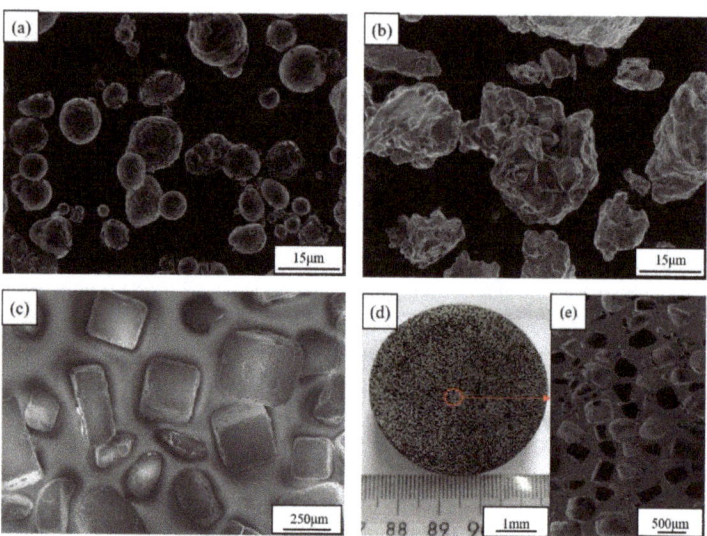

Figure 1. (a–c) SEM images of the raw powders: (a) 2024Al, (b) Zr, (c) NaCl. (d,e) the sample surface morphology of the sample.

Experimental procedures: First, 2024Al-Zr metal powder with Zr content of 5–30 wt.% was put into a planetary ball mill to mix the powder uniformly with a ball-to-ball ratio of 4:1 and a speed of 100 r/min for 3 h. In the second step, a 50–70 vol.% NaCl space holder was mixed uniformly with the above powder mixture with the same ball milling parameters as in the first step. The powder particle size was not changed before and after mixing. The final mixed powder was placed in the stainless-steel mould (inner diameter 40 mm, outer diameter 100 mm, height 100 mm), and the cold-pressed block with a height of 10–12 mm was obtained by applying constant pressure (400 MPa). The block was heated to 650 °C at a heating rate of 5 °C/min in an argon atmosphere for 1 h [21]. The obtained sample was placed in water at 50 °C to remove NaCl. The sample surface morphology of the sample is shown in Figure 1d,e. The preparation and sintering process curves of the sample are shown in Figure 2.

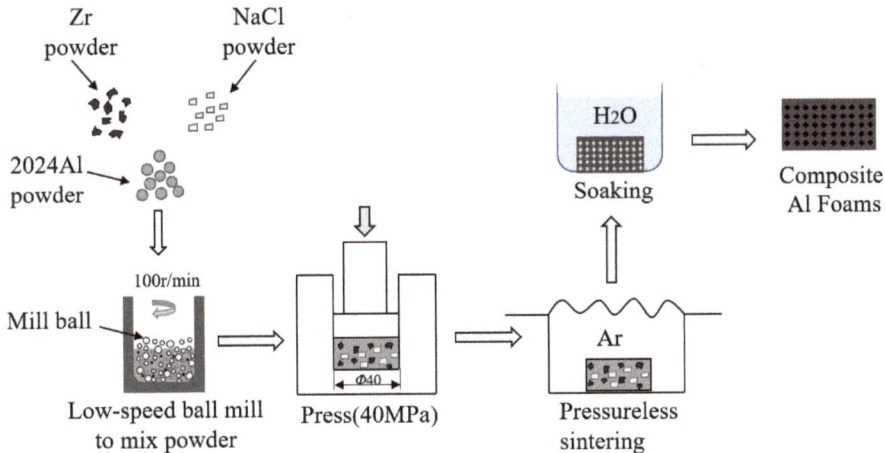

Figure 2. Schematic illustration showing the fabrication techniques for porous materials.

Characterisation of the porous composite: A rectangular compressed sample with a size of 10 × 10 × (10–12) mm was obtained through the wire electrical discharge machining. The porosity of the sample can be calculated by the following equation [22]:

$$P = 1 - \frac{m}{v\rho_s} \quad (1)$$

where m is the mass of the sample, v is the volume of the sample, and ρ_s is the theoretical density of the composite material, which can be calculated by the following equation [23]:

$$\rho_s = \frac{1}{\frac{m_{Al}}{\rho_{Al}} + \frac{m_{Zr}}{\rho_{Zr}}} \quad (2)$$

where m_{Al}, m_{Zr} are respectively mass fractions of 2024Al and Zr, the 2024Al is 2.78 g/cm^3 and Zr is 6.49 g/cm^3. In this study, to facilitate the description of the two types of pores studied and consequently the two sizes, we will use "intergranular pore" which corresponds to the pores between the particles (2024Al and Zr) in the matrix material and range from 1 to 10 μm. "Pores" correspond to the leached NaCl space holder which ranges from 100 to 500 μm. The intergranular porosity can be calculated in the following equation [24]:

$$P_m = P - P_n \quad (3)$$

where P_m is the intergranular porosity, P is the total porosity, P_n is the volume fraction of NaCl.

The compression test was carried out by the CMT-4204 material creep testing machine at room temperature. The previously mentioned rectangular compressed specimen of dimensions 10 × 10 × (10–12) mm was placed in the middle of the loading table at the strain rate of approximately 3.3 × 10^{-3} s^{-1}. When the material was compressed to the identified region, the compression stress–strain curve of the sample was obtained. The energy absorption capacity of the material is calculated by the following equation [25]:

$$w = \int_0^{e_0} \sigma \, de \qquad (4)$$

where w is the energy absorption capacity, e_0 is the densification strain and σ is the compressive stress. The compressive yield stress is defined as the first peak of the compressive stress–strain curve, and the plateau stress is set as the average stress in the range from 20% to 30% strain.

The micro-hardness of the pore wall of the sample was measured by a Vickers hardness tester with a load of 0.98 N (0.1 Kgf) along the vertical line. The hardness test was conducted at least ten test times, and the results were averaged. The phase composition of the composite was analysed through an Empyrean sharp shadow X-ray diffractometer at the scanning rate of 10°/min with Cu-Kα radiation; the obtained porous material was processed by wire cutting, mounting, polishing and cleaning, and the tissue was observed under an optical microscope. After it was determined that the surface of the material was free of scratches, the microstructure of the sample was observed by Zeiss Gemini 300 scanning electron microscope (SEM), and the phase composition was analysed by EDS.

3. Results and Discussion

3.1. Effect of Zr Content on Phase and Structure Morphology of Composites

Figure 3 presents the XRD pattern of the composites as an increase in the Zr content. As Zr content increases, the relative intensity of the Al peak decreases and the relative intensity of the Al$_3$Zr peak increases. The specimens were confirmed to be composed of α-Al(fcc), Al$_3$Zr(D0$_{23}$) and Al$_2$Cu(C$_{16}$) phases, indicating that Zr reacted with Al to form Al$_3$Zr under the designed sintering process. The result was consistent with the Al-Zr binary phase diagram [26]. In the case where the mass fraction of Zr is less than 53%, Al and Zr will form a single-phase Al$_3$Zr intermetallic compound under equilibrium conditions. Moreover, there is no new phase formed involving the elements Na and Cl, reflecting that NaCl has no effect on the composition of the composite after sintering. It can be employed as an excellent space holder.

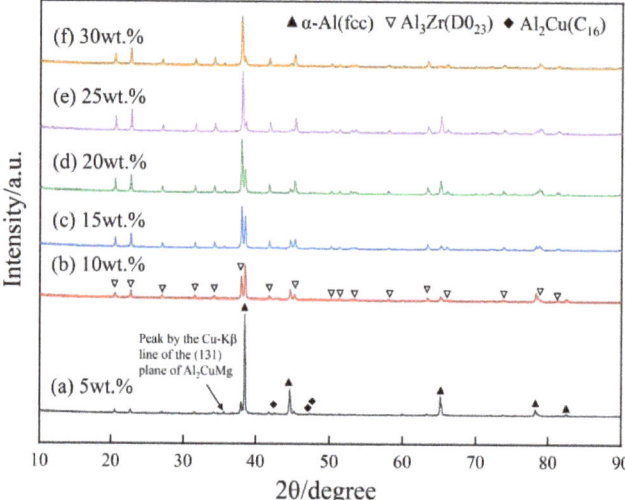

Figure 3. XRD patterns of the composites with different Zr contents.

Figure 4 shows the SEM image and results of the EDS quantitative analysis of composites with different Zr content. It can be concluded that the grey part is Al_3Zr and the morphology is mainly a fine block. In the case of Zr content 30 wt.%, the point scan result is Al_3Zr, which is consistent with the result of XRD, indicating that Zr and Al in this content will only form Al_3Zr and there is no other substance. An increase in Zr content also causes the appearance of Al_3Zr agglomerates with short rod morphology.

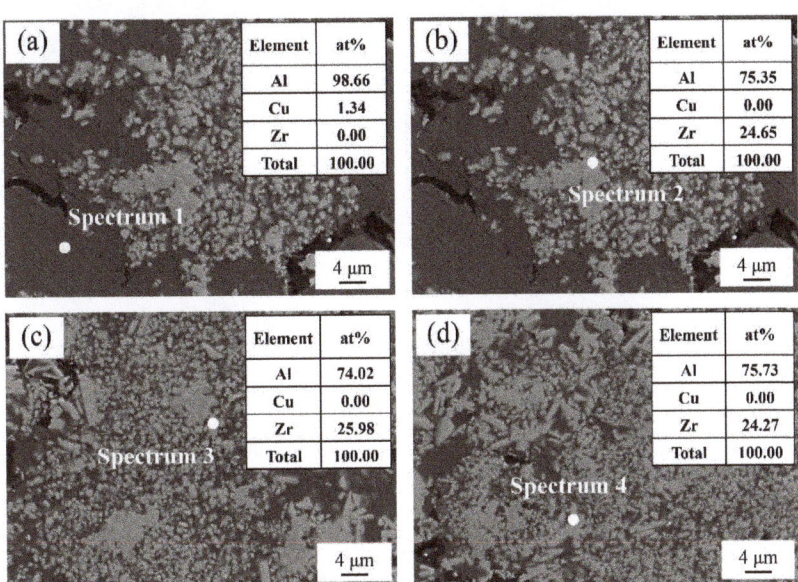

Figure 4. The points show the SEM images and results of EDS quantitative analysis of the $Al_3Zr/2024Al$ composites prepared with different Zr contents. (**a,b**) 10Zr; (**c**) 20Zr; (**d**) 30Zr.

Figure 5 is the SEM image of the pore wall of $Al_3Zr/2024Al$ porous composites with different Zr contents (space holder content is 60 vol.%). With the increase of Zr content, the number of Al_3Zr in the pore wall gradually increases, and the main components in the pore wall gradually change from 2024Al to Al_3Zr. As shown in Figure 5a,b, the Al_3Zr are mainly formed at the contact interface between Zr and 2024Al. In the region of the Al_3Zr phase, the microstructure was more uniform and regular compared with other regions, and there are fewer defects in the reaction area. However, obvious powder gaps and defects can be found between the mechanically bonded 2024Al powder particles, there is an obvious gap between the 2024Al powders, and there is much unevenness in the pore wall. When the Zr content continues to increase, the defect and gap in the material are declined and the distribution of Al_3Zr in the pore wall is more dispersed. In the case of Zr content 20 wt.%, the gap between 2024Al particles caused by mechanical bonding cannot be observed. It was supposed that the bonding mode of the material was metallurgical bonding. As the content of Zr further increases, a large number of 2024Al react with Zr. Almost the whole pore wall was occupied by Al_3Zr particles. It demonstrates that the material is mainly Al_3Zr. However, the amount of Al_3Zr possibly causes serious brittleness in the samples. The qualitative relationship between the content of Al_3Zr and the mechanical properties of the material will be investigated in the future.

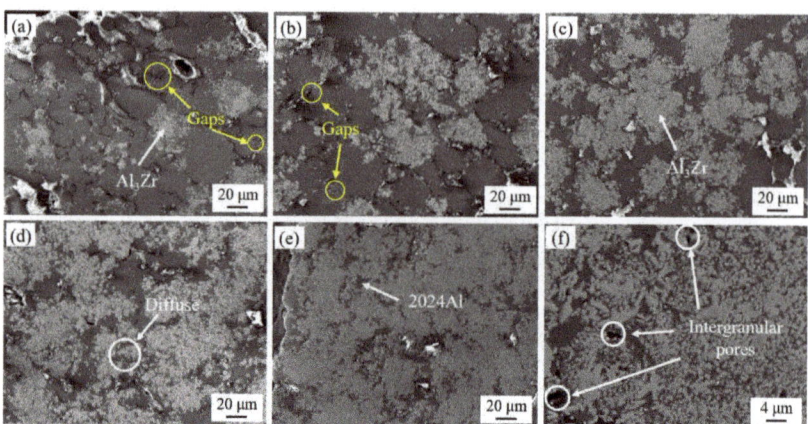

Figure 5. The SEM images of the composite pore wall were prepared with different Zr contents. (**a**) 5Zr; (**b**) 10Zr; (**c**) 15Zr; (**d**) 20Zr; (**e**) 25Zr; (**f**) 30Zr.

Figure 6 shows the SEM images of the pore walls of the porous composites. In the case of composition Zr5, there are obvious gaps (Intergranular pores) between the crystals in the pore wall. This phenomenon occurred since the 2024Al in the sample melted at the sintering temperature of 650 °C and flowed out due to the low interfacial tension, which resulted in the precipitation of some aluminium alloy on the outside of the sintered sample. In contrast, the powder is tightly connected, and the gap is small, and the surrounding defects are greatly reduced at the composition Zr20. It illustrates that the Al$_3$Zr formation under the action of surface tension will prevent the outflow of molten 2024Al, even liquid phase sintering also can keep the shape of cold pressing, and the reaction heat between particles will accelerate the bonding between particles.

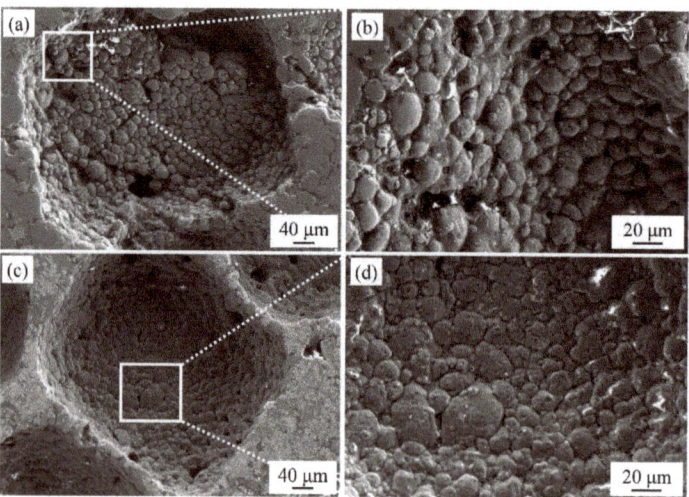

Figure 6. The SEM images of the pore walls of the porous composites with Zr content of (**a**,**b**) 5Zr and (**c**,**d**) 20Zr.

3.2. Effect of Zr Content on Mechanical Properties of Composites

Figure 7 shows the hardness of the pore wall of the composites with different Zr content. The hardness of the pore wall increases continuously with the increase of Zr content, Through the analysis of the hardness value, the highest hardness value (145.4 HV0.1) is obtained in the case of Zr content 30 wt.%, which is 71.6% higher than that of 84.73HV0.1 in the case of Zr content 5 wt.%. Based on the above analysis, the low pore wall hardness of composites is attributed to the weak bonding of powder particles with low content of Al$_3$Zr. With the increase of Zr content, the main composition of the pore wall changes to Al$_3$Zr, leading to higher hardness.

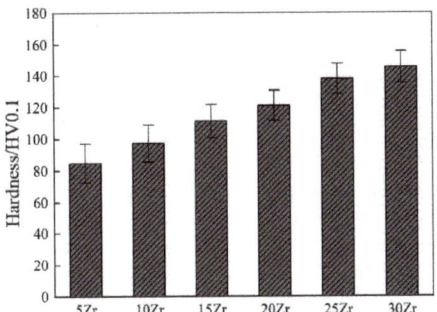

Figure 7. The hardness value of pore wall with different Zr content.

Figure 8 shows the compressive stress–strain and energy absorption capacity curves of the composites with different Zr content (space holder content is 60 vol.%). The curve in Figure 8a was divided into three regions: (1) Elastic region: the elastic deformation occurs, and the compression process is similar to dense materials, and the compression curve shows that the stress increases linearly with the increase of strain. The yield stress of the material is the value when the stress reaches the highest point for the first time. Moreover, the slope of the material with different Zr content in the elastic region is similar, indicating that the composition of the material has little effect on the elastic stage, and the main difference in the elastic region is the different yield stress of the material; (2) plateau region: most of the curves in this region is wavy or meandering. This phenomenon can be explained by the fact that the pore wall collapsed and deformed under the compressive load, and some pores ruptured rapidly with the increase of compressive strain, which eventually led to the stress weakening [27]. When the compression load is distributed to the uncompressed cavity, the support of the pore wall will make the stress increase again. So, the degree of curve bending is determined by the combined action of macroscopic and microstructural instability of porous materials [28,29]; (3) densified region: the pores in this area have been compacted, the material is in a dense state, and the compressive strain makes the stress rise rapidly.

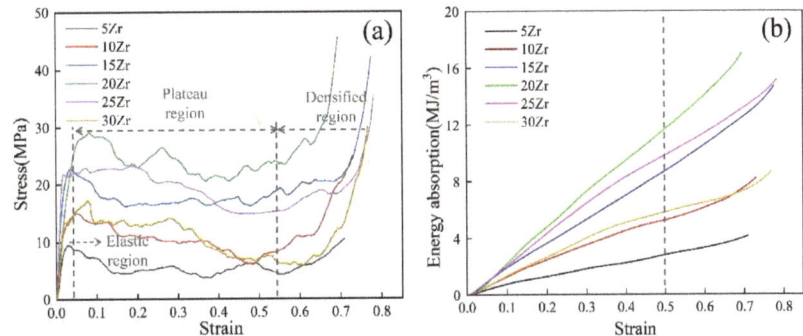

Figure 8. (**a**) Compressive stress–strain and (**b**) energy absorption capacity curves of composite materials with different Zr contents.

Compared with the upper limit of 50% strain, the curves in Figure 8b show that the energy absorption capacity of the material increases at first and then decreases with the increase of Zr content. In the case of Zr content 20 wt.%, the energy absorption capacity reaches the highest of 11.68 MJ/m^3, which is 315.6% higher than that of 5Zr (2.81 MJ/m^3). This phenomenon can be explained by the fact that the energy absorption capacity is mainly related to the area under the stress–strain curve in the case of samples with similar porosity. Due to the good stiffness and elastic modulus of Al$_3$Zr, the performance of the material in the plateau stress region is improved as well as the energy absorption performance.

Figure 9 presents the yield stress and plateau stress of the composites with different Zr content. It can be seen that the numerical gap between yield stress and plateau stress decreases gradually. This phenomenon can be explained by the fact that the density of porous materials increased with the increase of Zr content, and the microstructure of the material gradually stabilised. Under the influence of these comprehensive factors, the plateau stress value gradually approached yield stress. The figure also shows that yield stress and plateau stress increase at first and then decrease with the increase of Zr content, and reached their highest point at a Zr content of 20 wt.%. The yield stress was 28.11 MPa, which was 198.1% higher than 5Zr (9.43 MPa), and the plateau stress was 24.87 MPa, 358.8% higher than 5Zr (5.42 MPa). Among them, the stress rise of the plateau is more than the yield stress, which is due to the low density and many defects in the case of Zr content 5 wt.%, the pores will collapse quickly after entering the plateau region, resulting in a rapid decrease in the platform stress.

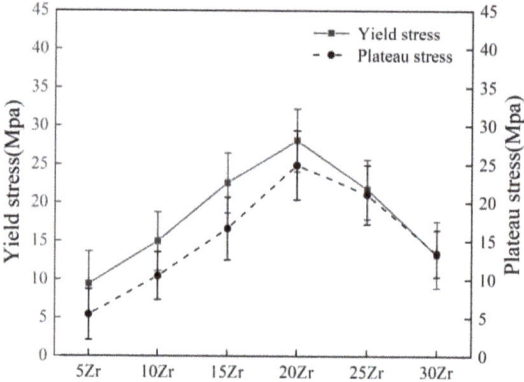

Figure 9. The yield stress and plateau stress of composites with different Zr contents.

In the case of Zr content 25 wt.% and 30 wt.%, there is mainly brittle phase Al$_3$Zr in the pore wall, cracks generated at defects will propagate easily, which could decline the material properties.

Figure 10 presents the SEM images of compression fracture of composites with different Zr contents. There are no obvious fracture characteristics in the area around the white box in Figure 10a, indicating that the white box area is one of the starting points for the fracture failure of the pore wall under the compression load. Moreover, some pore walls in the material are not in a strictly closed state, which causes the regional stress concentration, collapse and deformation in the compression process. At the fracture of Figure 10c, there are continuous dimples of different sizes, indicating that the fractured part is coherent. This may be due to the crack propagating in all directions because the thickness and structure of the pore wall are uniform, the fracture occurs in the whole area. However, it is difficult to directly observe the initial position of the fracture. Unlike the morphology of the sample in Figure 10c, there is a smooth area surrounded by cracks at the fracture of the pore wall in Figure 10e. Because the thickness distribution of the pore wall around this area is different and this irregular porous structure will cause stress concentration. The stress concentration in the thinner part of the pore wall will be greater than that in the thicker pore wall, Zettl [30] also concluded that the initial defect is the preferred medial part of the pore wall and the damage begins in the thinner part of the pore wall.

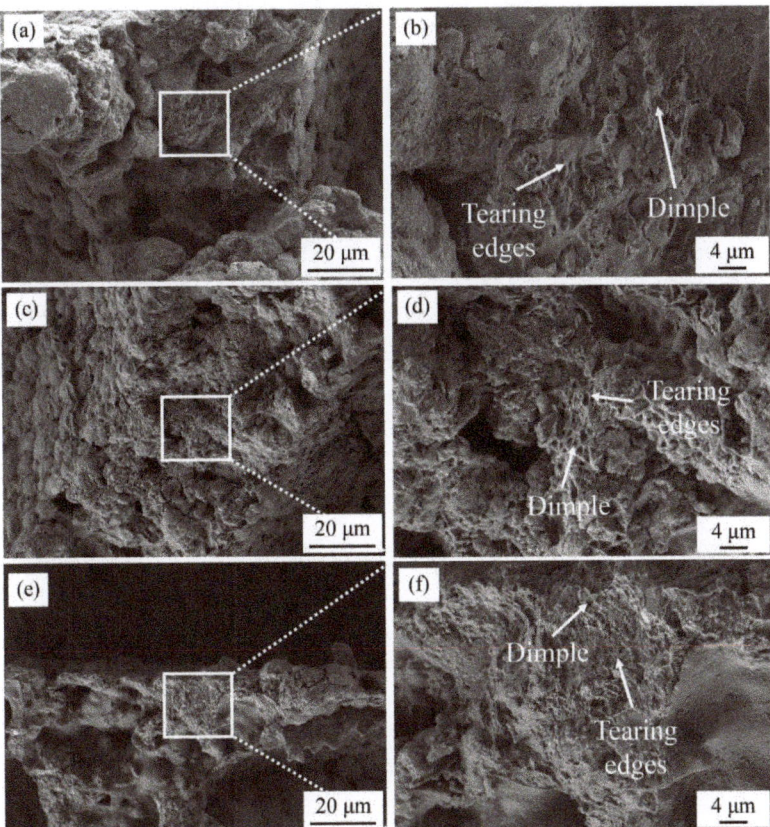

Figure 10. The SEM image of compression fracture of composites with different Zr content (**a,b**) 10Zr; (**c,d**) 20Zr; (**e,f**) 30Zr.

From the SEM of the pore walls, it can be seen that the fracture surfaces of 10Zr and 20Zr have a large number of dents and torn edges, indicating that the pore walls are ductile fractures. The dents of the fractures of 30Zr are larger and shallower than those of 20Zr, which indicates that brittle fractures occurred in the pore walls. Based on the explanation of the morphology and properties of porous materials with different Zr content, it can be considered that there are two fracture modes in the pore wall of $Al_3Zr/2024Al$ porous composites during compression. One is the fracture between ductile phase 2024Al, which is mainly displayed in Figure 10b. There are a large number of tear edges formed by the fracture between 2024Al. When the crack develops near the Al_3Zr particles, the fine Al_3Zr will hinder the crack extension and form dimples. As a result, the compression properties of the composites are enhanced. With the increase of the Al_3Zr number, the resistance of crack extension also increases, and the number of dimples in the fracture surface of the pore wall also increases. Another one is the brittle fracture of Al_3Zr, the interior of the material is mainly brittle phase Al_3Zr when too much Al_3Zr is formed. Due to the lack of ductile phase as support, the crack will develop rapidly after compression fracture, which leads to the fracture of some agglomerated Al_3Zr particles earlier than the ductile phase and makes the dimples on the fracture surface larger and shallower. Finally, this brittle structure causes the decline of material properties.

According to the previous explanation of pore wall morphology, the reaction of Zr and 2024Al from 5Zr to 20Zr forms finely dispersed Al_3Zr particles that hinder crack growth by reducing the stress concentration. The reaction also improved the density and reduced the number of defects on the pore wall. Under the combined effect of these factors, the composite performance is significantly improved. However, when the Zr content continues to increase, the composite is mainly brittle phase Al_3Zr, the crack will quickly spread to the surrounding parts under a compression load, thereby degrading the properties of the sample. Therefore, Al_3Zr should be composited with toughness to compensate for its brittleness, to obtain porous composites with excellent compressive and energy-absorbing properties.

3.3. Effect of Space Holder Content on Morphology and Properties of Composites

Figure 11 shows the morphology of the samples with different space holder contents (20 wt.% Zr content). Figure 11a,c,e show the macroscopic morphology of the specimens with space holder contents of 50 vol.%, 60 vol.% and 70 vol.%, where the morphology can remain intact after the removal of the space holder by impregnation. However, in the case of space holder content greater than 70 vol.%, some areas of the sintered specimens were difficult to maintain integrity due to the clumping of the space holder during cold pressing and collapse or detachment during impregnation, so porous specimens with space holder content greater than 70 vol.% were not explored in this paper. Figure 11b,d,f shows the enlarged shapes of specimens with 50 vol.%, 60 vol.% and 70 vol.% space holders. After removing the NaCl space holder by immersion, the macroscopic pores basically maintain the original shape of the NaCl particles, and with the increase of the space holder, the pores are more frequently connected, and the pore wall thickness gradually decreases.

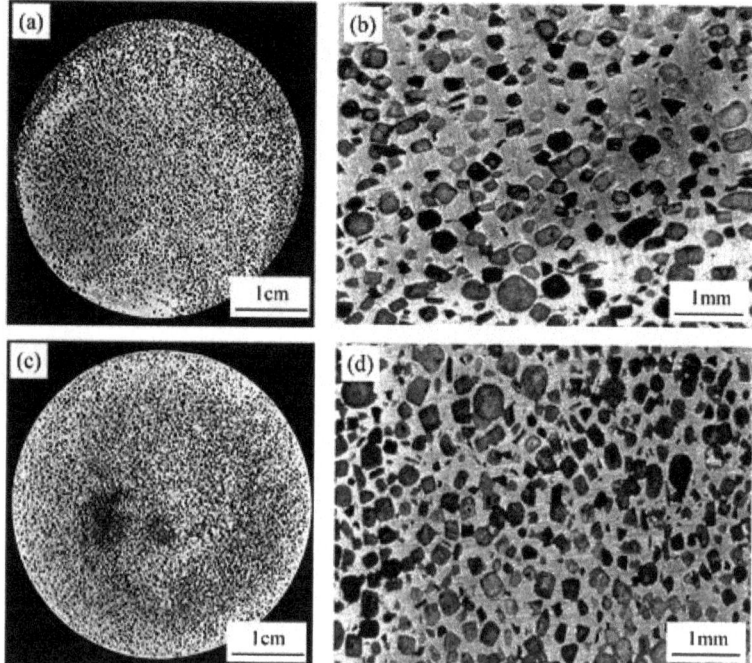

Figure 11. Cont.

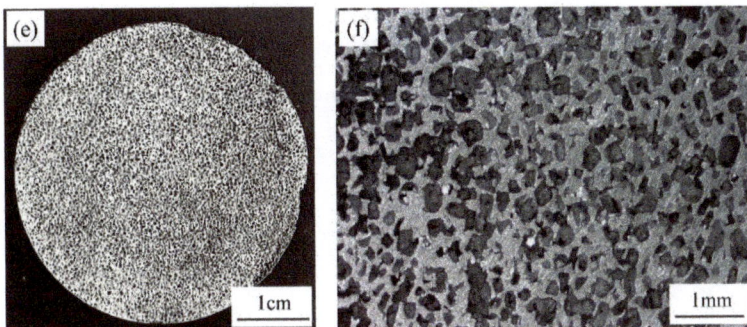

Figure 11. Macroscopic morphologies of porous materials with different contents of NaCl space holder (**a,b**) 50 vol.%; (**c,d**) 60 vol.%; (**e,f**) 70 vol.%.

This section introduces the relative density data according to the literature [30] investigation to better study the effects of different space holder contents on material properties [30]. The relative density is the ratio of the apparent density of the porous material to the theoretical density when the material is densified. Table 2 shows the mass, volume, apparent density and theoretical density of the 20Zr samples with different space holder percentages.

Table 2. The mass, volume, apparent density and theoretical density of the 20Zr samples with different space holder percentages.

Space Holder Percent/%	Mass/g	Volume/cm^3	Apparent Density/(g/cm^3)	Theoretical Density/(g/cm^3)
50	1.61	1 × 1 × 1.1	1.46	3.14
55	1.32	1 × 1 × 1.0	1.32	3.14
60	1.31	1 × 1 × 1.1	1.19	3.14
65	1.28	1 × 1 × 1.2	1.07	3.14
70	1.10	1 × 1 × 1.2	0.92	3.14

Relative density is also another manifestation of the density of the material, which can be calculated by the following equation [31]:

$$R = \frac{\rho_P}{\rho_s} \qquad (5)$$

where R is the relative density, ρ_P is the apparent density of porous materials and ρ_s is the theoretical density of materials when they are dense. Table 3 shows the porosity, intergranular porosity and relative density of the composites with different space holder contents. It illuminates that with the increase of space holder content, the number of intergranular pores decreases gradually, and the porosity of the material is closer to the volume fraction of the space holder. This is due to the reduction of metal powder, the powder gap caused by cold pressing and the Kirkendall voids formed by the in situ reaction is also reduced [32], and the pores of porous materials are mainly formed by space holders.

Figure 12 shows the compressive stress–strain and energy absorption capacity curves with different relative densities (20Zr). It can be found that with the increase of porosity, the elastic strain limit and the slope of the curve decrease, and the yield stress of materials also decreases gradually. The plateau region in the curves is longer and flatter with the decreases in relative density. It is because porous materials with lower relative density (or higher porosity) have more pore structures. When the pore wall collapses and deforms under a compression load, there is enough space inside to bear the deformation of the pore wall. From the energy absorption capacity curve, it can be concluded that the energy absorption performance of the porous material has a positive correlation with the relative density, because when the relative density is higher, there are more matrix materials in the porous material, and the compressive performance of the material will be better. Therefore, with the decrease in relative density, the energy absorption capacity also decreases.

Table 3. The porosity, intergranular porosity and relative density of the composites sintering at different space holder percentages.

Space Holder Percent/%	Porosity/%	Intergranular Porosity Porosity/%	Relative Density
50	53.51	3.51	0.47
55	57.96	2.96	0.42
60	62.10	2.10	0.38
65	65.92	0.92	0.34
70	70.70	0.70	0.29

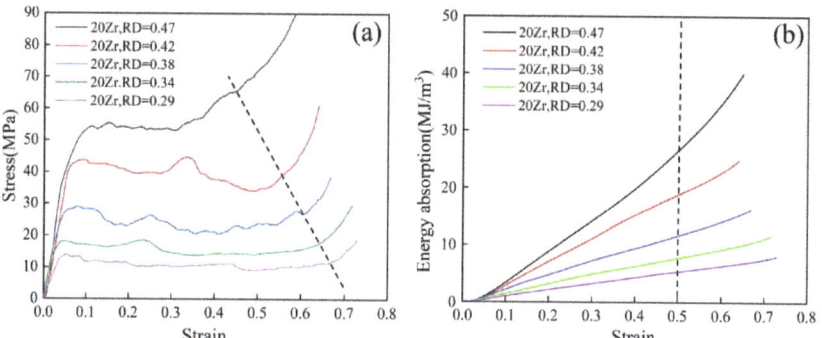

Figure 12. (a) Compressive stress–strain and (b) energy absorption capacity curves of composite materials with different relative densities.

Figure 13 shows the yield stress and plateau stress of the composites with different relative densities. It can be seen that the performance range of porous materials increases with the increase of relative density. This rule is mainly because with the increase of porosity, the space of the pore wall in the same area is smaller, and the thickness of the pore wall is thinner, the supporting capacity to the load decreases greatly, so the performance of the porous material decreases gradually. However, the decline in performance brings a larger specific surface area and lower density.

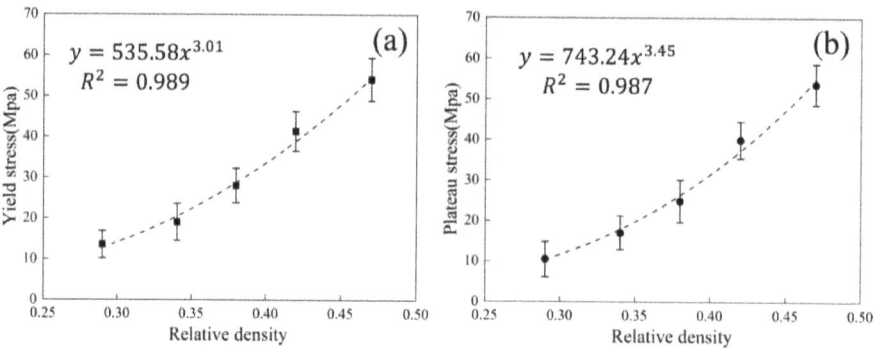

Figure 13. (a) Yield stress and (b) plateau stress of composites at different relative densities.

By comparing the prepared Al$_3$Zr/2024Al porous composites with other references, the properties of the materials are significantly improved, and the corresponding results are shown in Table 4. It is worth noting that the properties of porous composites prepared by in situ synthesis are better than those of pure aluminium, indicating that in situ synthesis has a unique advantage in solving the poor properties of porous aluminium.

Table 4. Compression properties of different porous materials.

Composition	Porosity/%	Yield Stress/MPa	Plateau Stress/MPa	Reference
Al$_3$Zr/2024Al	53.51	54.15	53.63	Present study
	62.10	28.02	24.94	
	70.70	13.52	10.51	
Al$_3$Ti/Al	~70	13	10	[32]
CNT/Al	~60	22.24	33	[22]
SiC/Al	~40	58	-	[33]
Al$_2$O$_3$/Al	~60	15	10	[34]
Al	~50	17.0	-	[35]
	~60	11.2	-	
	~70	4.6	-	

To better describe the effect of the relative density of porous materials on the compression properties of porous materials, Gibson and Ashby [6] assume that the pore wall is a dense material, and the compression of porous materials is mainly accomplished by the buckling and compression of the pore wall, and its strength is mainly determined by its relative density. A prediction model is proposed for this assumption, which can be calculated by the following equation [6]:

$$\sigma = 0.3\left(\Phi\frac{\rho_f}{\rho_d}\right)^{\frac{3}{2}} + 0.4(1-\Phi)\frac{\rho_f}{\rho_d} \qquad (6)$$

where σ is the compressive strength of the porous material, ρ_f is the actual density of the porous material and ρ_d is the theoretical density when the material is fully dense, Φ is the volume percentage of the pore wall of the material. Because the porous material prepared by the space holder method is a three-dimensional reticular structure, after the space holder is removed, the whole material is composed of a pore wall, so the $\Phi = 1$. Therefore, D.P. Mondal [36] has made some modifications to this model, and the result is the following equation [36]:

$$\sigma = C\left(\frac{\rho_f}{\rho_d}\right)^n \qquad (7)$$

where C and n are constant, determined by the properties of the porous material. This equation shows that the relationship between compression properties and the relative density of porous materials is exponential. Applying this equation to this experiment, the relationship between yield stress, plateau stress and relative density is as follows:

$$\sigma_y = 535.58\left(\frac{\rho_f}{\rho_d}\right)^{3.01} \qquad (8)$$

$$\sigma_p = 743.24\left(\frac{\rho_f}{\rho_d}\right)^{3.45} \qquad (9)$$

where σ_y is the yield stress, σ_p is the plateau stress. However, the pore wall of the actual porous material is difficult to be completely dense, and there will inevitably be some defects in the interior, and its pore structure, such as wall thickness and pore size cannot be the same, and the distribution of pores is also difficult to achieve uniform distribution. But these equations can be used to predict the compressibility of porous materials, which will become a part of the production process route design in actual production.

4. Conclusions

The effects of differences on the microstructure and properties of Al3Zr/2024Al porous composites were studied.

(1) As the Zr content increases from 5 wt.% to 30 wt.%, the Al_3Zr content gradually increases, the pore walls become denser and the number of defects decreases. The hardness of the material also increases with the increase of Al_3Zr content from 84.73 HV0.1 to 145.4 HV0.1. The pressure properties and energy absorption properties first increase and then decrease, and the best overall performance is achieved with a compressive strength of 28.11 MPa and an energy absorption capacity of 11.68 MJ/m^3 at a Zr content of 20 wt.%.

(2) The compressive fractures of materials with different Zr contents show that Al_3Zr can improve the compressive properties of the material by hindering the propagation of cracks in the pore wall, but when the Al_3Zr content is too much, the pore wall will undergo brittle fracture and the performance will decrease.

(3) As the space frame content increases from 50% to 70%, the relative density of the material gradually decreases from 0.47 to 0.29, and the yield strength and platform stress subsequently show a power function trend from 54.15 MPa and 53.63 MPa to 13.52 MPa and 10.51 MPa, respectively. Thus, the compressive properties and energy absorption capacity of the material gradually decrease.

Therefore, porous composites with excellent compression and energy absorption properties can be obtained by combining the brittle phase and ductile phase of Al_3Zr.

Author Contributions: Conceptualization and formal analysis, X.Z.; methodology and investigation, K.X.; data curation and writing—original draft preparation, W.Z.; writing—review and editing and supervision, W.L. All authors have read and agreed to the published version of the manuscript.

Funding: This research received no external funding.

Data Availability Statement: Data available on request due to restrictions eg privacy or ethics. The data presented in this study are available on request from the corresponding author. The data are not publicly available.

Conflicts of Interest: The authors declare that we have no known competing financial interests or personal relationships that might influence the work reported herein, and that the authors unanimously agreed to submit the manuscript to your journal.

References

1. Colombo, P.; Degischer, H.P. Highly Porous Metals and Ceramics. *Mater. Sci. Technol.* **2013**, *26*, 1145–1158. [CrossRef]
2. Esmaeelzadeh, S.; Simchi, A.; Lehmhus, D. Effect of Ceramic Particle Addition on the Foaming Behavior, Cell Structure and Mechanical Properties of P/M AlSi7 Foam. *Mater. Sci. Eng. A* **2006**, *424*, 290–299. [CrossRef]
3. Shishkin, A.; Mironov, V.; Zemchenkov, V.; Antonov, M.; Hussainova, I. Hybrid Syntactic Foams of Metal—Fly Ash Cenosphere—Clay. *Key Eng. Mater.* **2016**, *674*, 35–40. [CrossRef]
4. Shishkin, A.; Drozdova, M.; Kozlov, V.; Hussainova, I.; Lehmhus, D. Vibration-Assisted Sputter Coating of Cenospheres: A New Approach for Realizing Cu-Based Metal Matrix Syntactic Foams. *Metals* **2017**, *7*, 16. [CrossRef]
5. Shishkin, A.; Hussainova, I.; Kozlov, V.; Lisnanskis, M.; Leroy, P.; Lehmhus, D. Metal-Coated Cenospheres Obtained Via Magnetron Sputter Coating: A New Precursor for Syntactic Foams. *J. Miner. Met. Mater. Soc.* **2018**, *70*, 1319–1325. [CrossRef]
6. Gibson, L.J.; Ashby, M.F. *Cellular Solids: Structure and Properties*; Cambridge University Press: Cambridge, UK, 1997.
7. Parveez, B.; Jamal, N.A.; Maleque, A.; Yusof, F.; Jamadon, N.H.; Adzila, S. Review on Advances in Porous Al Composites and the Possible Way Forward. *J. Mater. Res. Technol.* **2021**, *14*, 2017–2038. [CrossRef]
8. Ning, J.; Li, Y.; Zhao, G. Simple Multi-Sections Unit-Cell Model for Sound Absorption Characteristics of Lotus-Type Porous Metals. *Phys. Fluids* **2019**, *31*, 077102.
9. Wang, N.; Maire, E.; Chen, X.; Adrien, J.; Li, Y.; Amani, Y.; Hu, L.; Cheng, Y. Compressive Performance and Deformation Mechanism of the Dynamic Gas Injection Aluminum Foams. *Mater. Charact.* **2019**, *147*, 11–20. [CrossRef]

10. Yang, D.-H.; Chen, J.-Q.; Wang, L.; Jiang, J.-H.; Ma, A.-B. Fabrication of Al Foam without Thickening Process through Melt-Foaming Method. *J. Iron. Steel Res. Int.* **2018**, *25*, 90–98. [CrossRef]
11. Kubelka, P.; Körte, F.; Heimann, J.; Xiong, X.; Jost, N. Investigation of a Template-Based Process Chain for Investment Casting of Open-Cell Metal Foams. *Adv. Eng. Mater.* **2022**, *24*, 2100608. [CrossRef]
12. Shi, X.; Xu, G.; Liang, S.; Li, C.; Guo, S.; Xie, X.; Ma, X.; Zhou, J. Homogeneous Deposition of Zinc on Three-Dimensional Porous Copper Foam as a Superior Zinc Metal Anode. *ACS Sustain. Chem. Eng.* **2019**, *7*, 17737–17746. [CrossRef]
13. Jain, H.; Mondal, D.; Gupta, G.; Kumar, R.; Singh, S. Synthesis and Characterization of 316l Stainless Steel Foam Made through Two Different Removal Process of Space Holder Method. *Manuf. Lett.* **2020**, *26*, 33–36. [CrossRef]
14. Inoguchi, N.; Kobashi, M.; Kanetake, N. Synthesis of Porous Al_3Ti/Al Composite and Effect of Precursor Processing Condition on Cell Morphology. *Mater. Trans.* **2009**, *50*, 2609–2614. [CrossRef]
15. Atturan, U.A.; Nandam, S.H.; Murty, B.; Sankaran, S. Deformation Behaviour of in-Situ Tib2 Reinforced A357 Aluminium Alloy Composite Foams under Compressive and Impact Loading. *Mater. Sci. Eng. A* **2017**, *684*, 178–185. [CrossRef]
16. Morere, B.; Shahani, R.; Maurice, C.; Driver, J. The Influence of Al_3Zr Dispersoids on the Recrystallization of Hot-Deformed Aa 7010 Alloys. *Metall. Mater. Trans. A* **2001**, *32*, 625–632. [CrossRef]
17. Gautam, G.; Mohan, A. Effect of Zrb2 Particles on the Microstructure and Mechanical Properties of Hybrid (Zrb2+ Al3zr)/Aa5052 Insitu Composites. *J. Alloys Compd.* **2015**, *649*, 174–183. [CrossRef]
18. Varin, R. Intermetallic-Reinforced Light-Metal Matrix in-Situ Composites. *Met. Mater. Trans. A* **2002**, *33*, 193–201. [CrossRef]
19. Gupta, R.; Daniel, B. Strengthening Mechanisms in Al3zr-Reinforced Aluminum Composite Prepared by Ultrasonic Assisted Casting. *J. Mater. Eng. Perform.* **2021**, *30*, 2504–2513. [CrossRef]
20. Pourkhorshid, E.; Enayati, M.H.; Sabooni, S.; Karimzadeh, F.; Paydar, M.H. Bulk Al–Al3zr Composite Prepared by Mechanical Alloying and Hot Extrusion for High-Temperature Applications. *Int. J. Miner. Metall. Mater.* **2017**, *24*, 937–942. [CrossRef]
21. Xun, K.; Dai, W.; Liu, N.; Zhang, W.; Long, W.; Zhou, X. Effect of Sintering Temperature on Microstructure and Compressive Properties of Al_3Zr/2024Al Porous Composites. *Trans. Mater. HT.* **2022**, *43*, 20–27.
22. Yang, K.; Yang, X.; Liu, E.; Shi, C.; Ma, L.; He, C.; Li, Q.; Li, J.; Zhao, N. Elevated Temperature Compressive Properties and Energy Absorption Response of in-Situ Grown Cnt-Reinforced Al Composite Foams. *Mater. Sci. Eng. A* **2017**, *690*, 294–302. [CrossRef]
23. Moreira, R.C.S.; Kovalenko, O.; Souza, D.; Reis, R.P. Metal Matrix Composite Material Reinforced with Metal Wire and Produced with Gas Metal Arc Welding. *J. Compos. Mater.* **2019**, *53*, 4411–4426. [CrossRef]
24. Shu, Y.; Suzuki, A.; Takata, N.; Kobashi, M. Fabrication of Porous Nial Intermetallic Compounds with a Hierarchical Open-Cell Structure by Combustion Synthesis Reaction and Space Holder Method. *J. Mater. Process. Technol.* **2019**, *264*, 182–189. [CrossRef]
25. Bafti, H.; Habibolahzadeh, A. Production of Aluminum Foam by Spherical Carbamide Space Holder Technique-Processing Parameters. *Mater. Des.* **2010**, *31*, 4122–4129. [CrossRef]
26. Tamim, R.; Mahdouk, K. Thermodynamic Reassessment of the Al–Zr Binary System. *J. Therm. Anal. Calorim.* **2018**, *131*, 1187–1200. [CrossRef]
27. Zhao, A.; Qiu, S.; Hu, Y. Effect of Parametric Variations on the Local Compression Deformation of Aluminum Foam Sandwich Panels. *Mater. Trans.* **2017**, *58*, 880–885. [CrossRef]
28. Luo, Y.; Yu, S.; Li, W.; Liu, J.; Wei, M. Compressive Behavior of SiCp/AlSi9Mg Composite Foams. *J. Alloys Compd.* **2008**, *460*, 294–298. [CrossRef]
29. Raj, R.E.; Daniel, B. Aluminum Melt Foam Processing for Light-Weight Structures. *Mater. Manuf. Process.* **2007**, *22*, 525–530. [CrossRef]
30. Zettl, B.; Mayer, H.; Stanzl-Tschegg, S.; Degischer, H. Fatigue Properties of Aluminium Foams at High Numbers of Cycles. *Mater. Sci. Eng. A* **2000**, *292*, 1–7. [CrossRef]
31. Jain, H.; Gupta, G.; Kumar, R.; Mondal, D.P. Microstructure and Compressive Deformation Behavior of Ss Foam Made through Evaporation of Urea as Space Holder. *Mater. Chem. Phys.* **2019**, *223*, 737–744. [CrossRef]
32. Suzuki, A.; Kosugi, N.; Takata, N.; Kobashi, M. Microstructure and Compressive Properties of Porous Hybrid Materials Consisting of Ductile Al/Ti and Brittle Al3ti Phases Fabricated by Reaction Sintering with Space Holder. *Mater. Sci. Eng. A* **2020**, *776*, 139000. [CrossRef]
33. Zhiwei, D.; Jian, C.; Binna, S. Preparation of Sic/Al Composite Foams by Spark Plasma Sintering and Dissolution and Its Compression Properties. *J. Mater. Metall.* **2019**, *18*, 121–126.
34. Alizadeh, M.; Mirzaei-Aliabadi, M. Compressive Properties and Energy Absorption Behavior of Al–Al_2O_3 Composite Foam Synthesized by Space-Holder Technique. *Mater. Des.* **2012**, *35*, 419–424. [CrossRef]
35. Yang, X.; Hu, Q.; Du, J.; Song, H.; Zou, T.; Sha, J.; He, C.; Zhao, N. Compression Fatigue Properties of Open-Cell Aluminum Foams Fabricated by Space-Holder Method. *Int. J. Fatigue* **2019**, *121*, 272–280. [CrossRef]
36. Mondal, D.; Majumder, J.D.; Jha, N.; Badkul, A.; Das, S.; Patel, A.; Gupta, G. Titanium-Cenosphere Syntactic Foam Made through Powder Metallurgy Route. *Mater. Des.* **2012**, *34*, 82–89. [CrossRef]

Article

Microstructure Comparison for AlSn20Cu Antifriction Alloys Prepared by Semi-Continuous Casting, Semi-Solid Die Casting, and Spray Forming

Shuhui Huang [1,2], Baohong Zhu [1,2,*], Yongan Zhang [1,2], Hongwei Liu [1,2], Shuaishuai Wu [1,2] and Haofeng Xie [1,2]

1   State Key Laboratory of Nonferrous Metals and Processes, GRINM Group Co., Ltd., Beijing 100088, China
2   GRIMAT Engineering Institute Co., Ltd., Beijing 101407, China
*   Correspondence: zhubh@grinm.com

Abstract: Antifriction alloys such as AlSn20Cu are key material options for sliding bearings used in machinery. Uniform distribution and a near-equiaxed granularity tin phase are generally considered to be ideal characteristics of an AlSn20Cu antifriction alloy, although these properties vary by fabrication method. In this study, to analyze the variation of the microstructure with the fabrication method, AlSn20Cu alloys are prepared by three methods: semi-continuous casting, semi-solid die casting, and spray forming. Bearing blanks are subsequently prepared from the fabricated alloys using different processes. Morphological information, such as the total area ratio and average particle diameter of the tin phase, are quantitatively characterized. For the tin phase of the AlSn20Cu alloy, the deformation and annealing involved in semi-continuous casting leads to a prolate particle shape. The average particle diameter of the tin phase is 12.6 µm, and the overall distribution state is related to the deformation direction. The tin phase of AlSn20Cu alloys prepared by semi-solid die casting presents both nearly spherical and strip shapes, with an average particle diameter of 9.6 µm. The tin phase of AlSn20Cu alloys prepared by spray forming and blocking hot extrusion presents a nearly equilateral shape, with an average particle diameter of 6.2 µm. These results indicate that, of the three preparation methods analyzed in this study, semi-solid die casting provides the shortest process flow time, whereas a finer and more uniform tin-phase structure may be obtained using the spray-forming process. The semi-solid die casting method presents the greatest potential for industrial application, and this method therefore presents a promising possibility for further optimization.

Keywords: AlSn20Cu alloy; microstructure; semi-continuous casting; semi-solid die casting; spray forming; antifriction alloys; bearings

1. Introduction

Sliding bearings are key components that are commonly used in machinery; antifriction alloys are the main materials employed in their manufacture. Bearing antifriction alloy materials generally have two metallographic structures. The first type of structure is based on a soft-phase matrix, where the hard phase is evenly distributed in the form of particles, such as in tin- and lead-based alloys. The second type of structure is based on a hard-phase matrix, where the soft phase is uniformly distributed in the form of particles, such as in aluminum–tin alloys and copper–lead alloys [1,2]. The Babbitt alloy belongs to the first type of antifriction alloy. It exhibits good compliance, compatibility, and embedment with other materials, although it has poor bearing capacity and heat resistance. It is prone to sticking and corrosion because of its lead content. The Babbitt alloy is therefore suitable for use under stable load working conditions, but not for use under heavy load conditions, and has been gradually phased out of industrial production. The second type of antifriction alloy for preparing self-lubricant bearings presents significant advantages. During working, the hard matrix structure of the bearing ensures that the bearing bush is not deformed, whereas the soft phase is easily worn out, forming a gap between the bearing bush and the bearing that contains the lubricant. As a representative example of the second type of antifriction alloy, aluminum–tin and copper–lead alloys are widely used in high-load mechanisms. However, preparation of this second type of antifriction alloy is more difficult than that of the first, where the main difficulty is related to controlling the distribution morphology of the soft phase in the matrix. For aluminum–tin alloys, the tin-phase morphology in the alloy is one of the most critical technical indicators for potential applications, and the fine, uniform, and nearly equiaxed Sn phase is the ideal application state for alloys [3–5].

Semi-continuous casting of aluminum–tin alloy ingots, followed by processing the aluminum–tin alloy bearing blanks through multiple deformations and annealing, is presently the most commonly used process. The morphology of the tin phase of alloys prepared in this manner is affected by the deformation, and the phase forms a flat strip along the deformation direction. Annealing treatments above the melting point of tin are used to obtain a granular tin phase, which causes tin liquefaction and overflow from the matrix [6–10]. Aluminum–tin alloys can be prepared by powder metallurgy [11–15]. However, this process is extremely complicated, involving multiple processes such as powder milling, powder mixing, wrapping, degassing, and sintering. In addition, if the sintering temperature is higher than the melting point of the tin phase (232 °C), the tin phase becomes reticular and segregates. However, if the sintering temperature is lower than the melting point of the tin phase, the aluminum matrix is not able to metallurgically bond. Although there are many reports on the preparation of aluminum alloys by spray forming, few studies describe aluminum–tin alloy preparation by spray forming, although these studies [16–19] confirmed the feasibility of preparing Al-Sn alloys using this approach. However, because spray forming involves the use of high-speed airflow as the driving force for deposition and forming, defects are inevitably introduced; few studies have considered the analysis and elimination of these defects. In contrast, the preparation of aluminum alloys by semi-solid die casting is a mature technology [20–22], although there have been no studies describing the preparation of aluminum–tin alloys by semi-solid die casting.

In this study, three methods: semi-continuous casting, semi-solid die casting, and spray forming are used to prepare aluminum–tin alloy billets for analyzing the variation of microstructure with fabrication method. Different processes are used for the subsequent processing (except semi-solid die casting) to obtain self-lubricating and antifriction bearing bush blanks. The microstructures of the materials obtained by these three processes are compared, providing valuable data that may be used in the improvement of aluminum–tin alloy processing technologies. This study also provides a meaningful experimental reference for the preparation of alloys such as copper–lead alloys, in which a low-melting-point phase is uniformly distributed in a high-melting-point matrix.

2. Materials and Methods

The composition of the alloy prepared in this experiment is 20.0 wt.% Sn, 2.0 wt.% Cu, with the remainder of the material comprising Al. The purity of all raw materials is above 99.9%. The software PANDAT (CompuTherm LLC, Middleton, WI, USA) was used to calculate the equilibrium phase diagram of the AlSn20Cu alloy and the results are shown in Figure 1. Figure 1a shows the phase transition with increasing Sn content in Al–2Cu matrix, and Figure 1b shows the phase transition with increasing Cu content in Al–20Sn matrix. Tin and aluminum cannot dissolve each other in the solid state, they both maintain the original phase structure. Copper is present in the aluminum matrix (FCC_Al) in the form of Al_2Cu (AlCu_Theta).

Spray forming is conducted using SF-500 equipment (GRIMAT Ltd., Beijing, China), which is a self-made machine. The schematic diagram of SF-500 equipment for spray forming is shown in Figure 2. Semi-continuous casting and semi-solid die casting was conducted using universal equipment. A metallographic microscope and scanning electron microscope (SEM) are used to observe the microstructures of the alloys. Image processing software (ImageJ software, National Institutes of Health & Laboratory for Optical and

Computational Instrumentation, Madison, WI, USA) is used to analyze the morphology of the Sn phase.

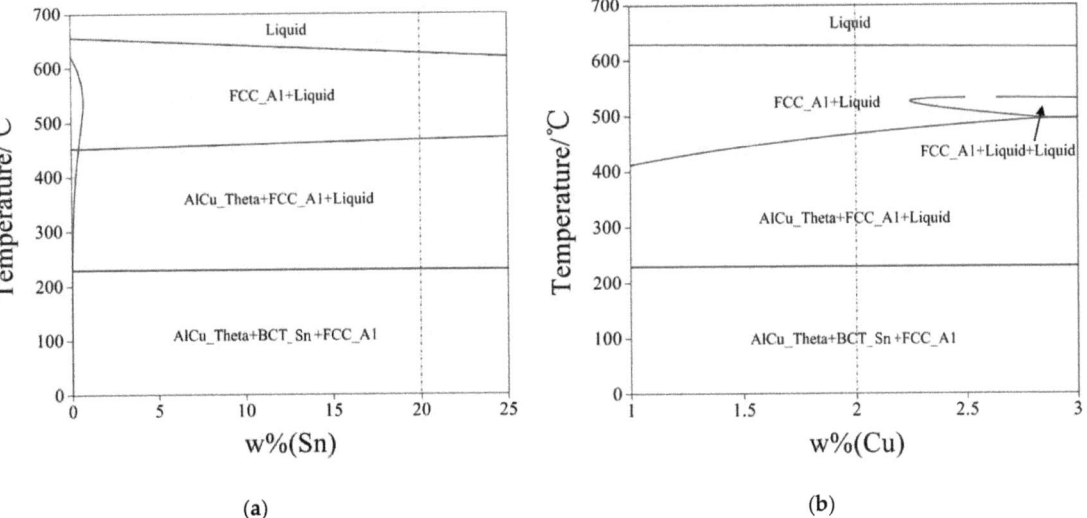

Figure 1. The equilibrium phase diagram of the AlSn20Cu alloy calculated by PANDAT: (**a**) the phase transition with increasing Sn content in Al–2Cu matrix, (**b**) the phase transition with increasing Cu content in Al–20Sn matrix.

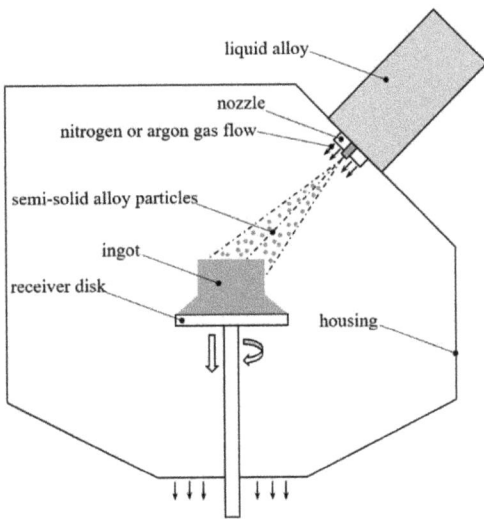

Figure 2. Schematic diagram of the spray-forming process.

The AlSnCu alloy ingot prepared by semi-continuous casting is cylindrical, which needs to be rolled and heat treated to form the plate for making the bearing bush. Gleeble compression experiments were used to study the process parameters of rolling deformation. The compression experiment temperature is 20, 100, 150, 202, 222, and 242 °C; the strain rate is 0.01, 0.1, and 1 s^{-1}; the deformation is 60%; and the quenching and cooling are realized within 3 s after deformation. The AlSnCu alloy ingot prepared by semi-solid die casting is flat, which reduces the production process of the bearing bush. The AlSnCu alloy ingots prepared by spray forming are also cylindrical. Because there are many defects in it, it needs to be reduced by densification process, and finally the ingot is sliced into plates.

3. Results

3.1. Semi-Continuous Casting

The process and ingot of an AlSn20Cu alloy prepared by semi-continuous casting is shown in Figure 3, for which the ingot diameter is ϕ150 mm. The metallographic and SEM photos of the semi-continuously cast AlSn20Cu alloy are shown in Figure 4. There are some porosity casting defects in the ingot. In addition to the tin phase that is distributed at the grain boundaries of the aluminum matrix, there are also granular tin phases inside the grains. Thus, the tin phase is distributed in a network in the aluminum matrix, resulting in poor deformation properties of the alloy.

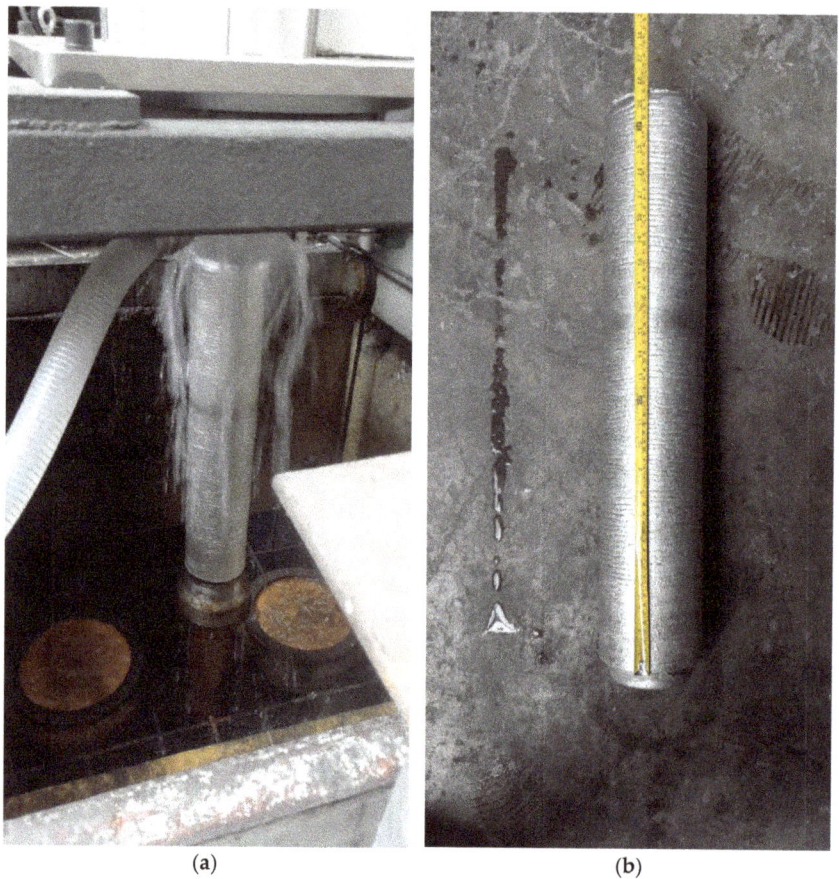

Figure 3. (**a**) Semi-continuous casting process, and (**b**) resulting AlSn20Cu alloy ingot.

Figure 5 shows the EDS analysis of semi-continuous casting AlSnCu alloy, and the order corresponds to the points marked in Figure 4b. From the results of EDS analysis, it can be seen that the matrix is mainly composed of aluminum, and a small amount of copper is solid solution in the matrix. The brightest phase is tin, and a small amount of gray phase is Al$_2$Cu. Subsequently, the phase of the alloy was analyzed by XRD, and the results are shown in Figure 6. XRD analysis can only see the diffraction characteristics of Al and Sn, but not the diffraction information of Al$_2$Cu. This also shows that the amount of Al$_2$Cu is very small, so in the subsequent quantitative analysis of the phase, the Al$_2$Cu phase can be ignored.

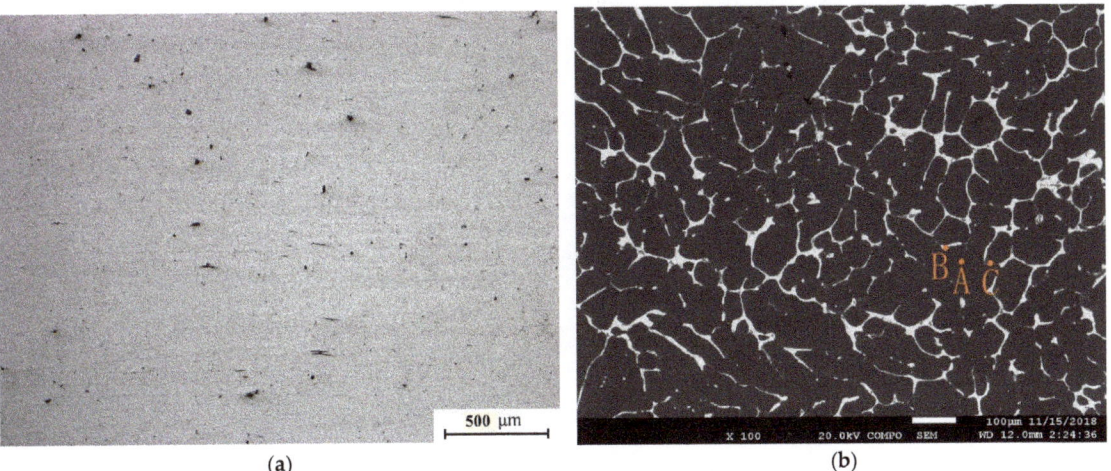

Figure 4. (a) Metallographic and (b) SEM images of the AlSn20Cu alloy fabricated using semi-continuous-casting.

The deformation properties of AlSn20Cu are studied using a Gleeble compression experiment. Macro-photographs of the post-compression test are shown in Figure 7.

When subject to deformation above the Sn melting point (232 °C), the Sn phase melts and is extruded from the matrix. Deformation below the melting point of the Sn phase results in a smooth sample surface. However, when deformation is too fast, deformation heat causes an increase in sample temperature, resulting in melting of the Sn phase. In this experiment, Sn phase melting occurs when deformation is conducted at a temperature and strain rate of 222 °C and 1 s^{-1}, respectively.

When the deformation temperature is below 150 °C, the alloy will crack, even at an extremely low deformation rate. In Figure 7, the red boxes mark samples with cracking, the blue boxes mark samples with tin-phase spillage, and the green boxes mark samples with neither cracking nor tin-phase spillage. Therefore, the blocking hot extrusion process will be conducted at a strain rate of 0.01–0.1 s^{-1} between 202 and 222 °C.

Figure 8 shows photos of the preparation of an antifriction bearing bush blank using a semi-continuous cast AlSn20Cu alloy. Figure 8a,b indicates a hot rolled billet at 210 °C for the first time and second time, respectively. Coating was applied to the surface of the sheet to prevent tin spillage, and the sheet was annealed at 250 °C for 10 h (Figure 8c). Figure 8d shows the final milled sheet after rolling.

The antifriction bearing bush blank is prepared by two hot rolling processes, one heat treatment and one cold rolling process. Its microstructure is shown in Figure 9. The initial network morphology of the tin phase is destroyed and transformed into a granular form. The particle shape shows directionality and is stretched along the deformation direction.

3.2. Semi-Solid Die Casting

According to the Al-Sn-Cu phase diagram in Figure 1, the temperature of the AlSn20Cu alloy solution is adjusted to approximately 720 °C. It is then transferred to a mechanical vibration platform. Under the action of mechanical vibration and stirring, when the alloy solution drops to 610 ± 5 °C, it is transferred to the die casting machine, for which the speed of the in-mold gate is fixed at approximately 40 m/s. Figure 10 shows the semi-solid die casting billet; its corresponding SEM image is shown in Figure 11. Compared with that produced using semi-continuous casting, the AlSn20Cu alloy prepared by semi-solid die casting does not require deformation and annealing, and a granular tin phase may be obtained. No defects are observed in the ingot, although some tin phases are observed in strips along the grain boundaries of the aluminum matrix.

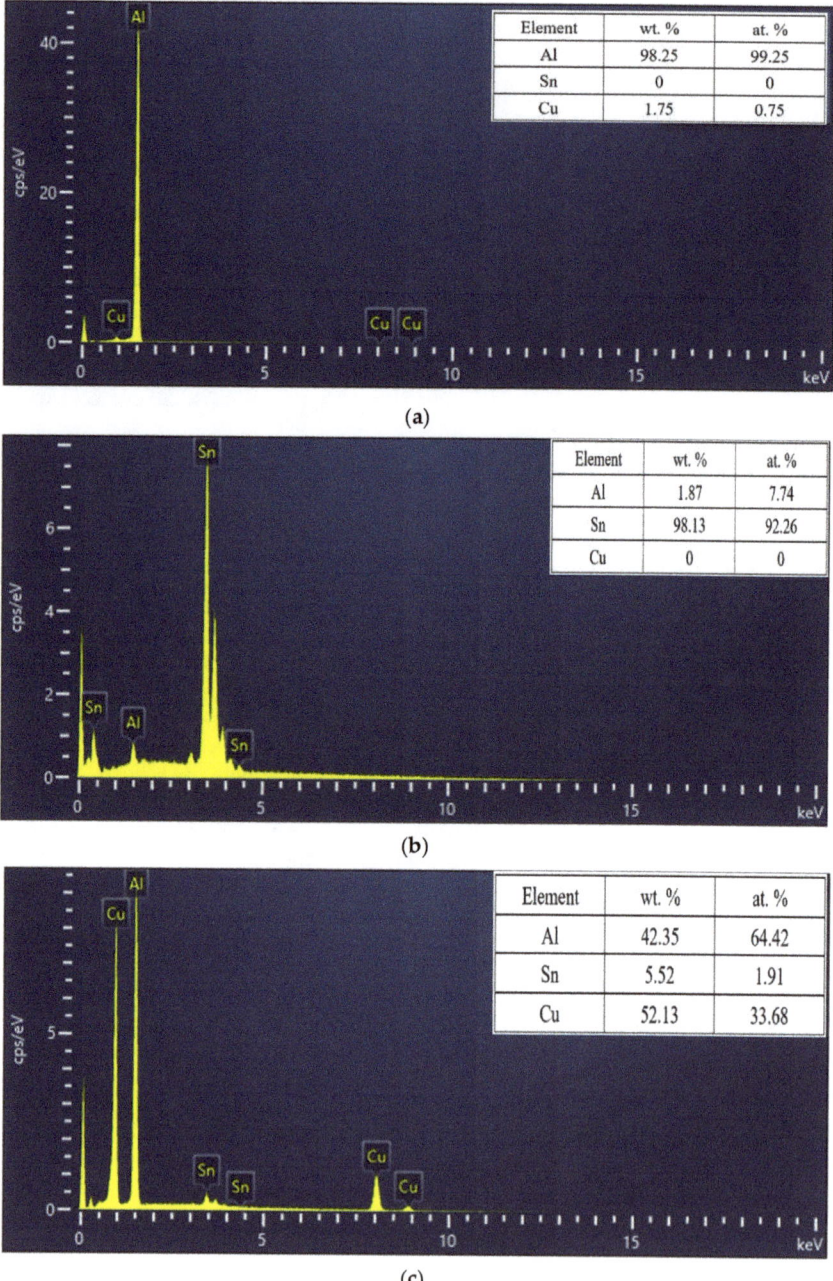

Figure 5. EDS analysis of semi-continuously cast AlSnCu alloy: the order corresponds to the points marked in Figure 4b. (**a**) point A, (**b**) point B, (**c**) point C.

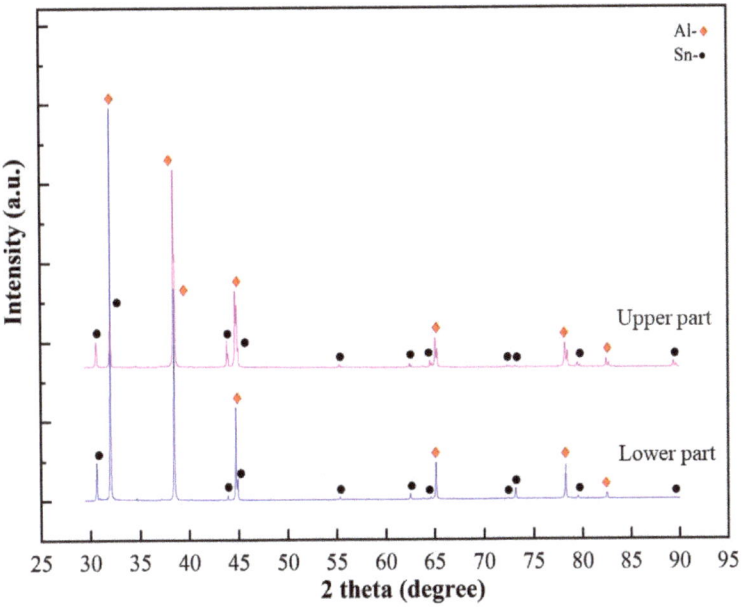

Figure 6. XRD analysis of semi-continuously cast AlSnCu alloy.

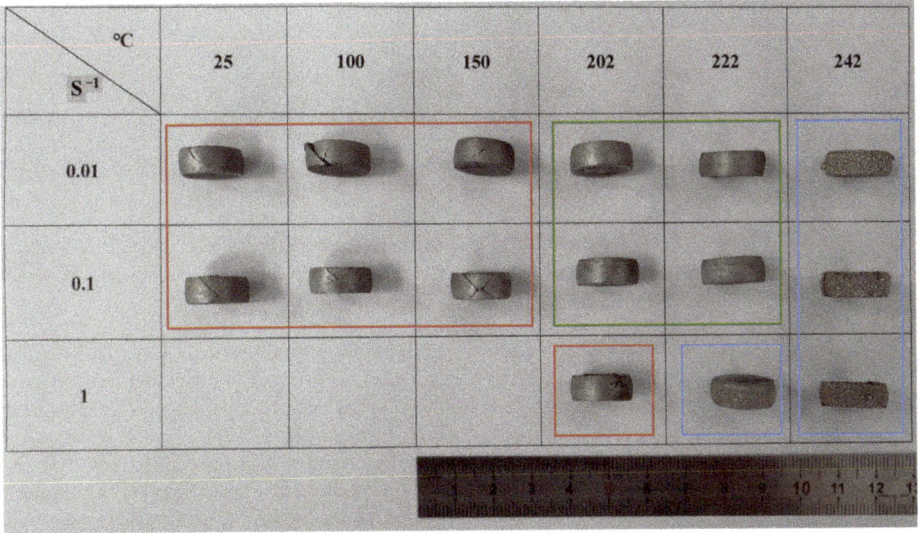

Figure 7. Photos of the alloy specimen after hot compression tests.

3.3. Spray Forming

The alloy ingot with an outer contour size of Φ190 mm × 500 mm prepared by spray forming is shown in Figure 12. Figure 13 shows a metallographic photo of the ingot produced using spray forming. It can be observed that, compared with the other ingot making processes, the number of defects in the ingot prepared by spray forming is larger, but the defect size is significantly smaller than that of ordinary casting. Figure 14 shows an SEM image of the ingot produced using spray forming.

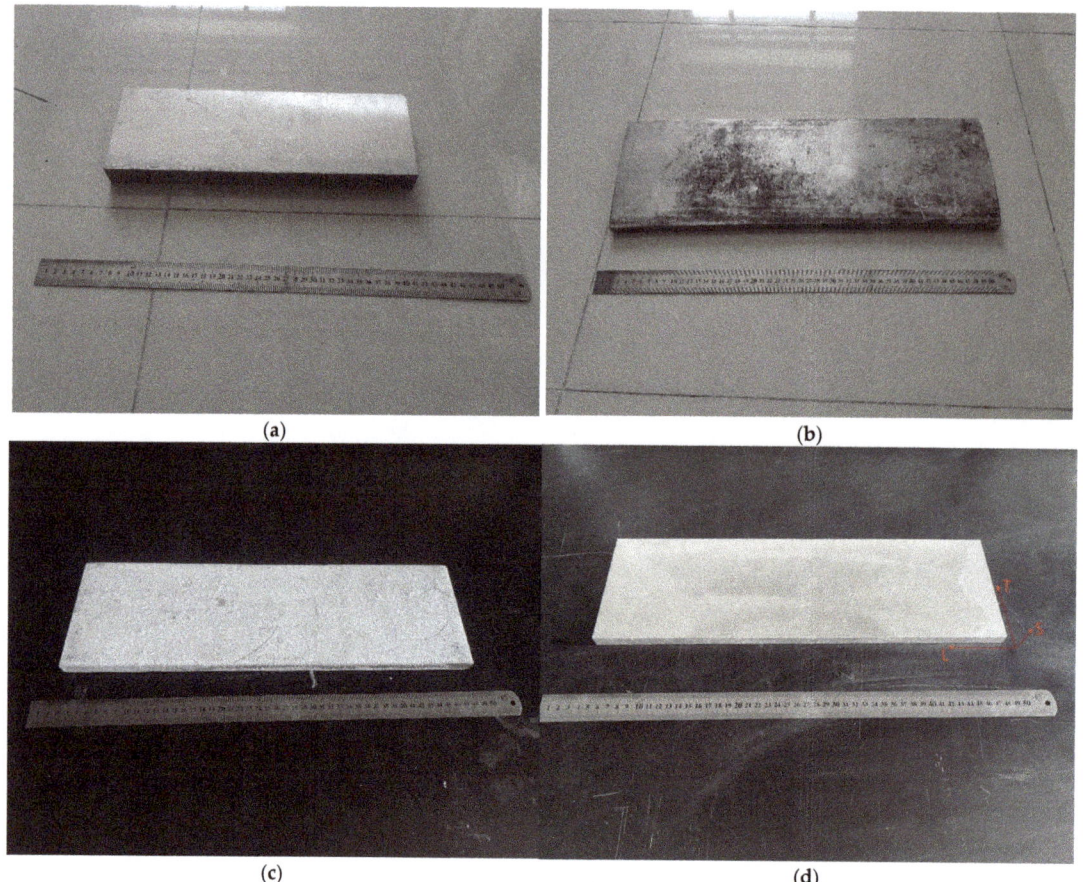

Figure 8. Preparation process for an antifriction bearing bush blank produced using a semi-continuous cast AlSn20Cu alloy: (**a**) first hot rolling, (**b**) second hot rolling, (**c**) annealing, and (**d**) final milled sheet after cold rolling.

High-speed ejected nitrogen or argon is used to drag the metal droplets towards the receiver desk during the spray-forming process, and the metal droplets rapidly cool during flight to form semi-solid particles, which are superimposed and deposited into the ingot. The particles inevitably entrain gases during the deposition process, and these gas particles cause defects in the ingot. The hot isostatic pressing densification process often used in powder metallurgy is not suitable for the elimination of gas-containing defects in spray-formed ingots, and the schematic diagram is shown in Figure 15. During hot isostatic pressing, the ingot is completely immersed in high pressure gas. The ingot is subjected to the same gas pressure in all directions, so the load applied by the hot isostatic pressing process only contains the spherical tensor of the stress. Under the action of the spherical tensor of stress, gas-containing defects can be compressed but not discharged. As mentioned earlier, large plastic deformation processing such as rolling may be used to eliminate defects in the original ingot, but this will also significantly change the shape of the tin phase, thereby destroying the equilateral granular tin in the original structure. To eliminate gas without introducing large deformation to change the morphology of the tin phase, a densification method by hot extrusion is proposed in this study, as shown in Figure 16.

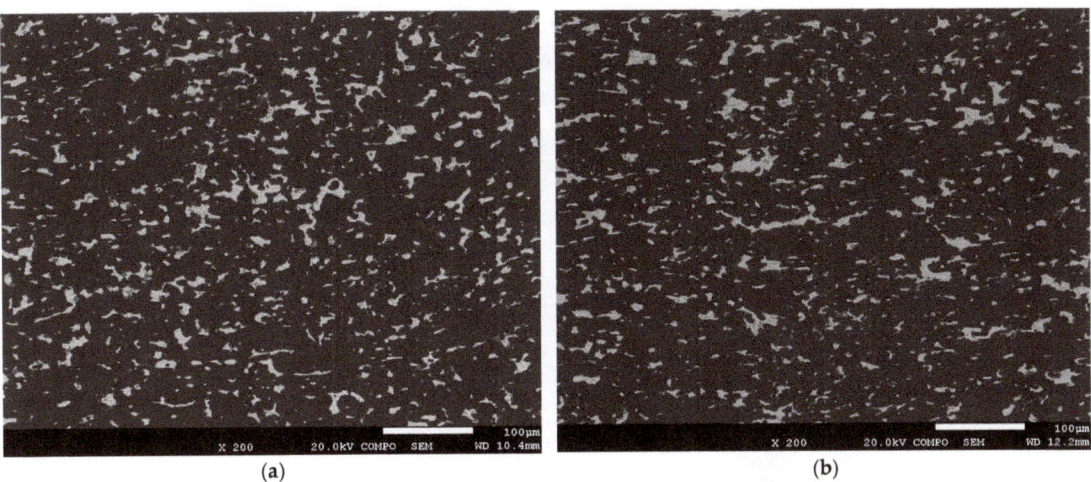

Figure 9. Scanning electron microscope (SEM) images of the final milled sheet: (**a**) S-T direction, and (**b**) L-S direction.

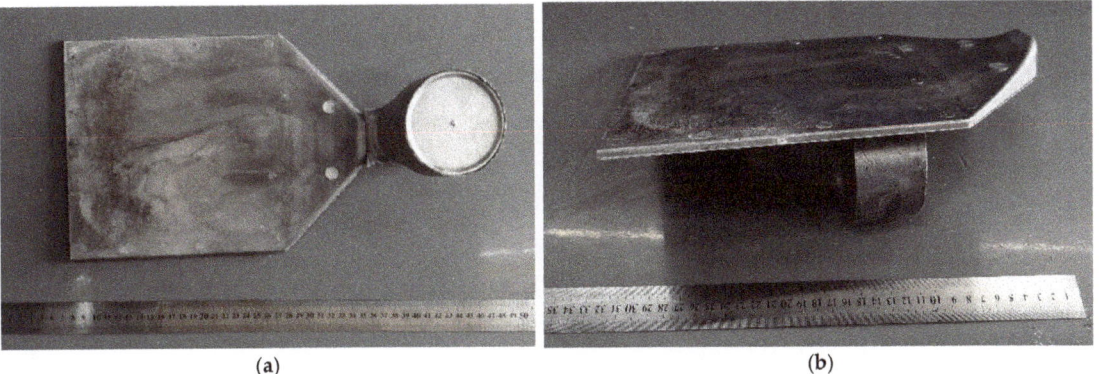

Figure 10. AlSn20Cu alloy billet produced using semi-solid die casting: (**a**) front view, and (**b**) side view.

Figure 17 shows the effects of traditional hot isostatic pressing and blocking hot extrusion on gas-containing defects. The load applied by the blocking hot extrusion process includes not only the stress ball tensor but also the stress deviator tensor. The gas is squeezed out of the defect and the defect is bridged. In blocking hot extrusion, the densification of the billet may be achieved with a small macroscopic deformation resulting from the small diameter of the extrusion barrel that limits the size of the billet.

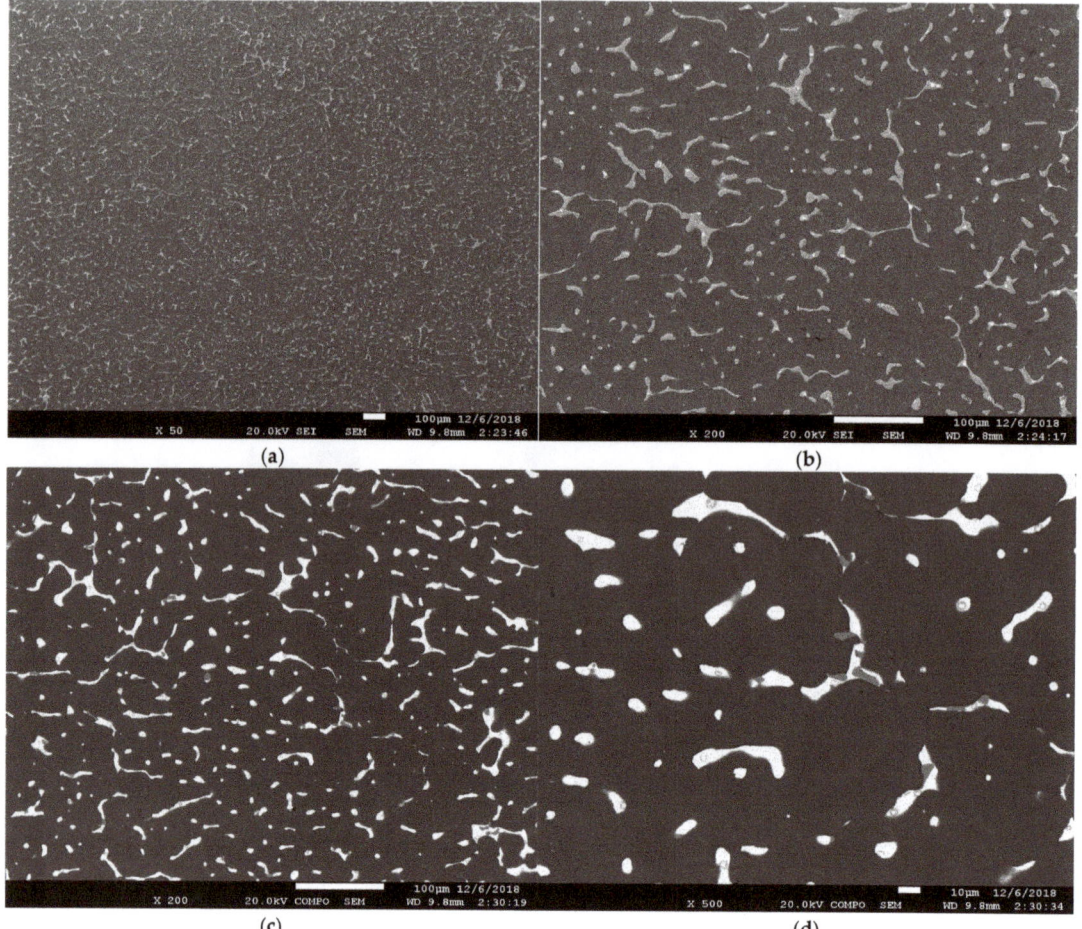

Figure 11. SEM images of a billet produced using semi-solid die casting: (**a**,**b**) secondary electron imaging, (**c**,**d**) backscattered electron imaging.

The original diameter of the extrusion billet is Φ175 mm, with an extrusion cylinder diameter of Φ180 mm. Graphite lubricant is applied to the surface to reduce friction and enhance the flow capacity of the material. The extrusion is conducted using a 1250 ton extruder. According to the Gleeble compression experiment results, the billet and extrusion cylinder are heated to 215 ± 5 °C. The ingot is continuously pressed three times using a load of more than 1000 tons. The billet is analyzed after blocking hot extrusion, as shown in the SEM images in Figure 18. No defects are found in the secondary electron images, and the Sn phase morphology in the backscattered electron images shows nearly equilateral granular morphology.

Figure 12. Photo of spray-formed alloy ingot.

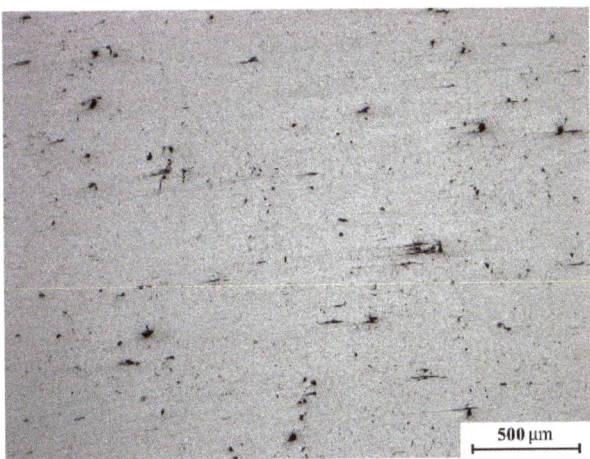

Figure 13. Metallographic photo of the spray-formed alloy ingot.

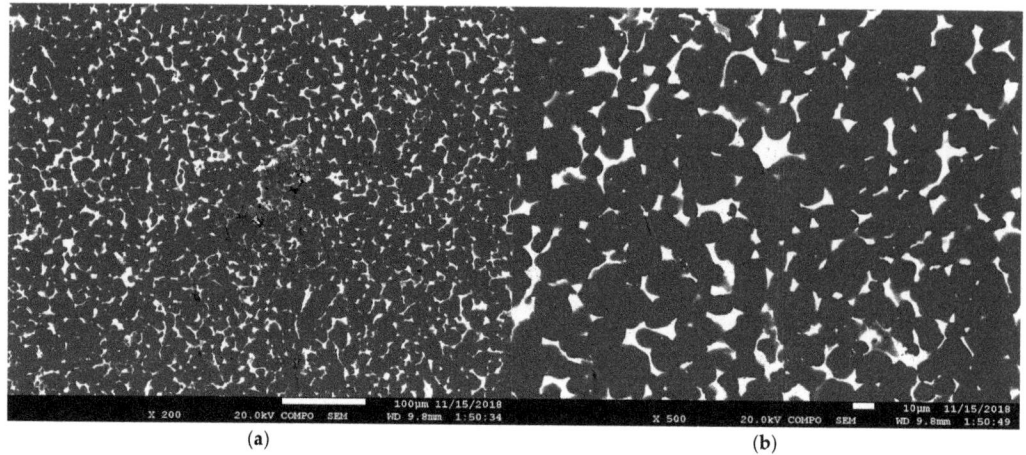

Figure 14. SEM image of the spray-formed alloy ingot. (**a**) magnified 200 times, (**b**) magnified 500 times.

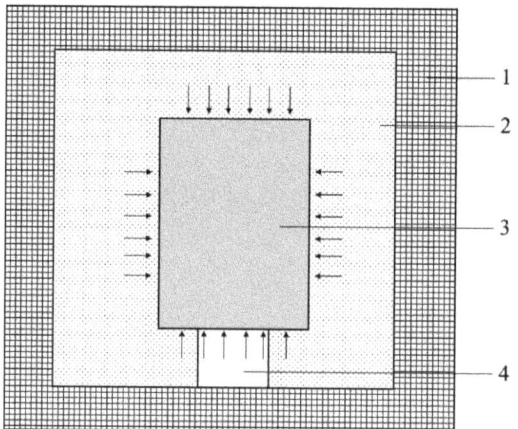

Figure 15. Schematic diagram showing the densification process of hot isostatic pressing. 1: hot isostatic pressing furnace, 2: gas, 3: billet, 4: holder.

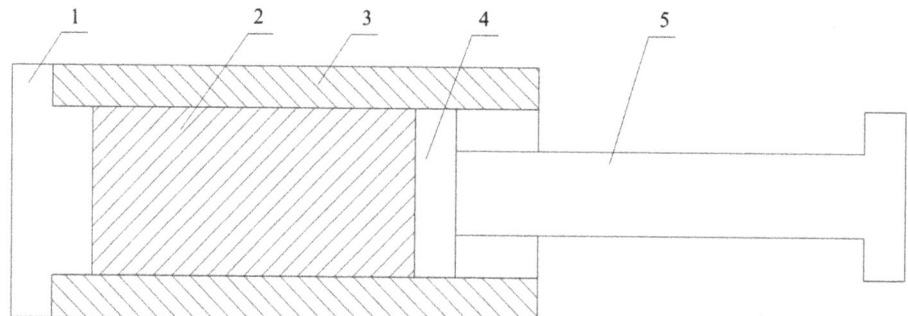

Figure 16. Schematic diagram showing the densification process of blocking hot extrusion. 1: Blocking extrusion die, 2: billet, 3: extrusion cylinder, 4: extrusion pad, and 5: extrusion rod.

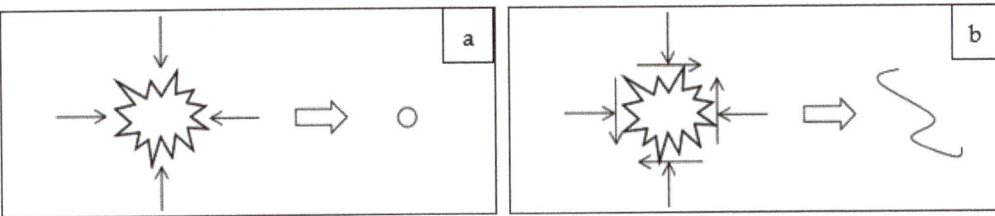

Figure 17. Schematic diagram showing the effect of (**a**) hot isostatic pressing, and (**b**) blocking hot extrusion on gas-containing defects in the alloy.

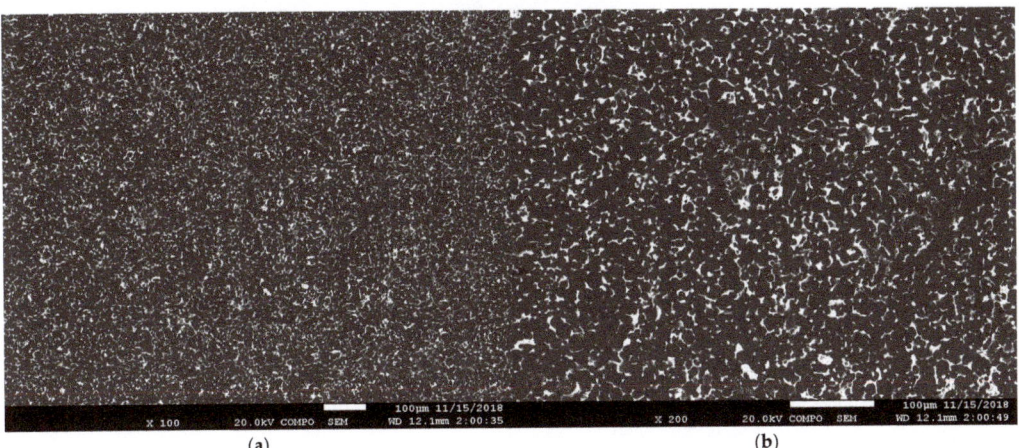

Figure 18. SEM images of a spray-formed ingot after densification. (**a**) magnified 100 times, (**b**) magnified 200 times.

4. Discussion

Cu exists in the AlSn20Cu alloy aluminum matrix as a solid solution element, and Sn is mostly insoluble in aluminum. The discontinuous distribution of the tin phase in the grain boundaries of the aluminum matrix results in poor plasticity of the alloy. As mentioned in Section 1, aluminum–tin antifriction alloys generally require a tin-phase morphology that is in the form of nearly equilateral particles. Because of the weak deformation capacity of the alloy, the acquisition of nearly uniformly distributed equilateral granular tin phase is the main focus of this study. According to engineering practice experience and the literature [5,18], the ideal distribution of the tin phase in an AlSn20Cu alloy in terms of bearing preparation is shown in Figure 19. The tin phase is uniformly distributed in a spherical shape in a matrix that comprises the aluminum alloy.

The densities of aluminum, tin, and copper are 2.70, 7.31, and 8.96 g/cm^3, respectively. Assuming that there is no volume change before and after the formation of a solid solution of aluminum and copper, the area fraction of the Sn phase in the cross-section of the AlSn20Cu alloy may be calculated as follows.

$$M_{Al}:M_{Sn}:M_{Cu} = 0.78:0.2:2 \tag{1}$$

$$M_{Al}:M_{Sn}:M_{Cu} = (\rho V)_{Al}:(\rho V)_{Sn}:(\rho V)_{Cu} \Rightarrow V_{Al}:V_{Sn}:V_{Cu} = 0.907:0.086:0.007 \tag{2}$$

$$V^{1/3} \propto S^{1/2} \Rightarrow S_{Al}:S_{Sn}:S_{Cu} = 0.803:0.166:0.031 \tag{3}$$

where M, V, and ρ are mass, volume and density of the elements in AlSn20Cu alloy, and S is area of the elements in a section. On the premise that the weight ratio of each element in the alloy is known, according to the density of each element, the volume fraction occupied by them can be calculated. In any section of the alloy block, it is obvious that the square root of the area occupied by each element is proportional to the cube root of the volume occupied by each element in the block. According to these calculations, the area fraction occupied by the tin phase in the microstructure section of the ideal-state AlSn20Cu should be 16.6%.

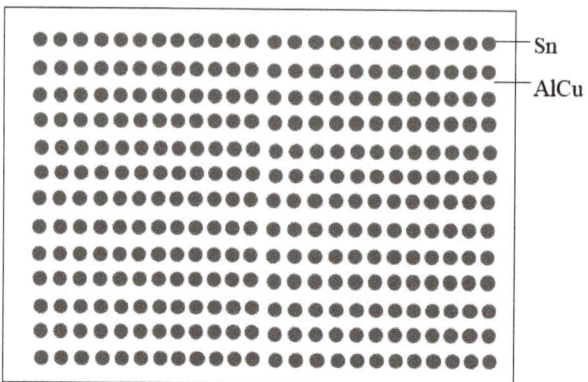

Figure 19. Ideal distribution of the tin phase in an aluminum–tin alloy.

The final microstructures of the AlSn20Cu alloys prepared using the three different processes are shown in Figures 4, 9, 11, 14 and 18. The same raw material is used in each process. The morphology of the Sn phase is quantitatively analyzed using image processing software, as shown in Table 1. The area fraction, particle size, and number density of tin-phase particles are analyzed, and the data in Table 1 are converted from pixel data.

The spray-forming process provides the highest ingot Sn content, and the Sn content area fraction of its cross-section reaches 13.6%, which is close to the ideal value of 16.6%. Semi-solid die casting provides the lowest ingot Sn content, at only 9.2%. The tin content in the aluminum alloy matrix is positively correlated with the cooling rate, and the faster the cooling rate, the higher is the tin content in the alloy. The smallest tin phase is observed in the ingot prepared by spray forming, with an average particle diameter of 6.5 µm. The tin phase in the ingot prepared by semi-continuous casting is the coarsest, with an average particle diameter of 13.1 µm. The agitation and vibration in the semi-solid die casting process have significant effects on the refinement of the tin phase.

Table 1. Result of quantitative analysis of Sn phase morphology.

Morphological Parameters of Tin Phase	Original State			Final State		
	Semi-Continuous Casting	Semi-Solid Die Casting	Spray Forming	Semi-Continuous Casting	Semi-Solid Die Casting	Spray Forming
Total area ratio (%)	11.4	9.2	13.6	8.2	9.2	13.8
Quantity density [number/(100µm)^2]	8.4	12.6	39.3	6.5	12.6	40.2
Average particle area (µm^2)	135.8	72.6	32.8	125.7	72.6	30.5
Average particle diameter (µm)	13.1	9.6	6.5	12.6	9.6	6.2

After semi-continuous casting, rolling deformation and annealing, a large amount of tin is lost, and the area fraction of Sn content in the cross-section decreases from 11.4 to 8.2%. After spray forming and hot extrusion, there is no significant change in tin content and morphology. A large amount of tin in the ingot prepared by semi-solid die casting is left in the final solidified cylindrical biscuit, which essentially corresponds to the macro-segregation of tin.

According to the Al-Sn-Cu pseudo-binary partial phase diagram (Figure 1), the aluminum–tin alloy first forms Al dendrites below 660 °C during the solidification process. When cooled below 232 °C, the Sn phase gradually solidifies along the grain boundaries of the aluminum dendrites, thereby forming an as-cast network structure. The cooling rate of semi-continuous casting is of the order of 10 K/s, although the Sn phase is still distributed in a network shape along the grain boundaries of the Al matrix. After multiple deformations and annealing treatments, granular Sn that is uniformly distributed in the aluminum matrix may be obtained, and the average particle diameter of the Sn phase reaches approximately 12.6 µm. The annealing temperature must be above the melting point of the Sn phase. After the Sn phase is liquefied, the individual particles tend to be spherical under the action of surface tension. During the annealing process, after the Sn phase on the surface of the billet is liquefied, it will inevitably overflow, which results in an uneven distribution of the tin phase and a waste of materials.

The cooling rate of semi-solid die casting is similar to that of semi-continuous casting, which is in the order of 10 K/s. However, the vibration stirring and high-speed filling employed in the semi-solid die casting process have a crushing effect on the dendrites of the aluminum matrix. When the temperature of the alloy solution is lowered to approximately 610 °C under the conditions of vibration and stirring, the dendrites formed by the aluminum matrix are continuously broken and tend to be granular; the solidified aluminum reaches more than half of the total content. This is then injected into the mold cavity by pressure. The semi-solid alloy is rapidly cooled below the melting point of the tin phase, and the alloy is completely solidified. The AlSn20Cu alloy prepared by semi-solid die casting has both the characteristics of a network and granular shape. If active cooling can be introduced through appropriate design of the mold structure and the cooling rate after alloy injection can be increased, a more equilateral tin-phase granular microstructure can be obtained.

High-speed ejected nitrogen or argon is used as a power to drag the metal droplets during the spray-forming process, and the cooling speed may reach the order of 10^3 k/s. High-velocity airflow not only disperses the droplets, but also prompts their rapid cooling, changing the solidification process from both a thermal and mechanical perspective. The dispersed droplets have a particle size of approximately 30–50 µm, and are rapidly cooled from 720 °C until solidified. The aluminum matrix is gradually deposited and solidifies before forming dendrites, whereas the tin phase also rapidly solidifies. The Sn phase of the spray-formed ingot is in the shape of polygonal particles, with an average particle diameter of 6.5 µm. There are certain porosity defects in the ingot, but the gas may be discharged through the blocking extrusion, thereby eliminating the defects. One characteristic of the blocking extrusion process is that the densification of the material is achieved with less deformation, which implies that the morphology of the Sn phase does not significantly change after the densification process.

The initial shape of the Sn phase is determined by both thermal and mechanical factors during preparation. Different cooling rates and external force conditions cause significant differences in the tin-phase morphology of the AlSn20Cu alloys prepared by the three processes used in this study. Compared to semi-continuous casting and semi-solid die casting, the alloy cools faster during spray forming, in which the high-speed airflow suppresses and destroys aluminum dendrites. Although the cooling rate of semi-solid die casting is similar to that of semi-continuous casting, the destruction of dendrites by the vibration and stirring process prevents the tin phase from exhibiting a network-like distribution.

The study at this stage mainly focuses on the influence of the preparation process on tin phase morphology in AlSn20Cu alloys, and their wear-reducing properties are only qualitatively predicted based on tin phase morphology. The authors will carry out the wear reduction experiments of AlSn20Cu alloys prepared by three processes in the following work, and illustrate the effect of the tin phase morphology on the wear reduction performance of the alloys with specific experimental data.

5. Conclusions

(1) For the AlSn20Cu alloy prepared by semi-continuous casting, the majority of the tin phase is distributed in a network along the grain boundaries of the aluminum matrix. After deformation and annealing treatment, the tin-phase morphology changes from that of a network to prolate particles. The average particle diameter and total area ratio of the tin phase are 12.6 μm and 8.2%, respectively. Although the annealing process results in a granular tin phase, it also leads to a situation in which the tin phase overflows from the aluminum matrix.

(2) The tin phase of AlSn20Cu alloy products prepared by semi-solid die casting forms two shapes: nearly spherical and strips. The average particle diameter and total area ratio of the tin phase are 9.6 μm and 9.2%, respectively. The cooling rate of the semi-solid die casting process used in this study is not sufficient to prevent serious macro-segregation of the tin.

(3) In the AlSn20Cu alloy prepared by spray forming, the tin phase is mostly equilateral, although there are some defects in the matrix. After hot extrusion at 215 °C, the defects are completely eliminated, and the tin-phase morphology remains almost unchanged. The average particle diameter and total area ratio of the tin phase are 6.2 μm and 13.8%, respectively.

(4) The initial shape of the Sn phase is determined by both thermal and mechanical factors during preparation. A finer and more uniform tin-phase structure may be obtained by using the spray-forming process. Preparing an AlSn20Cu alloy by semi-solid die casting requires the shortest time of the three studied methods, and this method therefore presents a promising possibility for further optimization.

Author Contributions: Supervision, B.Z.; writing—original draft preparation, S.H.; writing—review and editing, S.H., B.Z. and Y.Z., and H.L.; investigations, S.W. and H.X. All authors have read and agreed to the published version of the manuscript.

Funding: This research received no external funding.

Data Availability Statement: Not applicable.

Conflicts of Interest: The authors declare no conflict of interest.

References

1. Stuczynski, T. Metallurgical problems associated with the production of aluminium-tin alloys. *Mater. Des.* **1997**, *18*, 369–372. [CrossRef]
2. Lu, Z.C.; Gao, Y.; Zeng, M.Q.; Zhu, M. Improving wear performance of dual-scale Al-Sn alloys: The role of Mg addition in enhancing Sn distribution and tribolayer stability. *Wear* **2014**, *309*, 216–225. [CrossRef]
3. Bertelli, F.; Brito, C.; Ferreira, I.L.; Reinhart, G.; Nguyen-Thi, H.; Mangelinck-Noël, N.; Cheung, N.; Garcia, A. Cooling thermal parameters, microstructure, segregation and hardness in directionally solidified Al-Sn-(Si;Cu) alloys. *Mater. Des.* **2015**, *72*, 31–42. [CrossRef]
4. Belova, N.A.; Akopyan, T.K.; Gershman, I.S.; Stolyarova, O.O.; Yakovleva, A.O. Effect of Si and Cu additions on the phase composition, microstructure and properties of Al-Sn alloys. *J. Alloys Compd.* **2017**, *695*, 2730–2739. [CrossRef]
5. Bertelli, F.; Freitas, E.S.; Cheung, N.; Arenas, M.A.; Conde, A.; Damborenea, J.; Garcia, A. Microstructure, tensile properties and wear resistance correlations on directionally solidified Al-Sn-(Cu; Si) alloys. *J. Alloys Compd.* **2017**, *695*, 3621–3631. [CrossRef]
6. Xu, K.; Russell, A.M. Texture strength relationships in a deformation processed Al-Sn metal-metal composite. *Mater. Sci. Eng. A* **2004**, *373*, 99–106. [CrossRef]
7. Mirkovic, D.; Grobner, J.; Schmid-Fetzer, R. Liquid demixing and microstructure formation in ternary Al–Sn–Cu alloys. *Mater. Sci. Eng. A* **2008**, *487*, 456–467. [CrossRef]

8. Schouwenaars, R.; Ramírez, E.I.; Romero, J.; Jacobo, V.H.; Ortiz, A. Fracture of thin cast slabs of Al-Sn alloys during cold rolling. *Eng. Fail. Anal.* **2012**, *25*, 175–181. [CrossRef]
9. Hernández, O.; Gonzalez, G. Microstructural and mechanical behavior of highly deformed Al–Sn alloys. *Mater. Charact.* **2008**, *59*, 534–541. [CrossRef]
10. Mahdavian, M.M.; Khatami-Hamedani, H.; Abedi, H.R. Macrostructure evolution and mechanical properties of accumulative roll bonded Al/Cu/Sn multilayer composite. *J. Alloys Compd.* **2017**, *703*, 605–613. [CrossRef]
11. Liu, X.; Zeng, M.Q.; Ma, Y.; Zhu, M. Promoting the high load-carrying capability of Al-20 wt%Sn bearing alloys through creating nanocomposite structure by mechanical alloying. *Wear* **2012**, *294–295*, 387–394. [CrossRef]
12. Xu, K.; Russell, A.M.; Chumbley, L.S.; Laabs, F.C. A deformation processed Al-20%Sn in-situ composite. *Scr. Mater.* **2001**, *44*, 935–940. [CrossRef]
13. Patel, J.; Morsi, K. Effect of mechanical alloying on the microstructure and properties of Al–Sn–Mg alloy. *J. Alloys Compd.* **2012**, *540*, 100–106. [CrossRef]
14. Lu, Z.C.; Zeng, M.Q.; Gao, Y.; Zhu, M. Significant improvement of wear properties by creating micro/nano dual-scale structure in Al-Sn alloys. *Wear* **2012**, *296*, 469–478. [CrossRef]
15. Liu, X.; Zeng, M.Q.; Ma, Y.; Zhu, M. Wear behavior of Al-Sn alloys with different distribution of Sn dispersoids manipulated by mechanical alloying and sintering. *Wear* **2008**, *265*, 1857–1863. [CrossRef]
16. Lavernia, E.J.; Ayers, J.D.; Srivatsan, T.S. Rapid solidification processing with specific application to aluminium alloys. *Int. Mater. Rev.* **1992**, *37*, 1–44. [CrossRef]
17. Lavernia, E.J.; Gutierrez, E.M.; Szekely, J. Spray deposition of metals. *Mater. Sci. Eng. A* **1988**, *98*, 381–394. [CrossRef]
18. Lucchetta, M.C.; Saporiti, F.; Audebert, F. Improvement of surface properties of an Al-Sn-Cu plain bearing alloy produced by rapid solidification. *J. Alloys Compd.* **2019**, *805*, 709–717. [CrossRef]
19. Li, H.; Jiang, X.; Wang, X. Effects of Target Microstructure on Al-Cu Alloy Sputtering and Depositing Performance. *Rare Met.* **2009**, *33*, 442–445.
20. Zhu, Q. Semi-solid moulding: Competition to cast and machine from forging in making automotive complex components. *Trans. Nonferrous Met. Soc.* **2010**, *20* (Suppl. S3), sl042–sl047. [CrossRef]
21. Atkinson, H.V.; Liu, D. Microstructural coarsening of semi-solid aluminium alloys. *Mater. Sci. Eng. A* **2008**, *496*, 439–446. [CrossRef]
22. Tebib, M.; Morin, J.B.; Jersch, F.A. Semi-solid processing of hypereutectic A390 alloys using novel rheoforming process. *Trans. Nonferrous Met. Soc.* **2010**, *20*, 1743–1748. [CrossRef]

Article

Hot Deformation Behavior of Alloy AA7003 with Different Zn/Mg Ratios

Xu Zheng [1,2], Jianguo Tang [1,3], Li Wan [1], Yan Zhao [1], Chuanrong Jiao [1] and Yong Zhang [1,3,*]

1 School of Materials Science and Engineering, Central South University, Changsha 410083, China
2 Guangxi Key Laboratory of Materials and Processes of Aluminum Alloys, ALG Aluminium Inc., Nanning 530031, China
3 Key Laboratory of Non-Ferrous Metals Science and Engineering, Ministry of Education, Changsha 410083, China
* Correspondence: yong.zhang@csu.edu.cn

Abstract: The hot-deformation behavior of three medium-strength Al-Zn-Mg alloys with different Zn/Mg ratios was studied using isothermal-deformation compression tests; the true strain and true stress were recorded for constructing series-processing maps. A few constitutive equations describe the relationship between flow stress and hot-working parameters. The microstructures were characterized using an electron backscatter diffraction (EBSD) detector and transmission electron microscope (TEM). The results show that the optimized deformation parameters for ternary alloy AA7003 are within a temperature range of 653 K to 813 K and with strain rates lower than 0.3 S^{-1}. The microstructures show that materials with a lower Zn/Mg ratio of 6.3 could lead to a problematic hot-deformation capability. Alloys with a higher Zn/Mg ratio of 10.8 exhibited better workability than lower Zn/Mg ratios. The Al_3Zr dispersoids are effective in inhibiting the recrystallization for alloy AA7003, and the Zn/Mg ratios could potentially affect the drag force of the dispersoids.

Keywords: deformation maps; Al-Zn-Mg alloys; Al_3Zr dispersoids; Zn/Mg ratios; recrystallization

1. Introduction

Al-Zn-Mg ternary alloys have strong work-hardening capabilities at room temperature. In order to achieve the desired microstructures and mechanical properties, it is required that they be processed at elevated temperatures [1,2]. An appropriate thermal–mechanical-processing (TMP) route should be carefully selected, as dynamic recrystallization or cracking may happen during the deformation process.

Prasad developed dynamic material modeling (DMM) to calculate the processing maps using a set of flow stress data as a function of temperatures and strain rates over a wide strain range [3–6]. The calculated processing maps can optimize the hot-processing parameters and determine flow instability regimes that should be avoided during processing. Several published papers have demonstrated that the processing maps have been successfully applied for steels [7], zirconium alloys [8], and aluminum alloys [9–11].

For example, Lin et al. built up the hot deformation and processing map for a typical aluminum alloy AA7075. They proposed that the optimum hot-working domain for this high-strength alloy should be within the temperature range of 623–723 K and strain rate range of 0.001–0.05 S^{-1} [12]. Xiao et al. also demonstrated similar optimum processing parameters for their studied alloy AA7050 [11]. However, Lu et al. also showed that the optimal hot-working processing parameters for alloy AA7075 sheet are within the temperature range of 695–723 K and the strain rate range of 0.05–1 S^{-1} [13]. Their strain rate conditions for a given alloy are different. Zhao et al. claimed that the initial structures for Al-Zn-Mg-Cu before deformation could cause a significant difference. Their results demonstrated that the recrystallization mechanism might be different for different grains. Therefore, the alloys with different microstructures should be deformed accordingly [14]. Luo et al.

studied the deformation behavior of alloy 7A09 during the isothermal-compression test. Their results showed that the maximum power dissipation efficiency was about 0.34 for the studied alloy deformed at 713 K and a strain rate of 0.01 S^{-1} [15]. In comparison, Liu studied the isothermal-compression process of alloy AA7085. Their results demonstrated that dynamic recrystallization could happen if the alloy deformed at a temperature higher than 673 K with higher-strain-rate conditions. Therefore, the alloy should be deformed at a temperature of 673 K and a strain rate of 1 S^{-1} [16]. Yang et al. demonstrated that the optimized deformation parameters for alloy AA7085 are within a temperature range of 663–723 K and at strain rates lower than 0.1 S^{-1} [10].

However, Bylya et al. showed validation of the simulation results compared with processing maps for alloy AA2099. The author claimed that the underlying mechanisms of instability regions for processing maps are unclear. Therefore, more meaningful processing maps might be generated using more complex testing scenarios [17].

In this article, we develop processing maps for alloy AA7003 with different Zn/Mg ratios. This alloy is known as a medium-strength alloy, and it can be easily processed during manufacturing. M. Kumar et al. demonstrated that a medium-strength alloy AA7020 exhibited the desired workability at temperatures above 423 K and was sensitive to temperature and strain rate [18–20]. However, there is always the requirement, from an industrial point of view, that the alloy be extruded as fast as possible. Then, the question would be whether the alloy AA7003 can be deformed at a relatively faster or lower temperature range? Can they be easily deformed with slight change in alloying compositions? It is also reported that such medium-strength Al-Zn-Mg alloys suffer a strong natural-ageing effect after quenching [21–23]. By altering the Zn/Mg ratios, the natural ageing effect can be inhibited, but on the other side, the deformation capabilities for different Zn/Mg ratios should be evaluated systematically. The microstructure characterization of a series of processing maps can indicate different workability of variable alloy compositions.

2. Materials and Experiments

The materials used in the present study were deliberately designed AA7003 alloys with three different Zn/Mg ratios, but the total content of Mg + Zn was the same (~6.6 wt.%). The measured chemical compositions are shown in Table 1.

Table 1. Chemical composition of the tested 7003 alloys (wt.%).

Fe	Si	Zn	Mg	Cu	Mn	Ti	Zr	Al	Zn/Mg
0.14	0.06	5.73	0.91	0.16	0.02	0.03	0.20	Bal.	6.3
0.14	0.04	5.95	0.72	0.17	0.02	0.03	0.19	Bal.	8.3
0.14	0.05	6.05	0.56	0.18	0.03	0.03	0.23	Bal.	10.8

The cylinder samples with a dimension of Ø10 × 12 mm were cut along the longitude direction of commercially direct-chilled casting ingots. The homogenization was carried out at 733 K for 48 h. The isothermal-deformation compression tests were carried out on a computer servo-controlled Gleeble-1500 thermo-simulation machine. All samples were lubricated with graphite paste at both ends to reduce friction and increase thermal conductivity. The samples were heated to target temperatures at a constant heating rate of 1 K/s and held at setting temperatures for 5 min before compression. Five different temperatures (653, 693, 733, 773, and 813 K) were chosen to cover the real industrial-manufacturing situations. Samples were deformed at constant true strain rates of 0.01, 0.1, 1, and 10 S^{-1} over a selected temperature range. All samples were deformed to a strain of about 0.9. True stress–true strain curves were recorded during the compression test.

Microstructures were characterized using a ZEISS EVOMA10 scanning electron microscope with an OXFORD electron backscatter diffraction (EBSD) detector. Samples were electrolytically polished in a 10% perchloric acid solution mixed with 90% ethanol. Transmission electron microscope (TEM) samples were 3 mm discs punched from an 80 μm-thick foil. The specimens were further polished using twin-jet electropolishing in a solution of

80% methanol and 20% nitric acid at a temperature below 248 K. A Tecnai G2 F20 TEM then examined the samples operated at 200 kV.

3. Results and Discussion

3.1. True Stress–Strain Curves

Before the stress–strain curves are presented, the initial microstructures are shown in Figure 1. It is shown that all three alloys demonstrated large grains with casting dendrites inside. The grain size is very close for the alloy with a Zn/Mg ratio of 6.3 (as shown in Figure 1a) and a Zn/Mg ratio of 8.3 (as shown in Figure 1b). The grain size for the alloy with a Zn/Mg ratio of 6.3 is relatively more prominent than the other two alloys. The measured average grain sizes are 200 ± 35 µm, 156 ± 28 µm, 175 ± 40 µm, respectively. However, this is not conclusive as the as-cast microstructure varies from place to place. It is also noticed that some eutectic phases can be observed. This could be due to the high alloying content for the studied alloys.

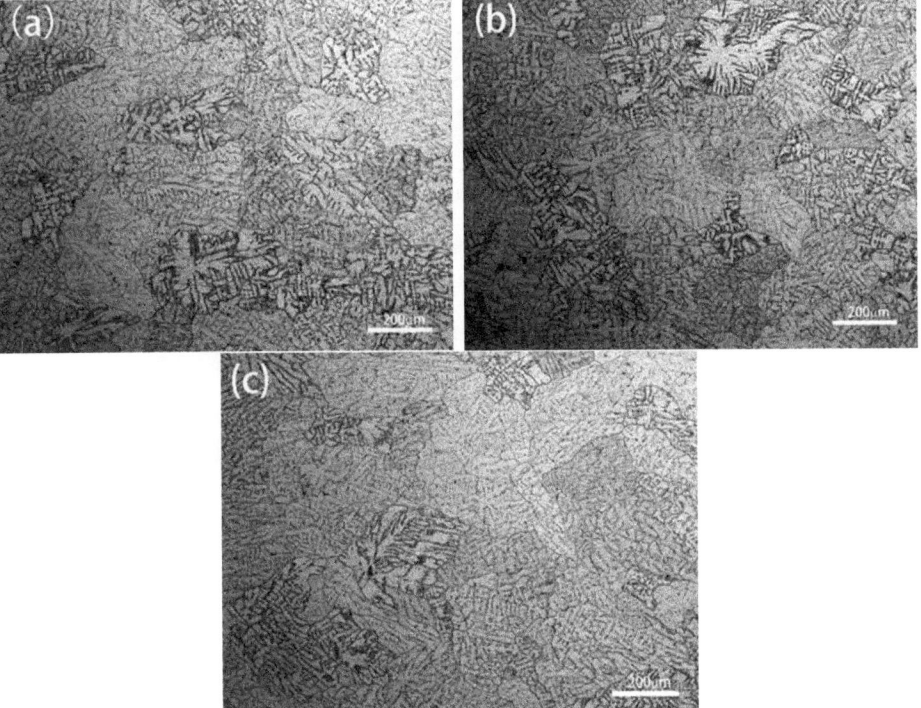

Figure 1. The initial microstructure of the studied alloy with (**a**) Zn/Mg ratio of 6.3, (**b**) Zn/Mg ratio of 8.3, and (**c**) Zn/Mg ratio of 10.8.

There are five different temperature conditions, four different strain rates, and three different alloys. In total, 60 curves need to be present. For the readers' convenience, the true stress–strain curves for the alloy with Zn/Mg = 6.3 were chosen to demonstrate the data process routine, while the results of other alloys are calculated using the same methodology.

Figure 2 shows the true stress–strain curves during the hot-compression testing of alloy AA7003 with a Zn/Mg ratio of 6.3 under the strain rate condition of (a) 0.01 s^{-1}, (b) 0.1 s^{-1}, (c) 1 s^{-1}, and (d) 10 s^{-1} for different temperatures. The true stress–strain curves need to be corrected as the friction could either leading to inhomogeneous deformation or temperature rising. As a result, the deformed samples start to form a barrel shape. The detailed friction correction method and temperature correction method can be found from Ebrahimi [9] and Z-P Wan et al. [20]. The authors here use the Ebrahimi method to correct the stress–strain curves; the basic equations are:

$$\frac{P_{ave}}{\sigma} = 8b\frac{R}{H}\left\{\left[\frac{1}{12} + \left(\frac{H}{R}\right)^2\frac{1}{b^2}\right]^{3/2} - \left(\frac{H}{R}\right)^3\frac{1}{b^3} - \frac{m}{24\sqrt{3}}\frac{\exp(-b/2)}{\exp(-b/2)-1}\right\} \quad (1)$$

$$b = \frac{4m/\sqrt{3}}{(R/H) + \left(2m/3\sqrt{3}\right)} \quad (2)$$

$$m = \frac{(R/H)b}{\left(4/\sqrt{3}\right) - \left(2b/3\sqrt{3}\right)} \quad (3)$$

where m is the constant friction factor in the compression test, σ is the corrected true stress, Pave is the uncorrected external pressure applied to specimens in compression (the measured stress), b is the barrel parameter, and R and H are the radius and height of samples during compression. Temperature correction was also carried out so the raw data and corrected data are shown in Figure 2. It is shown in Figure 2a that, for a given temperature condition, the true stress values initially increase rapidly with increasing true strain values. It reaches its maximum values after a small number of strain values and remains constant for the rest of the strain values before it descends to tremendous strain values. Comparing stress values at different temperatures demonstrates that the alloy is more resistant to deformation at lower temperatures. This trend is generally observed in other strain rate conditions, as shown in Figure 2b–d. When comparing different strain rate conditions for a given temperature, the true stress increases with increasing strain rate conditions (as shown in the same color in different figures). It is also interesting to find out that the true stress is relatively stable at low-strain-rate conditions (as shown in Figure 2a,b). The strain–stress curves become more fluctuated at relatively high-strain-rate conditions (as shown in Figure 2c,d). Moreover, the stress curves exhibit a significant drop at high-strain-rate conditions. This could be an indication of the instability of the studied alloys.

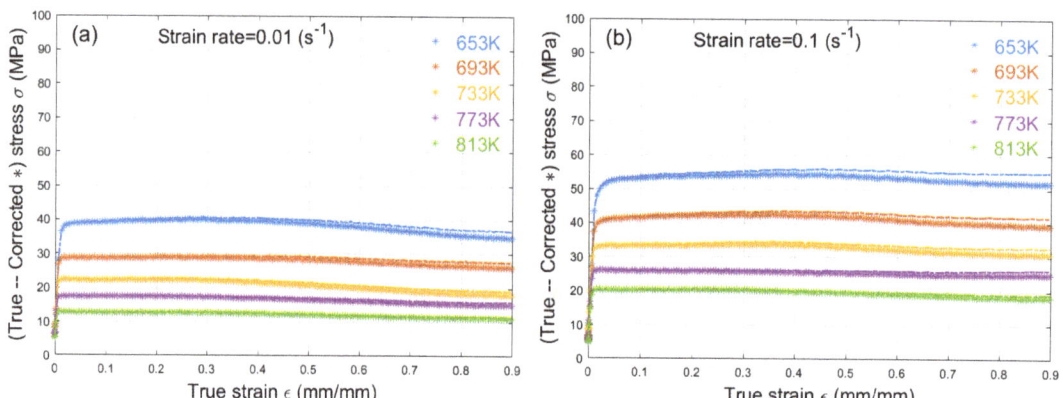

Figure 2. Cont.

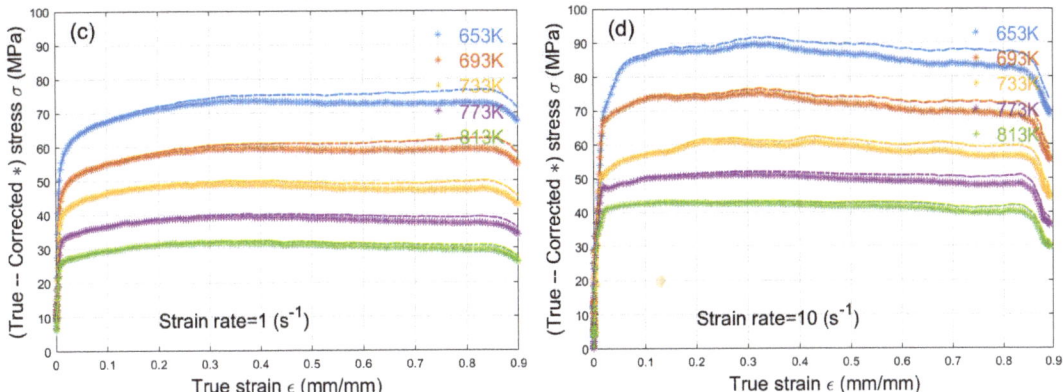

Figure 2. Raw and corrected (friction and temperature) stress vs. true strain curves of alloy AA7003 with Zn/Mg ratio of 6.3 during hot-compression testing at different strain rates for different temperatures. (**a**) strain rate of 0.01 s^{-1}, (**b**) strain rate of 0.1 s^{-1}, (**c**) strain rate of 1 s^{-1} and (**d**) strain rate of 10 s^{-1}.

Table 2 summarizes the detailed flow stress values (in MPa) of alloy with a Zn/Mg ratio of 6.3 at different temperatures and strain rates for various strains.

Table 2. Flow stress values (in MPa) of alloy with Zn/Mg ratio of 6.3 at different temperatures and strain rates for various strains.

Strain	Strain Rate, s^{-1}	Temperature (K)				
		653	693	733	773	813
0.2	0.01	39.73	28.90	22.16	17.42	12.55
	0.1	53.86	42.07	33.35	25.86	20.28
	1	71.13	57.64	48.13	38.34	30.98
	10	87.47	73.90	60.46	50.75	42.62
0.4	0.01	39.54	28.46	21.26	16.93	12.31
	0.1	54.23	42.22	33.43	25.62	19.63
	1	73.42	59.38	48.63	39.05	31.34
	10	87.91	73.14	60.25	50.46	42.17
0.6	0.01	37.91	27.58	19.83	16.08	11.69
	0.1	53.22	40.74	31.95	24.98	18.93
	1	72.62	58.57	47.28	38.24	30.27
	10	84.68	70.09	57.51	49.02	41.49
0.8	0.01	35.55	26.34	18.25	15.16	11.23
	0.1	51.92	39.65	30.99	24.86	18.11
	1	72.76	59.11	47.05	37.28	29.36
	10	82.38	68.74	56.43	48.04	39.94

3.2. Constitutive Equations

The constitutive equations describe the relationship between flow stress and hot-working parameters, such as strain, deformed temperatures, and strain rates. Sellars and McTegart developed a hyperbolic sine model in 1960 [24]. This model has been widely used to describe the workability of different alloys. Workability depends on the initiated microstructures, chemical compositions, and processing histories. For example, the annealed materials (O temper) exhibit better workabilities than the deformed materials (H temper). It should also be noted that the friction between the sample's edges and the dies, such as adiabatic heating during the deformation, may significantly influence the true stress–true strain curves. This could be corrected by introducing lubricant during the deformation process or empirically corrected by estimation [25,26]. Since we used graphite paste on both ends of the samples, the friction effect can be negligible in this case.

In the 1960s, Sellars and McTegart proposed that the isothermal stress–strain relation is based on Arrhenius equations [24]:

$$\dot{\varepsilon} = f(\sigma) \exp\left(-\frac{Q}{RT}\right) \quad (4)$$

where the strain rate unit is s^{-1}; R is gas constant, Q is activation energy for hot deformation, unit in kJ/mol; T is isothermal temperature, unit in Kelvin; and $f(\sigma)$ is the strain-related equation, also called Zener–Hollomon parameters in some publications.

It turns out that the alloy may exhibit different deformation mechanisms within other strain regions. Therefore, many publications proposed describing the data using different subfunctions for different intervals:

$$\text{When } \alpha\sigma < 0.8,\ \dot{\varepsilon} = A_1 \sigma^{n_1} \exp\left(-\frac{Q}{RT}\right) \quad (5)$$

$$\text{When } \alpha\sigma > 1.2,\ \dot{\varepsilon} = A_2 \exp(\beta\sigma) \exp\left(-\frac{Q}{RT}\right) \quad (6)$$

$$\text{For all other } \alpha\sigma,\ \dot{\varepsilon} = A[\sinh(\alpha\sigma)]^n \exp\left(-\frac{Q}{RT}\right) \quad (7)$$

where n_1, β, and α are the material's constants, and $\alpha = \beta/n_1$. Taking the natural logarithm on both sides of the equation yields:

$$\ln \dot{\varepsilon} = \ln A_1 + n_1 \ln \sigma - Q/RT \quad (8)$$

$$\ln \dot{\varepsilon} = \ln A_2 + \beta\sigma - Q/RT \quad (9)$$

$$\ln \dot{\varepsilon} = \ln A + n \ln[\sinh(\alpha\sigma)] - Q/RT \quad (10)$$

According to Equation (5), the slope value of the linear relationship between $\ln \dot{\varepsilon}$ and $\ln(\sigma)$ is the n_1 value. According to Equation (6) at different temperatures, the slope value of the linear relationship between $\ln \dot{\varepsilon}$ and σ is β. The α can be calculated accordingly. The detailed plots are shown in Figure 3a–c. The activation energy Q value for hot deformation can be extrapolated by linear fitting for 1/T and $\ln[\sinh(\alpha\sigma)]$ at a given strain rate condition.

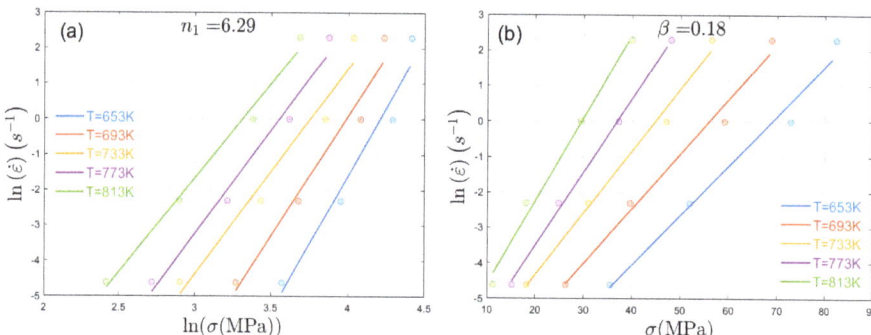

Figure 3. Cont.

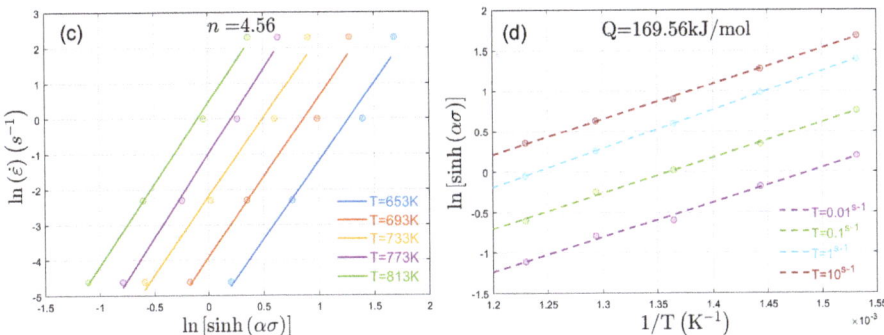

Figure 3. (**a**) linear fitting for $\ln(\sigma)$ and $\ln(\dot{\varepsilon})$ at different temperatures; (**b**) linear fitting for σ and $\ln(\dot{\varepsilon})$ at different temperatures; (**c**) linear fitting for $\ln[\sinh(\alpha\sigma)]$ and $\ln(\dot{\varepsilon})$; and (**d**) linear fitting for $1/T$ and $\ln[\sinh(\alpha\sigma)]$ at given strain rate conditions.

Applying this methodology to other strain conditions, one can calculate the activation energy Q values for studied alloys (as shown in Figure 4). It was demonstrated that an alloy with a Zn/Mg ratio of 10.8 exhibited deficient activation energy compared to the other two studied alloys. This could be an indication that this alloy is easy to be deformed. The activation energy for the other two alloys with close Zn/Mg ratios (Zn/Mg ratio of 6.3 and Zn/Mg ratio of 8.3) demonstrated very similar values. However, their values show a different trend with increasing strain.

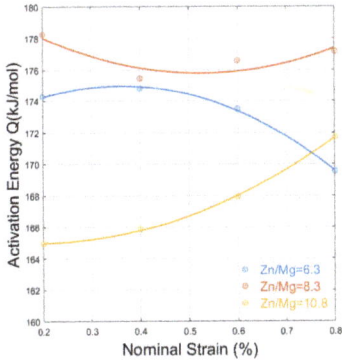

Figure 4. Calculated Activation energy value under different nominal strain conditions for three studied alloys with different Zn/Mg ratios.

3.3. Processing Maps

The constitutive equation may be helpful in the interpretation of strain–stress curves. However, the processing map could be a straightforward method to describe the workability of the studied alloys. Although it is an explicit representation of the response of studied alloys, it has been widely applied in many process parameter selections.

According to Prasad and Srivatsana [27,28], the input energy causing deformation at a given temperature could be dissipated by heat or the so-called "conduction entropy" and microstructural changes induced by dislocation movement.

$$P = G + J = \int_0^{\dot{\varepsilon}} \sigma d\dot{\varepsilon} + \int_0^{\sigma} \dot{\varepsilon} d\sigma \tag{11}$$

where the first integral (G content) is dissipated energy as temperature arises, while the second integral (J co-content) is energy dissipated due to microstructural changes.

$$m = \frac{\partial J}{\partial G} = \frac{\dot{\varepsilon}\partial \sigma}{\sigma \partial \dot{\varepsilon}} = \frac{\partial \ln \sigma}{\partial \ln \dot{\varepsilon}} \quad (12)$$

The strain rate sensitivity (m) is given by Equation (9). This strain rate sensitivity value defines the relationship between $\ln \dot{\varepsilon}$ and $\ln \sigma$. According to Prasad [4,5,8], the m value is generally between 0 and 1 for aluminium alloys. Equation (8) states that:

$$\Delta J/\Delta P = \frac{m}{m+1} \quad (13)$$

The efficiency of power dissipation (η) is, therefore, defined by:

$$\frac{\Delta J/\Delta P}{(\Delta J/\Delta P)_{linear}} = \frac{2m}{m+1} = \eta \quad (14)$$

The denominator in Equation (11) indicates that the G-content is equal to J co-content in the ideal dissipation system. The efficiency of power dissipation (η) describes how close the current system is compared to the ideal dissipation system, since the J-content is more related to the microstructural changes. Therefore, the efficiency of power dissipation (η) essentially describes the microscopic deformation mechanism of the materials within the range of applied temperatures and strains. The efficiency of power dissipation (η) changes with temperature and strain rate to form a power dissipation map, representing the microstructure change in the studied materials. Since various failures (such as void formation and cracking propagation) or metallurgical changes (such as dynamic recovery, dynamic recrystallization, etc.) in the plastic deformation process dissipate input energy, with the help of microstructural characterization, the power dissipation diagram can be used to analyze different deformation mechanisms under other deformation conditions. It is necessary, first, to determine the processing instability zone of the studied alloys. According to Prasad [5,8,29], the instability criteria are given by:

$$\xi(\dot{\varepsilon}) = \frac{\partial \ln\left(\frac{m}{m+1}\right)}{\partial \ln \dot{\varepsilon}} + m < 0 \quad (15)$$

Figure 5 shows the constructed processing maps for the studied alloy with different Zn/Mg ratios at different strain conditions. The processing map is an instability diagram overlapped with an energy dissipation diagram at the current strain condition. As shown in Figure 5, the yellow shaded region represents the instability regions. When comparing with other publications, we have presented a series of processing maps that show a systematic change with different conditions. In general, the instability regions for studied alloys are within the lower-temperature and higher-strain-rate conditions. This is consistent with the idea that alloy AA7003 is easier to be deformed when compared to other high-strength 7xxx series alloys [5,9,26]. It is interesting that the studied alloys represent different workability with different Zn/Mg ratios, i.e., different rows in Figure 5. The alloys with a lower Zn/Mg ratio exhibit more significant instability regions. Therefore, it is concluded that alloy AA7003 with a higher Zn/Mg ratio could have better formability than lower Zn/Mg ratios. When comparing different strain conditions, i.e., different columns, it is clear that the higher-strain condition exhibits more significant instability regions. This also agrees with our shared knowledge that higher-strain conditions could lead to void formation and cracking propagation. The result also agrees with Figure 4 that the calculated activation energy value for the alloy with a Zn/Mg ratio of 10.8 is significantly lower than the other two alloys.

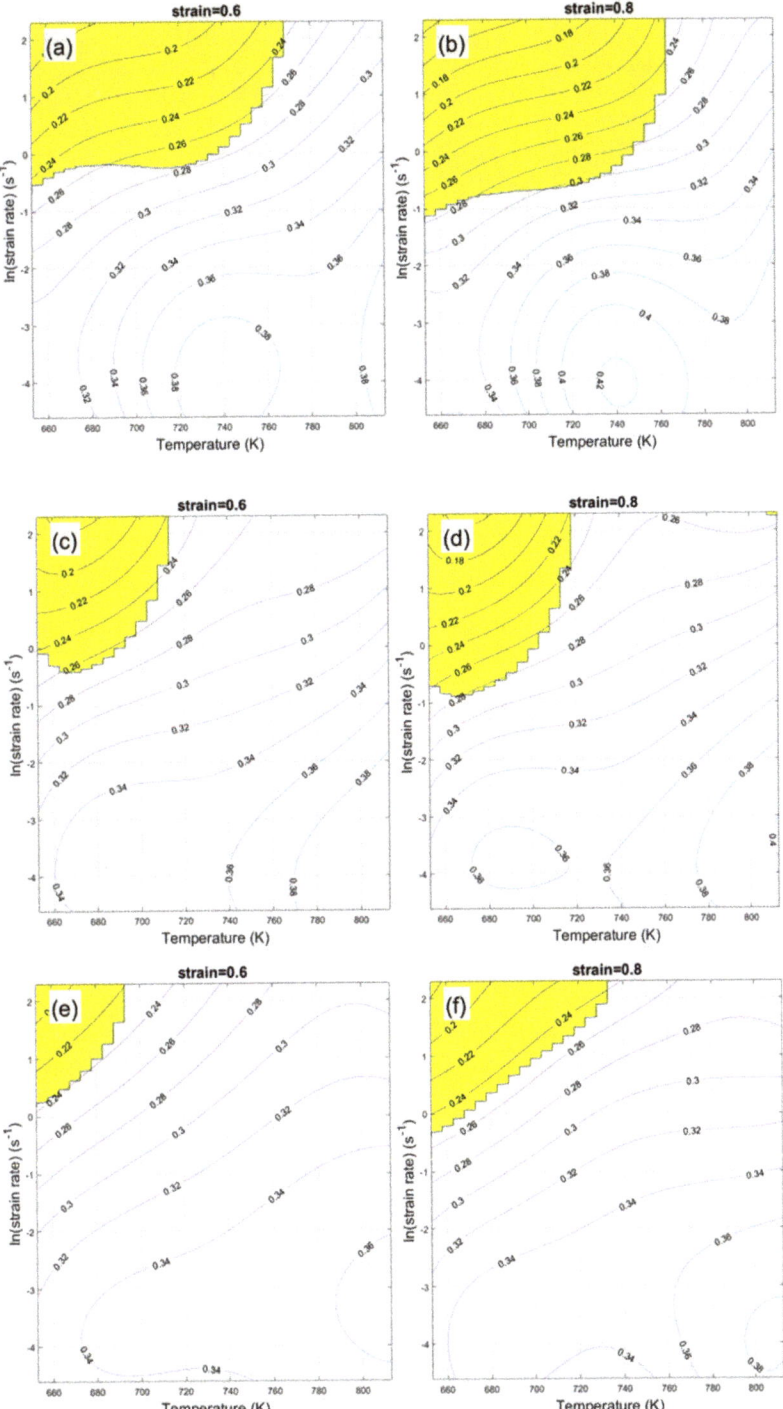

Figure 5. The processing maps for alloys with (**a**,**b**) Zn/Mg = 6.3, (**c**,**d**) Zn/Mg = 8.3, (**e**,**f**) Zn/Mg = 10.8, at strain of 0.6 (left column) and 0.8 (right column), respectively, while the shaded region represents the instability regions.

The contour lines in Figure 5 represent the efficiency of power dissipation (η) at the current strain condition. As discussed earlier, the maximum efficiency is 100%, while G-content is equal to J co-content in the ideal dissipation system. In the current dissipation system, the maximum efficiency (η) is about 50%, and the high efficiency of power dissipation (η) is always shown in the lower-right corner of each diagram. This indicates that the input energy is more likely to be dissipated at higher temperatures and low-strain-rate conditions without cracking or instability. Therefore, it is concluded that optimized deformation parameters for ternary alloy AA7003 are within a temperature range of 653–813 K and with strain rates lower than 0.3 S^{-1}.

4. Discussions

As discussed earlier, the J co-content is energy dissipated due to microstructural changes. The alloy with a Zn/Mg ratio of 8.3 was chosen for typical microstructure characterization. The electron backscatter diffraction (EBSD) technique was selected to analyze deformed microstructures, recrystallized microstructures, and substructures quantitatively. The specific microstructural characterization is shown in Figure 6. When the grain's misorientation angle (θc) exceeds 15°, it is classified as a deformed microstructure. Grains consisting of subgrains whose internal misorientation is below 15°, but whose misorientation from subgrain to subgrain is above 2°, are classified as substructures. All the remaining grains are classified as recrystallized. In Figure 6, the colored maps present different microstructures, i.e., blue stands for recrystallized structures, yellow stands for substructures, and red stands for deformed structures. It is shown in Figure 6a that the primary remaining microstructures are deformed structures that coexist with a small number of substructures and few recrystallized structures. When deformed at a higher temperature, i.e., 733 K in Figure 6b, it is shown that more substructures and more recrystallized structures were found. It is shown in Figure 5b that the primary microstructures are substructures. Additionally, the recrystallized structures are significantly increased. Figure 6c shows that when deformed at 813 K, the recrystallized structures become the dominated microstructures.

The statistical analysis of EBSD mappings at a different temperature and at a strain rate of 0.1 s^{-1} is shown in Figure 7. It is demonstrated in Figure 7a that the area fraction of recrystallized grains significantly increased with deformation temperatures, while the frequency of deformed microstructures declined dramatically. This indicates that recrystallization occurs rapidly for the alloy with Zn/Mg = 6.3. The frequency of recrystallized structures showed a moderate decrease for the alloy with Zn/Mg = 8.3. However, it is shown in Figure 7b that the substructures arose significantly at different temperature conditions. For the alloy that contained the highest Zn/Mg ratio (as shown in Figure 7c), the substructures exhibited a remarkable frequency at a temperature of 733 K and 813 K. It is also shown that the amount of recrystallized structures increased slightly with temperature. It also should be noticed that the deformed microstructures decreased with rising temperatures within all three alloys. In general, the EBSD results show that the alloy with Zn/Mg = 10.8 exhibited a significantly small amount of recrystallization, while the other two alloys exhibited a moderate amount of recrystallization. This could be an indication that the dynamic recrystallization is the main factor to dissipate the deformation energy. It is also shown that the alloy with Zn/Mg = 10.8 retained a large number of substructures. This could also be an indication that this alloy can be further deformed without cracking.

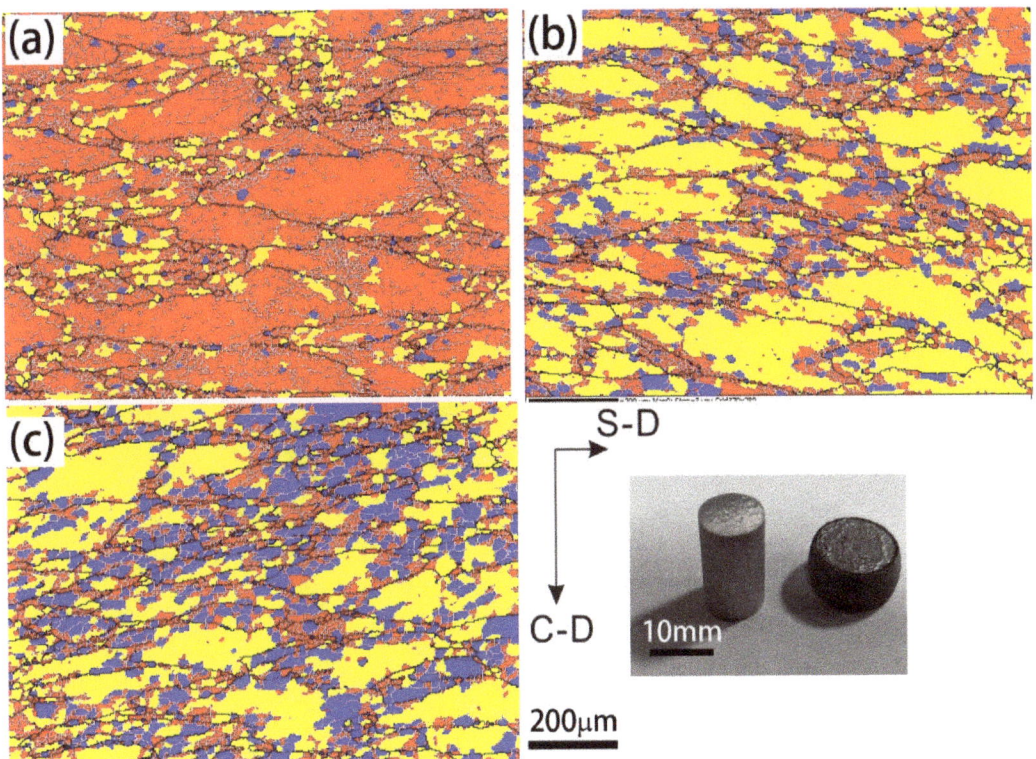

Figure 6. The typical EBSD mappings for the alloy with Zn/Mg = 6.3 deformed at (**a**) 653 K, (**b**) 733 K and (**c**) 813 K at a strain rate of 0.1 S^{-1}, with the different colors representing different microstructures, i.e., yellow stand for substructures, blue stand for recrystallized structures and red stand for deformed structures; the bottom right image shows the compressed samples before and after deformation.

Many reports focus on the effects of dispersoid particles [30,31]. It is proven that these particles effectively inhibit the dislocation movement and grain boundary movements [32–35]. In the current research work, ~0.2% Zr was added to the studied alloys. The typical particles are known to be Al$_3$Zr dispersoids. It is shown in Figure 8a that there are a significant number of spherical Al$_3$Zr particles; they have an L1$_2$ crystalline structure (as shown by the selected area diffraction pattern (SADP)) and are confirmed to be coherent with the Al matrix. The SADP also indicates that these particles exhibit a simple cubic/cubic orientation relationship with the Al matrix, where the diffraction pattern from Al$_3$Zr particles is located at {100} planes.

It is shown in Figure 8a that the size of Al$_3$Zr dispersoids is within the range of 20–50 nm. In fact, due to the same homogenization treatment and Zr addition, all three studied alloys have a similar size distribution of Al$_3$Zr dispersoids. However, it should be noted that these dispersoids are heterogeneously distributed within grains. Some regions (as shown in Figure 8b) have very-low-number density while others have relatively high-number density (as shown in Figure 8a). Interestingly, these particles constantly interact with dislocations or grain boundaries. It is shown in Figure 8b that the dislocations bypass a single spherical Al$_3$Zr dispersoid by bowing around. According to classical deformation theory, an Orowan loop is left afterward. The classical deformation theory considers the shear stress or line tension caused by dispersoids themselves. However, it should be noted that the studied alloys exhibit different thermal–mechanical behaviors, given they contain a similar amount of Zr addition. Therefore, it is concluded that the main alloying content, such as Zn/Mg ratios, could also affect the dynamic recrystallization process. In the current study, about the reaction with recrystallizations, three scenarios can be discerned: the low-Zn/Mg-ratio alloy being deformed with difficulty can cause recrystallization quickly, the medium-Zn/Mg-ratio alloy has a moderate trend of recrystallization, and the high-Zn/Mg-ratio alloy being deformed easily can lead to very low recrystallization. The underlying mechanism is not yet fully understood. More detailed work will be conducted in the future.

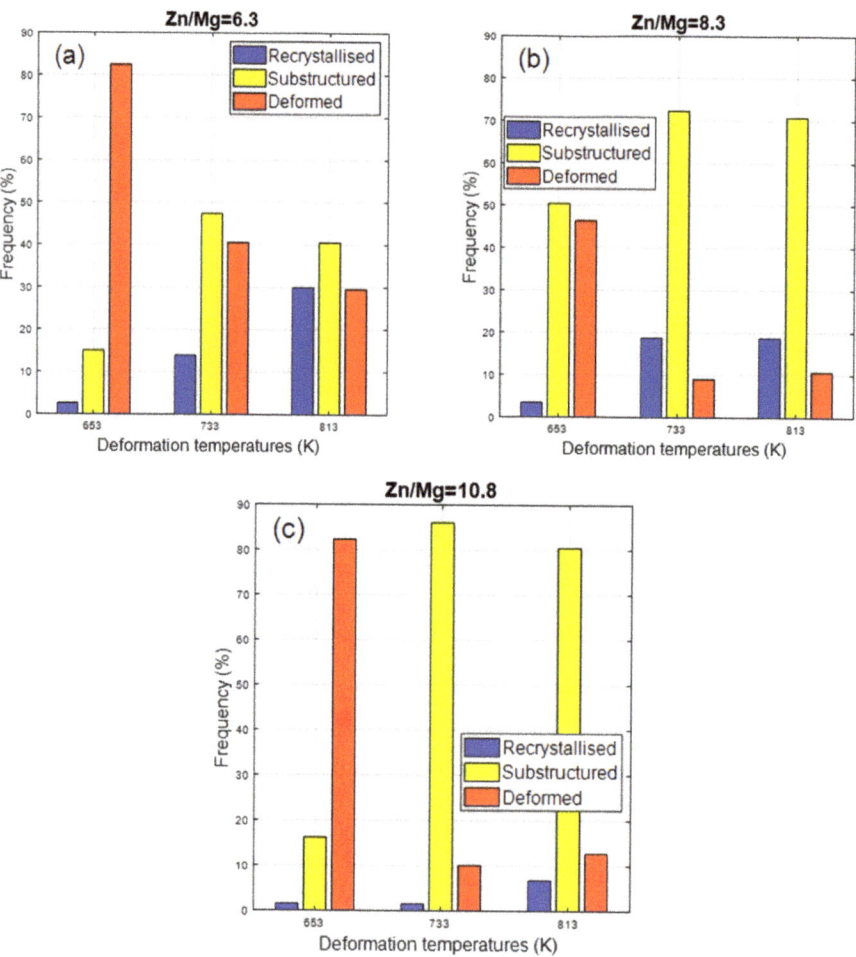

Figure 7. Statistical analysis of different area fractions of different microstructures at various deformation temperatures for (**a**) alloy with Zn/Mg = 6.3, (**b**) alloy with Zn/Mg = 8.3, and (**c**) alloy with Zn/Mg = 10.8.

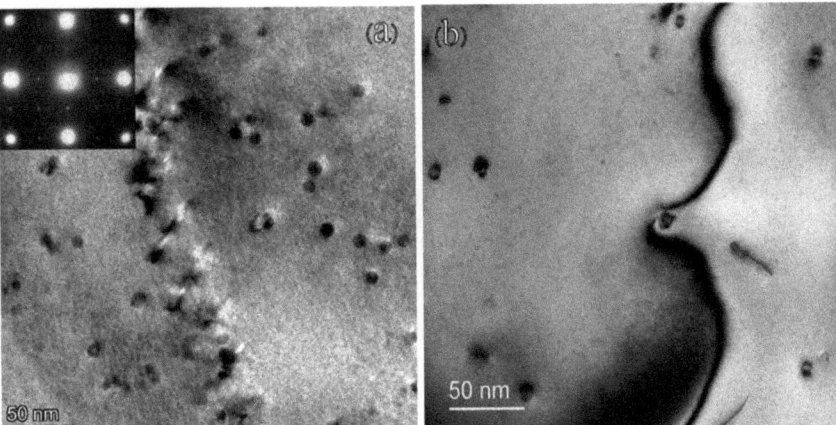

Figure 8. TEM observation of the alloy with Zn/Mg = 6.3 deformed at 733 K at a strain rate of 0.1 S^{-1}, (**a**) Distribution of Al_3Zr dispersoids within grains (an alloy with Zn/Mg = 6.3), viewing from $\{100\}_{Al}$, (**b**) an Al_3Zr dispersoid tangled with dislocation movement during deformation.

5. Conclusions

The hot-deformation behavior of three Al-Zn-Mg alloys with different Zn/Mg ratios was studied. The main conclusions are:

(1) When comparing the processing maps for AA7003 alloys with different Zn/Mg ratios, alloys with a low Zn/Mg ratio of 6.3 led to a problematic hot-deformation capability. In contrast, alloys with a higher Zn/Mg ratio of 10.8 exhibited better workability than lower Zn/Mg ratios.

(2) The optimized deformation parameters for ternary alloy AA7003 were within a temperature range of 653–813 K and at strain rates lower than 0.3 S^{-1}.

(3) When comparing the microstructures after hot deformation, alloy AA7003 with a lower Zn/Mg ratio of 6.3 had a more negligible fraction of substructures but higher frequency of recrystallized structures. In comparison, the alloy with a higher Zn/Mg ratio of 10.8 had a high fraction of substructures and low frequency of recrystallization.

(4) The Al_3Zr dispersoids were effective in inhibiting the recrystallization for alloy AA7003; three scenarios can be discerned when considering the interaction between dispersoids and recrystallization: the low-Zn/Mg-ratio alloy being deformed with difficulty can cause recrystallization easily, the medium-Zn/Mg-ratio alloy has a moderate trend of recrystallization, and the high-Zn/Mg-ratio alloy contains a minor fraction of recrystallization and, therefore, leads to easy deformability.

Author Contributions: Writing—original draft preparation, X.Z.; Writing—review and editing, J.T.; Materials supplier, L.W.; Data collection, Y.Z. (Yan Zhao) and C.J.; Project administration, Y.Z. (Yong Zhang). All authors have read and agreed to the published version of the manuscript.

Funding: This research was funded by the National Key Research and Development Program of China (No. 2016YFB0300901) and Guangxi Science & Technology Program (Guike AA22068075).

Data Availability Statement: Data is contained within the article.

Acknowledgments: The authors would like to thank Foshan Sanshui Fenglu Aluminium Co., Ltd. China, for providing the materials.

Conflicts of Interest: The authors declare no conflict of interest.

References

1. Hatch, J.E. *Aluminum: Properties and Physical Metallurgy*; Aluminum Association Inc.: Arlington County, VA, USA; ASM International: Almere, The Netherlands, 1984.
2. Polmear, I.J. *Light Alloys: Metallurgy of the Light Metals*, 4th ed.; Butterworth-Heinemann: Melbourne, VIC, Australia, 2006.
3. Prasad, Y.V.R.K.; Rao, K.P.; Sasidhara, S.; Staff, A.I. *Hot Working Guide: A Compendium of Processing Maps*; ASM International: Almere, The Netherlands, 2015.
4. Padmavardhani, D.; Prasad, Y. Characterization of hot deformation behavior of brasses using processing maps: Part II. β Brass and α-β brass. *Metall. Trans. A* **1991**, *22*, 2993–3001. [CrossRef]
5. Prasad, Y.; Gegel, H.; Doraivelu, S.; Malas, J.; Morgan, J.; Lark, K.; Barker, D. Modeling of dynamic material behavior in hot deformation: Forging of Ti-6242. *Metall. Trans. A* **1984**, *15*, 1883–1892. [CrossRef]
6. Padmavardhani, D.; Prasad, Y. Characterization of hot deformation behavior of brasses using processing maps: Part I. α Brass. *Metall. Trans. A* **1991**, *22*, 2985–2992. [CrossRef]
7. Łukaszek-Sołek, A.; Krawczyk, J.; Śleboda, T.; Grelowski, J. Optimization of the hot forging parameters for 4340 steel by processing maps. *J. Mater. Res. Technol.* **2019**, *8*, 3281–3290. [CrossRef]
8. Chakravartty, J.K.; Prasad, Y.V.R.K.; Asundi, M.K. Processing map for hot working of alpha-zirconium. *Metall. Trans. A* **1991**, *22*, 829–836. [CrossRef]
9. Ke, B.; Ye, L.; Tang, J.; Zhang, Y.; Liu, S.; Lin, H.; Dong, Y.; Liu, X. Hot deformation behavior and 3D processing maps of AA7020 aluminum alloy. *J. Alloys Compd.* **2020**, *845*, 156113. [CrossRef]
10. Qunying, Y.; Wenyi, L.; Zhiqing, Z.; Guangjie, H.; Xiaoyong, L. Hot Deformation Behavior and Processing Maps of AA7085 Aluminum Alloy. *Rare Met. Mater. Eng.* **2018**, *47*, 409–415. [CrossRef]
11. Xiao, D.; Peng, X.; Liang, X.; Deng, Y.; Xu, G.; Yin, Z. Study on Hot Workability of Al-5.87Zn-2.07Mg-2.28Cu Alloy Using Processing Map. *JOM* **2017**, *69*, 725–733. [CrossRef]
12. Lin, Y.C.; Li, L.-T.; Xia, Y.-C.; Jiang, Y.-Q. Hot deformation and processing map of a typical Al–Zn–Mg–Cu alloy. *J. Alloys Compd.* **2013**, *550*, 438–445. [CrossRef]
13. Lu, J.; Song, Y.; Hua, L.; Zheng, K.; Dai, D. Thermal deformation behavior and processing maps of 7075 aluminum alloy sheet based on isothermal uniaxial tensile tests. *J. Alloys Compd.* **2018**, *767*, 856–869. [CrossRef]
14. Zhao, J.; Deng, Y.; Xu, F.; Zhang, J. Effects of Initial Grain Size of Al-Zn-Mg-Cu Alloy on the Recrystallization Behavior and Recrystallization Mechanism in Isothermal Compression. *Metals* **2019**, *9*, 110. [CrossRef]
15. Luo, J.; Li, M.Q.; Ma, D.W. The deformation behavior and processing maps in the isothermal compression of 7A09 aluminum alloy. *Mater. Sci. Eng. A* **2012**, *532*, 548–557. [CrossRef]
16. Liu, W.; Zhao, H.; Li, D.; Zhang, Z.; Huang, G.; Liu, Q. Hot deformation behavior of AA7085 aluminum alloy during isothermal compression at elevated temperature. *Mater. Sci. Eng. A* **2014**, *596*, 176–182. [CrossRef]
17. Ferragut, R.; Somoza, A.; Tolley, A.; Torriani, I. Precipitation kinetics in Al–Zn–Mg commercial alloys. *J. Mater. Process. Technol.* **2003**, *141*, 35–40. [CrossRef]
18. Kumar, M.; Sotirov, N.; Chimani, C.M. Investigations on warm forming of AW-7020-T6 alloy sheet. *J. Mater. Process. Technol.* **2014**, *214*, 1769–1776. [CrossRef]
19. Ke, B.; Ye, L.; Zhang, Y.; Liu, X.; Dong, Y.; Wang, P.; Tang, J.; Liu, S. Enhanced strength and electrical conductivities of an Al-Zn-Mg aluminum alloy through a new aging process. *Mater. Lett.* **2021**, *304*, 130586. [CrossRef]
20. Ke, B.; Ye, L.; Zhang, Y.; Tang, J.; Liu, S.; Liu, X.; Dong, Y.; Wang, P. Enhanced mechanical properties and corrosion resistance of an Al-Zn-Mg aluminum alloy through variable-rate non-isothermal aging. *J. Alloys Compd.* **2022**, *890*, 161933. [CrossRef]
21. Bloem, C.; Salvador, M.; Amigo, V.; Vergara, M. Aluminium 7020 Alloy and Its Welding Fatigue Behaviour. Available online: http://www.intechopen.com/books/aluminium-alloys-theory-andapplications/aluminium-7020-alloy-and-its-welding-fatigue-behaviour (accessed on 2 August 2022).
22. Wan, L.; Deng, Y.-L.; Ye, L.-Y.; Zhang, Y. The natural ageing effect on pre-ageing kinetics of Al-Zn-Mg alloy. *J. Alloys Compd.* **2019**, *776*, 469–474. [CrossRef]
23. Zhang, Y.; Milkereit, B.; Kessler, O.; Schick, C.; Rometsch, P.A. Development of continuous cooling precipitation diagrams for aluminium alloys AA7150 and AA7020. *J. Alloys Compd.* **2014**, *584*, 581–589. [CrossRef]
24. Sellars, C.M.; McTegart, W.J. On the mechanism of hot deformation. *Acta Metall.* **1966**, *14*, 1136–1138. [CrossRef]
25. Raja, N.; Daniel, B.S.S. Microstructural evolution of Al-7.3Zn-2.2Mg-2Cu (Al7068) alloy in T6 condition during isothermal compression using 3-dimensional processing map. *J. Alloys Compd.* **2022**, *902*, 163690. [CrossRef]
26. Khomutov, M.; Pozdniakov, A.; Churyumov, A.; Barkov, R.; Solonin, A.; Glavatskikh, M. Flow Stress Modelling and 3D Processing Maps of Al4.5Zn4.5Mg1Cu0.12Zr Alloy with Different Scandium Contents. *Appl. Sci.* **2021**, *11*, 4587. [CrossRef]
27. Srivatsana, T.S.; Guruprasad, G.; Vasudevan, V.K. The Quasi Static Deformation and Facture Behavior of Aluminum Alloy 7150. *Mater. Des.* **2008**, *29*, 742–751. [CrossRef]
28. Prasad, K.S.; Murali, D.S.K.; Prasad, N.E.; Mukhopadhyay, A.K. Influence of microstructure on the subzero temperature tensile properties of heat treated AA7010 plates. *Trans. Inian Inst. Met.* **2010**, *63*, 799–805. [CrossRef]
29. Prasad, K.S.; Gokhale, A.A.; Mukhopadhyay, A.K.; Banerjee, D.; Goel, D.B. On the formation of faceted Al3Zr (b') Precipitates in Al-Li-Cu-Mg-Zr Alloys. *Acta Mater.* **1999**, *47*, 2581–2592. [CrossRef]

30. Xiao, T.; Deng, Y.; Ye, L.; Lin, H.; Shan, C.; Qian, P. Effect of three-stage homogenization on mechanical properties and stress corrosion cracking of Al-Zn-Mg-Zr alloys. *Mater. Sci. Eng. A* **2016**, *675*, 280–288. [CrossRef]
31. Wang, D.; Xiao, Z. Revealing the Al/L12-Al3Zr inter-facial properties: Insights from first-principles calculations. *Vacuum* **2022**, *195*, 110620. [CrossRef]
32. Jia, Z.-h.; CouziniÉ, J.-P.; Cherdoudi, N.; Guillot, I.; Arnberg, L.; ÅSholt, P.; Brusethaug, S.; Barlas, B.; Massinon, D. Precipitation behaviour of Al3Zr precipitate in Al–Cu–Zr and Al–Cu–Zr–Ti–V alloys. *Trans. Nonferrous Met. Soc. China* **2012**, *22*, 1860–1865. [CrossRef]
33. Tsivoulas, D.; Robson, J.D. Heterogeneous Zr solute segregation and Al3Zr dispersoid distributions in Al–Cu–Li alloys. *Acta Mater.* **2015**, *93*, 73–86. [CrossRef]
34. Guo, Z.; Zhao, G.; Chen, X.G. Effects of two-step homogenization on precipitation behavior of Al3Zr dispersoids and recrystallization resistance in 7150 aluminum alloy. *Mater. Charact.* **2015**, *102*, 122–130. [CrossRef]
35. Zhang, Y.; Bettles, C.; Rometsch, P.A. Effect of recrystallisation on Al_3Zr dispersoid behaviour in thick plates of aluminium alloy AA7150. *J. Mater. Sci.* **2014**, *49*, 1709–1715. [CrossRef]

Fabrication of Mg/Al Clad Strips by Direct Cladding from Molten Metals

Gengyan Feng [1,*], Hisaki Watari [2] and Toshio Haga [3]

[1] Graduate School of Advanced Science and Technology, Tokyo Denki University, Ishizaka, Hatoyama-machi, Hiki-gun, Saitama 350-0394, Japan
[2] Division of Mechanical Engineering, Tokyo Denki University, Ishizaka, Hatoyama-machi, Hiki-gun, Saitama 350-0394, Japan
[3] Department of Mechanical Engineering, Osaka Institute of Technology, 5-16-1 Omiya Asahi-ku, Osaka 535-8585, Japan
* Correspondence: 20udm02@ms.dendai.ac.jp

Abstract: This work describes the fabrication of AZ91D/A5052 clad strips by direct cladding from molten metals using a horizontal twin roll caster. Subsequently, the effects of roll speed, pouring sequence, and solidification length on the AZ91D/A5052 clad strips were investigated. The AZ91D/A5052 clad strips with a thickness of 4.9 mm were successfully cast at a roll speed of 9 m/min and with a 5 mm roll gap. The cladding ratio of AZ91D/A5052 was about 1:1. The single-roll casting results showed that the experimental solidification constants of AZ91D and A5052 were 62 mm/min$^{0.5}$ and 34 mm/min$^{0.5}$, respectively. The twin-roll casting results showed that the effect of rolling speed on the surface condition of A5052 was greater than that of AZ91D. In addition, the high melting point A5052 alloy poured into the lower nozzle could solve the remelting problem of the low melting point AZ91D. Moreover, extending the upper solidification distance could reduce the generation of intermetallic compounds. The EDS analysis results showed no voids at the bonding interface, while three intermetallic compound layers were also found at the bonding interface of AZ91D/A5052 strips, namely α-Mg + Mg$_{17}$Al$_{12}$, Mg$_{17}$Al$_{12}$, and Al$_3$Mg$_2$. This study could be instructive for dissimilar sheet metal bonding.

Keywords: magnesium alloy; aluminum alloy; twin-roll caster; molten metals; cladding

1. Introduction

Automotive light-weighting has been a hot research topic in recent years to improve fuel efficiency and reduce vehicle exhaust emissions [1–4]. Traditional automotive lightweighting technology has been used to replace steel materials with a single lightweight alloy [5]. However, with the development of new energy vehicles, the characteristics of single-metal materials can no longer meet the new demands of automotive industry development. On this basis, metal additive manufacturing technologies have also started to receive attention [6–9]. Among these, the fabrication of clad sheets has become a current research hotspot [10–14], as they will exhibit the characteristics of each metal alloy, and can realize the complementary advantages and disadvantages of different metal alloys.

As the lightest metal among the practical structural metals, magnesium (Mg) alloys have gained attention as they offer excellent specific strength, vibration damping performance, and recyclability. However, drawbacks such as low corrosion resistance and high production costs have limited their application in the automotive industry [15]. As one of the practical lightweight metals, aluminum (Al) alloys have been widely used in the transportation and aerospace fields due to their active chemical properties, which can easily react with oxygen in the air to produce a dense oxide film and improve their corrosion resistance. Composite sheets prepared by using Mg and Al alloys can improve the corrosion resistance of Mg alloys, while also endowing the composite sheets with the physical properties of two metal sheets [16].

Many preparation methods for Mg/Al composite sheets have been reported, such as vacuum diffusion bonding [17], in situ hot press bonding [18], cold rolling [19], hot rolling [20–24], explosion welding [25–29], and solid-liquid composites [30,31]. The vacuum diffusion bonding method consists of bonding Mg and Al sheets by heating, pressurizing, and cooling in a vacuum environment. Because oxides on the metal surface can hinder diffusion bonding, it is necessary to clean the oxides before experimentation; however, this method is tedious and has high equipment requirements [17]. Jin investigated an in situ hot press bonding technique to prepare Mg/Al composite sheets. This technique required only four steps, and no vacuum environment or protective gas was needed to avoid oxidation of the metal material. The results showed that the Mg/Al composite sheet had good interfacial strength, while the intermetallic compounds that were generated at the interface had little effect on the Mg/Al composite sheet in compression, but caused clad–core delamination under tensile conditions [18]. Cold and hot rolling is a combination of pressure on Mg and Al alloy sheets using rolling equipment. The advantage of this approach is the low production cost and the fact that the preparation process does not need to be performed under a vacuum. In addition, most of the Mg/Al composite sheets are rolled by hot rolling because of the difficulty of processing Mg alloys at room temperature. Generally, rolled Mg/Al composite sheets also must be annealed. Research has shown that the best bonding strength of Mg/Al composite plates can be achieved under an annealing temperature of 200 °C and annealing time of 1 h [21]. Cao et al. investigated the effect of secondary rolling on the interfacial bonding strength and mechanical properties of Al/Mg/Al composite sheets. The results showed that secondary rolling significantly improved the interfacial bond strength. After annealing, the elongation of the composite plate was as high as 21% and the interfacial strength was maintained at 12 MPa [24]. Explosion welding uses the high impact forces generated by the explosion to strongly bond Mg and Al alloy sheets. Wang et al. improved the explosion welding technique by adding an additional thin aluminum plate as a buffer layer between the Mg and Al sheets, while the other sides of the magnesium sheets were fixed with steel plates to avoid deformation. The welding process was also simulated using smoothed-particle hydrodynamics (SPH) simulations, and the bonding interface showed a regular wave shape, which was in agreement with the experimental results [28]. However, explosion welding poses certain safety risks and cannot be produced on a large scale. Solid-liquid composite uses twin-roll casting technology, which differs from traditional sheet production methods in that it significantly reduces production costs due to the use of a twin-roll caster, enabling continuous casting from molten metals. Park et al. successfully fabricated three-layer A5052/AZ31/A5052 clad strips by twin-roll casting and post-treatment, where the thinness of the A5052 alloy sheet resulted in an uneven surface condition of the composite sheet after casting [31]. In addition, Haga et al. successfully cast clad strips consisting of Al alloys in one step using a single-roll caster equipped with a scraper and an unequal diameter twin-roll caster. A theoretical formula for controlling the clad ratio by solidification length is also proposed. Because the second molten metal was located directly over the free solidification surface of the first molten metal, oxidation of the first metal as a result of contact with air was avoided [32,33]. However, relatively little research has been conducted on the liquid–liquid lamination of dissimilar metals.

In this study, AZ91D/A5052 clad strips were fabricated by direct cladding from molten metals using horizontal twin-roll casters, and the effects of experimental parameters such as the roll speed, pouring sequence, and solidification length on the surface conditions of the AZ91D/A5052 strips were investigated. Then, the microstructure of the fabricated strips was investigated using optical microscopy and electron microscopy.

2. Experimental Procedures

2.1. Materials and Methods

An AZ91D Mg alloy (Mg-9mass%Al-1mass%Zn alloy), with excellent specific strength, was chosen as the base material, and an A5052 Al alloy (Al-2.5mass%Mg alloy) with good corrosion resistance and plasticity was chosen as the covering material. Table 1 shows the chemical compositions of the AZ91D and A5052 alloys. Figure 1 shows a schematic illustration of the cladding processes of the two molten metals, which utilized a horizontal twin-roll caster (HTRC). The two molten metals were poured into the upper and lower nozzles separately, cooled rapidly when they made contact with the rolls, and then bonded between the rolls. The rolls were composed of copper and were 300 mm in diameter and 150 mm in width. In addition, both the upper and lower nozzles had a width of 150 mm.

Table 1. Chemical compositions of the AZ91D and A5052 alloys [mass%].

Materials	Mg	Al	Si	Fe	Cu	Mn	Zn	Cr
AZ91D	Rest	9	0.05	0.002	0.01	0.3	1	-
A5052	2.5	Rest	0.09	0.14	0.01	0.01	-	0.25

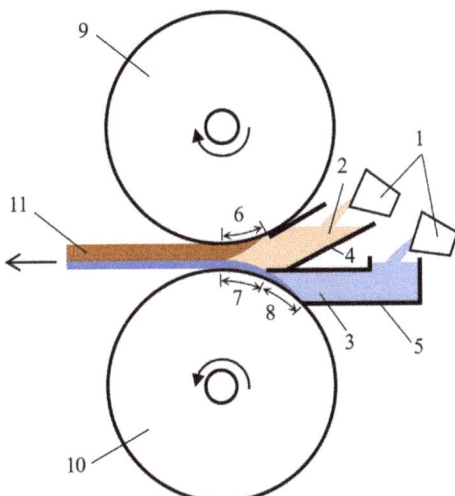

Figure 1. Schematic showing the cladding processes of Mg/Al using a horizontal twin-roll caster (1. crucible; 2. molten metal A; 3. molten metal B; 4. upper nozzle, 5. lower nozzle; 6. upper solidification length; 7. cooling length; 8. solidification length; 9. upper roll; 10. lower roll; 11. clad strip).

2.2. Control of Clad Ratio

In this study, because molten metal B in the lower nozzle solidifies before the molten metal A in the upper nozzle, the cladding ratio of the cladding material could be expressed by Equation (1), where the roll gap was the shortest distance between the surfaces of the twin rolls:

$$\text{clad ratio} = \frac{\text{roll gap} - \text{thickness of solidification layer of molten metal B}}{\text{thickness of solidification layer of molten metal B}}, \quad (1)$$

The thickness of the solidification layer of molten metal B was controlled by the solidification length and roll speed, and the thickness of the solidification layer is given by Equation (2) [32]:

$$d = K\sqrt{t} = K\sqrt{L/V}, \quad (2)$$

where d is the thickness of the solidification layer, K is the experimental solidification constant, t is the solidification time, L is the solidification length, V is the roll speed, and the K values were obtained from single-roll casting experiments. The schematic diagram of the single-roll casting experiment is shown in Figure 2, and the experimental conditions are shown in Table 2. The length of the molten metal in the lower nozzle in contact with the roll was defined as the solidification length (L), and the length of the molten metal in contact with the roll after completion of solidification was defined as the cooling length (Lc). The pouring temperature was the temperature of the molten metal inside the crucible during pouring, while the casting temperature was the temperature of the molten metal inside the nozzle. Because the temperature of the molten metal decreased as a result of the pouring process, the pouring temperature was increased by 5 °C to ensure that the temperature inside the nozzle reached the liquidus temperature of the metal material [34]. Because it was difficult to measure the temperature of the bonding interface in the twin-roll composite casting, the surface temperature of the solidified metal immediately out of the melt pool was measured in the single-roll casting experiment. The temperature at point A was approximately equal to the temperature of the metal bonding surface on the lower roll side before the two metals were compounded. The temperature of the metal contact surface on the upper roll side was its liquidus temperature. In addition, the continuous twin-roll caster used in this experiment was not equipped with water cooling. To investigate the variation of the surface temperature of the cast rolls, the surface temperature of point B during the casting process was measured.

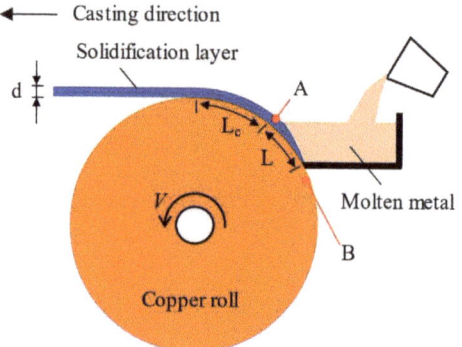

Figure 2. Schematic diagram of single-roll casting.

Table 2. Experimental conditions of the single-roll casting experiments for the cast AZ91D and A5052 alloys.

Materials	AZ91D	A5052
Solidus temperature [°C]	430	607
Liquidus temperature [°C]	595	649
Pouring temperature [°C]	600	654
Solidification length [mm]	50	50
Cooling length [mm]	50	50
Roll speed [m/min]	6–36	6–36

2.3. Cladding for Mg/Al Clad Strips by Twin-Roll Caster

The casting conditions for the Mg/Al clad strips are shown in Table 3. The experimental results of single-roll casting showed that the effect of roll speed on the thickness of the solidified layer was small when the roll speed exceeded 12 m/min. Therefore, in this experiment, to study the effect of rolling speed on the Mg/Al clad strip, casting experiments were conducted at rolling speeds ranging from 6–12 m/min, and different pouring sequences were also investigated. In addition, the settings of cooling length and solidification length were kept the same as for single-roll casting, and both were set to 50 mm. The roll gap was set at 5 mm. Because flame retardant gases such as SF_6 will destroy the ozone layer and cause environmental pollution, in this experiment, we did not use a flame retardant gas such as SF_6. To inhibit the oxidation and combustion of Mg alloys, 0.5 mass% flux (S.K.No.101, TACHIGAWACAST, Japan) was added during the dissolution and refining stages, and an oxide film formed on the surface of the crucible at the end of refining. This oxide film could prevent direct contact between the molten Mg alloy and the oxygen in the air, preventing further oxidation and combustion. In addition, the oxide film was removed from the surface before the Mg alloy was poured. In this study, the molten metal from the upper nozzle was poured directly onto the surface of the metal solidified by the lower nozzle, avoiding direct contact between the lower nozzle metal and the air. This replaced the traditional vacuum environment of hot rolling bonding. The aim of this work was to maximize the control of strip production costs.

Table 3. Casting conditions for the Mg/Al clad strips.

Materials	AZ91D	A5052
Pouring temperature [°C]	600	654
Pouring sequence [upper nozzle/lower nozzle]	A5052/AZ91D, AZ91D/A5052	
Upper solidification length [mm]	50, 100	
Solidification length [mm]	50	
Cooling length [mm]	50	
Roll gap [mm]	5	
Roll speed [m/min]	6–12	
Roll surface temperature [°C]	22	

2.4. Microstructure of the Bonding Interface

Subsequently, the bonding interface microstructure of the AZ91D/A5052 clad strip was observed using an OLYMPUS BX60M optical microscope. To clearly observe the diffusion layer of the bonding interface, the observation surfaces of the specimens were abraded in turn using #400, #800, #1500, and #2000 sandpaper. The specimens were then polished to a mirror finish using 6.0 and 0.25 µm diamond polishing solutions, in order. Then, the polished surface was sequentially chemically etched with a citric acid solution (10 g of $C_6H_8O_7$ + 90 g of H_2O) and sodium hydroxide solution (1 g of NaOH + 100 mL of H_2O). Then, the composition of the intermetallic compounds at the bonding interface was analyzed by ultra-low acceleration voltage scanning electron microscopy (JSM-7100F).

3. Results and Discussion

3.1. Calculation of the Experimental Solidification Constants

The temperature measurement results at point A showed that the surface temperature of AZ91D was between approximately 491 °C and 520 °C, while the surface temperature of A5052 was between approximately 600 °C and 626 °C. Because the metal surface was in a semi-solidified state, the thermocouple did not perfectly measure most of the surface temperature of the metal. Therefore, the measured temperature could only be used as a reference. In addition, during the experiment, the temperature at point B increased from the initial 22 °C to 30 °C. Compared to the temperature of the molten metal, the effect of this temperature difference was negligible.

The experimental results showed that AZ91D and A5052 could be cast continuously using a single roll, and the average thickness of the solidification layer was measured using a micrometer. The relationship between the roll speed and solidified layer thickness is shown in Figure 3a. With a higher roll speed, the thickness of the solidification layer was thinner [35]. The effect of the roll speed on the solidified layer thickness was more obvious when the roll speed was below 12 m/min, which indicated that the control of the solidified layer thickness of molten metal B, and consequently the cladding ratio of the cladding material, was easier to achieve under roll speed conditions below 12 m/min. The relationship between the square root of the solidification time and the solidification layer thickness is shown in Figure 3b. According to the figure, the experimental solidification constants of AZ91D and A5052 were 62 mm/min$^{0.5}$ and 34 mm/min$^{0.5}$, respectively.

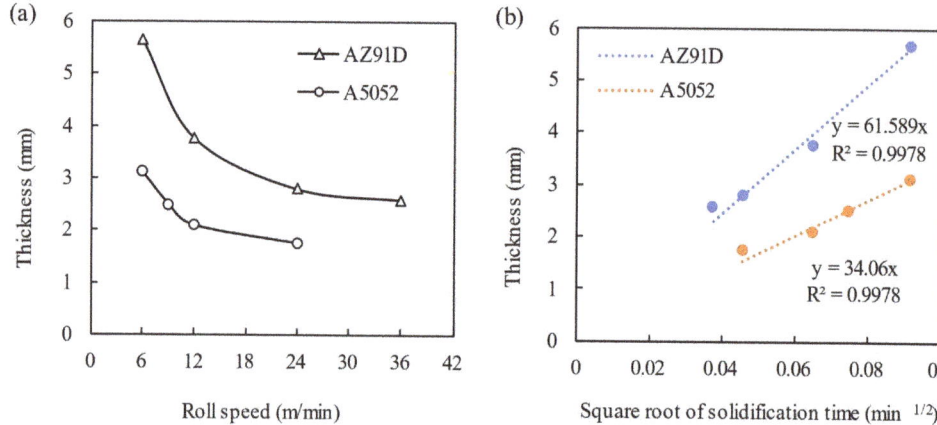

Figure 3. (a) Relationship between the roll speed and solidification layer thickness; (b) relationship between the square root of the solidification time and the thickness of the solidification layer.

The duration of the experiment was difficult to quantify. Currently, we have observed the approximate time from the pouring to the end of casting by video, which was not accurate. This made it difficult to quantify the duration of the experiment because the amount of molten metal poured at the beginning potentially did not reach the height of the melt pool, and the amount of molten metal in the melt pool slowly decreased after the end of pouring, resulting in thin ends at the front and back of the continuous sheet, with uniformity in the middle. In addition, calculating the solidification time was reasonable. For example, when the roll speed was 6 m/min, the solidification time was 0.5 s for a solidification length of 50 mm, assuming no relative sliding of the solidified molten metal and casting roll.

3.2. Effects of Roll Speeds and Pouring Sequences on the Surfaces of the Clad Strips

The experimental results under different roll speeds and pouring sequences are shown in Table 4. The first pouring sequence involved pouring the high melting point A5052 material into the upper nozzle and the low melting point AZ91D material into the lower nozzle. As a result, the remelting of AZ91D on the lower roll side occurred under different roll speed conditions, as shown in Figure 4, which also shows the experimental results of the first pouring sequence at a roll speed of 9 m/min. The upper roll contact surface of the strip showed a good surface condition; however, the lower roll contact surface was uneven. The front portions of the clad strips showed that it was possible to form AZ91D sheets without the A5052 covering. After A5052 covered the AZ91D layer, the underlying AZ91D was dissolved and mixed together, forming grey intermetallic compounds. In addition, two types of heat transfer were associated with AZ91D in this experiment: one with A5052, and the other with the lower copper roll. Because the solidus temperature of A5052 was higher than the liquidus temperature of AZ91D, heat could only be transferred from the A5052 side to the AZ91D side. When the heat absorbed by AZ91D from A5052 was greater than that absorbed by the copper roll, the basic conditions for AZ91D solidification were not present. This resulted in the remelting phenomenon shown in Figure 4b.

Table 4. Experimental results under different roll speeds and pouring sequences.

Pouring Sequence [Upper Nozzle/Lower Nozzle]	Roll Speed [m/min]		
	6	9	12
A5052/AZ91D	×	×	×
AZ91D/A5052	×	Δ	×

×: bad, Δ: not good.

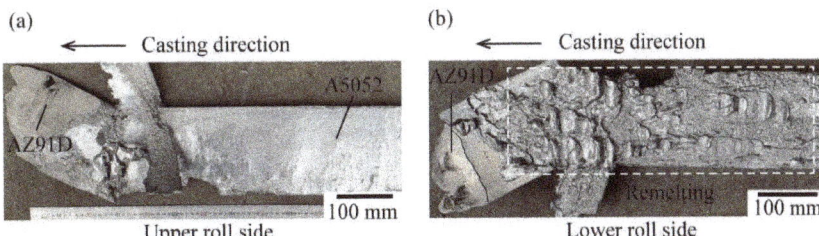

Figure 4. Surfaces of the A5052/AZ91D clad strips at a roll speed of 9 m/min. (**a**) Upper roll side, (**b**) Lower roll side.

The surfaces of the AZ91D/A5052 strips cast using the second pouring sequence at different rolling speeds are shown in Figure 5. The upper roll contact surface (AZ91D side) of the AZ91D/A5052 clad strip contained numerous ripple marks and some black oxides at a roll speed of 6 m/min. This black oxide could be easily removed by acid washing. When the roll speed was increased to 9 or 12 m/min, the ripple marks disappeared, and the surface condition was good with a smooth surface. However, we observed that the roll speed had a significant influence on the lower roll contact surface (A5052 side) of the AZ91D/A5052 clad strips. When the roll speed was 6 m/min, most of the area on the A5052 side was AZ91D and only a small portion of the area was A5052, and cracks were observed. The A5052 side showed a bright Al alloy surface at a roll speed of 9 m/min; however, intermetallic compounds appeared on both the surface and edges of the strip. Specifically, at 12 m/min, the A5052 side was directly covered with intermetallic compounds. We could speculate about these compounds. Because Mg and Al were in direct contact inside the nozzle, the interface output compound likely consisted of a liquid Mg alloy, an Al-Mg intermetallic compound, and a mixture of both.

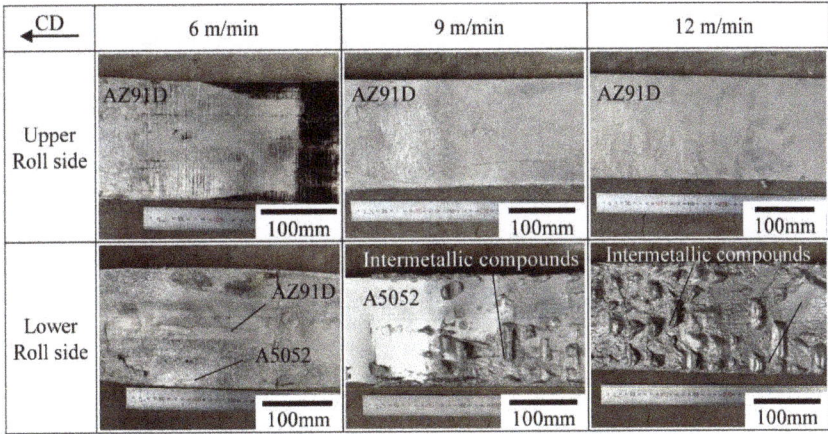

Figure 5. Surface of the AZ91D/A5052 strips cast using the second pouring sequence with different roll speeds.

The large size of these intermetallic compounds and the rapid solidification of the alloy were not very well connected. In the twin-roll experiment, the upper and lower sides of the sheet were close to the upper and lower roll faces, respectively, and could achieve rapid solidification. Therefore, the surface condition was good; however, the middle part of the sheet was far from the upper and lower rolls and did not solidify well. This led to incomplete solidification in the middle part of the sheet, and excess liquid metal or metal compounds were then squeezed out under pressure of the double rolls and flowed around the sheet. The large size could be explained by Figure 6. The excess liquid metal or metal mixture at the bonding interface flowed to the lower surface of the sheet and accumulated between the individual drums of the drum conveyor under the influence of its own gravity. After cooling and solidification, large-size intermetallic compounds formed.

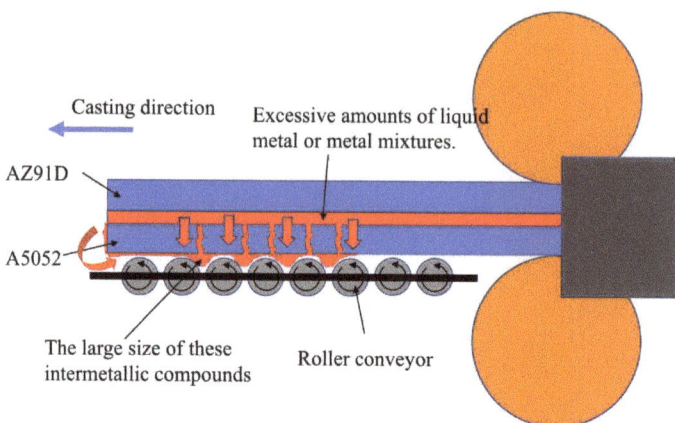

Figure 6. Schematic diagram of intermetallic compound output.

Figure 7 shows the casting process at different roll speeds. At low roll speeds, there was a significant increase in the thickness of the solidification layer in both molten metals. The increased thickness of the solidification layer of the Al alloy caused the lower nozzle to become blocked and the solidified Al alloy layer could not be smoothly brought out. Thus, only a small portion of the Al alloy was bound. However, the increased thickness of the solidification layer of the Mg alloy created a downward squeezing force. This inhibited the formation of the Al alloy in the lower nozzle, and this was consistent with the experimental results shown in Figure 5 (6 m/min). At high roll speeds, the thickness of the solidification layer in both molten metals was significantly reduced. Theoretically, with a constant roll gap, most of the intermediate gap would be filled by the molten Mg and Al alloys in a semi-solidified state. In this study, the bonding interface between the Mg alloy and Al alloy occurred at a high temperature, because the higher the temperature, the higher the energy of the atoms, the easier the migration, the higher the diffusion coefficient, and the faster the diffusion. Therefore, the mutual diffusion movement of the Mg and Al atoms was very strong, which would intensify the generation of intermetallic Mg and Al compounds. The experimental single-roll casting results showed that the surface temperature of A5052 before bonding was higher than 600 °C. In addition, the AZ91D in the upper nozzle was in a liquid state and the temperature of the bonding interface was the liquidus temperature (595 °C); thus, the average temperature of the bonding interface was higher than 595 °C. Because the temperature of the bonding interface was higher than 595 °C, the excess molten Mg alloy and intermetallic compounds remained in a liquid state and were extruded to both sides of the strip under the pressure of the twin rolls, and finally flowed along the edge of the Al layer to the lower roll contact surface of the strip via gravity. In addition, more intermetallic compounds were produced at the bonding interface at high roll speeds. This also resulted in more compounds flowing to the lower roll contact surface of the Mg/Al strips, which was consistent with the experimental results shown in Figure 5 (12 m/min).

Figure 5 shows that the surface condition of the Mg/Al clad strips was closest to being successful at a roll speed of 9 m/min with a 5 mm roll gap. The generation of intermetallic compounds was caused by excessive contact between the molten AZ91D and A5052 at the bonding interface. According to Equation (2), extending the solidification length at a certain roll speed increased the thickness of the solidification layer, which in turn reduced contact between the molten AZ91D and A5052. Theoretically, this could reduce the output of intermetallic compounds, which we verified experimentally. The upper solidification length was extended from the original 50 mm to 100 mm, and the experiments were carried out. Figure 8 shows the surface of the AZ91D/A5052 clad strips at an upper solidification length of 100 mm, showing the good overall surface condition of the AZ91D/A5052 clad strip. The picture of the lower roll side showed that the intermetallic compounds were only present at the edges of the AZ91D/A5052 clad strips. The effectiveness of extending the upper solidification length in reducing the generation of intermetallic compounds was verified. In this study, the surface of A5052 was smooth and had the metallic luster of the Al alloy, and the average thickness of the AZ91D/A5052 clad strip manufactured at a roll speed of 9 m/min and solidification length of 100 mm was 4.9 mm.

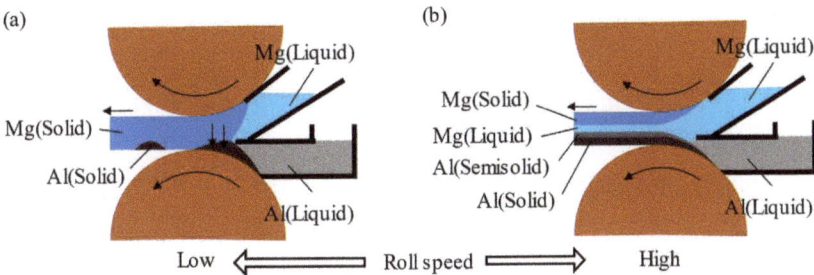

Figure 7. Casting process at different roll speeds. (**a**) Low speed, (**b**) High speed.

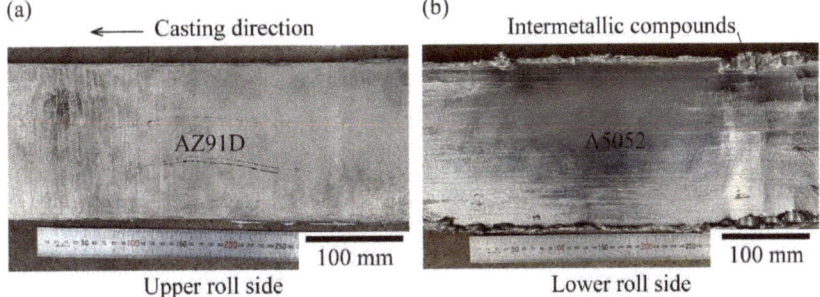

Figure 8. The surface of the AZ91D/A5052 clad strip with an upper solidification length of 100 mm. (**a**) Upper roll side, (**b**) Lower roll side.

3.3. Microstructure of the Bonding Interface

Figure 9a shows the cross-section of the AZ91D/A5052 clad strips cast by the horizontal twin-roll caster. Figure 9b shows the microstructure of region A of the bonding interface in Figure 9a. We clearly observed that the bonding interface was free of voids, with two diffusion layers at the bonding interface of AZ91D and A5052. The thickness of the total diffusion layer on the bonding interface was about 1 mm, which was much greater than the thickness of the diffusion layer obtained by hot rolling [21,22]. Because the solidus temperature of A5052 was higher than the liquidus temperature of AZ91D, when the molten Mg alloy and Al alloy were combined, the temperature at the bonding interface still remained above 595 °C. This accelerated the diffusion of Mg and Al elements on the bonding surface. In addition, the thicknesses of the AZ91D layer and A5052 layer were 2.1 mm and 1.8 mm, respectively, and the cladding ratio was approximately 1:1. According to Figure 3a, the theoretical solidification thickness of A5052 at a roll speed of 9 m/min was 2.4 mm, and the thickness of the A5052 layer in the clad strip was reduced by 25%. This was because the surface of A5052 in the lower nozzle was in a semi-solidified state after the end of solidification and before bonding with the molten AZ91D [35]. When the molten AZ91D was poured onto the semi-solidified surface of A5052, the thickness of A5052 on the lower roll side was smaller than the theoretical value due to the gravity of the molten AZ91D and rolling force of the upper roll. Combining the results in Figure 10 and Table 5, it was clear that the two diffusion layers in the microstructure consisted of the α-Mg+$Mg_{17}Al_{12}$ eutectic layer near the Mg side and the $Mg_{17}Al_{12}$+Al_3Mg_2 compound layer near the Al side. Because the lowest magnification image of the laboratory optical microscope did not show the complete cross-section image of the clad strip, the image in Figure 9a was obtained using a normal camera under the illumination of a vertical light source. In addition, the two etching solutions used in this experiment could not etch the $Mg_{17}Al_{12}$ and Al_3Mg_2 layers, which were the two intermetallic compound layers that also showed a mirror effect under polishing and reflected under the irradiation of a vertical light source, causing the photographed $Mg_{17}Al_{12}$ and Al_3Mg_2 layers to show dark colors.

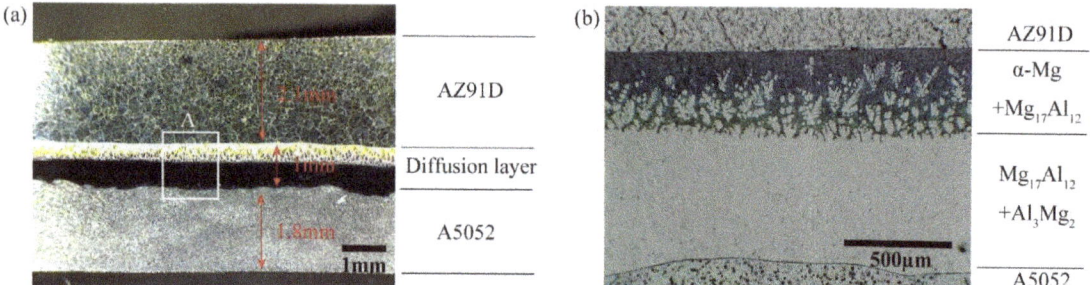

Figure 9. (**a**) Cross-section of the AZ91D/A5052 clad strips cast by the horizontal twin-roll caster; (**b**) microstructure of region A of the bonding interface in (**a**).

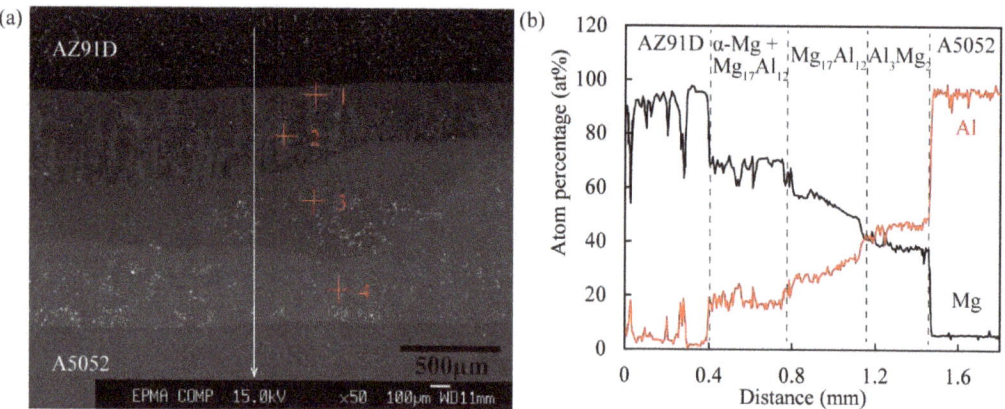

Figure 10. Microstructure and EDS analysis results on the bonding interface of the Mg/Al clad strips: (**a**) SEM image; (**b**) EDS line scan result corresponding to (**a**).

Table 5. Results of EDS point scan analysis at different positions of the interface corresponding to Figure 10a.

Point	Mg	Al	Possible Phase
1	90.15	9.49	α-Mg
2	66.70	33.30	α-Mg+$Mg_{17}Al_{12}$
3	56.86	43.14	$Mg_{17}Al_{12}$
4	43.32	56.68	Al_3Mg_2

The trace element content in AZ91D and A5052 was low and most of the trace elements would solidly dissolve into α-Mg and α-Al, respectively, where the content did not change significantly. In addition, the focus of this study was on the content changes of the major elements Mg and Al, and the composition of the compounds. Figure 10 shows an SEM image of the bonding interface of the A5052/AZ91D clad strip and the EDS line scan results of the bonding interface. The SEM image and EDS line scan results indicated that the diffusion layer was divided into three layers. The chemical composition of each layer was analyzed by EDS point scanning and the results are shown in Table 5. The results showed that the layers near the AZ91D side were α-Mg and $Mg_{17}Al_{12}$, the middle was $Mg_{17}Al_{12}$, and the layer near the A5052 side was Al_3Mg_2. The composition of the intermetallic compounds was consistent with the results of previous studies [36,37]. Moreover, reducing the generation of intermetallic compounds has been an effective means of improving the interfacial bond strength of composite sheets [21]. For example, Ni or Zn intermediate layers were inserted between Mg alloys and Al alloys to avoid direct contact between the Mg and Al alloys, which in turn reduced the generation of Al-Mg intermetallic compounds [38–40]. Therefore, the future direction of this study will be to insert an Ni foil in the middle of the upper and lower nozzles to reduce the generation of intermetallic compounds and improve the bonding strength of the composite strip.

4. Conclusions

In this work, the continuous casting performance of AZ91D and A5052 was verified by single-roll casting experiments, and the experimental solidification constants of AZ91D and A5052 were obtained by empirical equations and fitted curves, with values of 62 mm/min$^{0.5}$ and 34 mm/min$^{0.5}$, respectively. Next, the surface temperatures of the metals during single-roll casting were measured to provide data to support the subsequent temperature change discussion of the bonding interface during double-roll composite casting. In addition, for continuous casting of a small number of metal ingots, the surface temperature variations of the casting rolls could be neglected.

Furthermore, the AZ91D/A5052 clad strips were successfully fabricated by direct cladding from molten metals using a horizontal twin-roll caster. The optimal experimental parameters were investigated by conducting tests under different experimental conditions of roll speed, pouring sequence, and solidification length. The results showed that the AZ91D/A5052 clad strip with a thickness of 4.9 mm was successfully cast at a rolling speed of 9 m/min and rolling gap of 5 mm. We found that the roll speed affected the surface state of A5052 on the lower roll side more than AZ91D on the upper roll side. In addition, the casting sequence of AZ91D/A5052 (upper nozzle metal/lower nozzle metal) made it easier to form the strip compared to A5052/AZ91D, and the intermetallic compounds produced at the bonding interface of AZ91D and A5052 could be controlled by the upper solidification length. The cross-section (Figure 9) showed that the AZ91D/A5052 strip had a cladding ratio of approximately 1:1, while three intermetallic compound layers were found at the bonding interface. Finally, the use of a horizontal twin-roll caster allowed for the direct compounding of dissimilar metals.

Author Contributions: Writing—original draft preparation, G.F.; conceptualization, T.H. and H.W.; methodology, G.F. and H.W.; investigation, G.F.; resources, H.W.; validation, G.F. and H.W.; data curation, G.F. and H.W.; writing—review and editing, G.F. and H.W.; All authors have read and agreed to the published version of the manuscript.

Funding: This research received no external funding.

Institutional Review Board Statement: Not applicable.

Informed Consent Statement: Not applicable.

Data Availability Statement: Not applicable.

Conflicts of Interest: The authors declare no conflict of interest.

References

1. Taub, A.I.; Luo, A.A. Advanced lightweight materials and manufacturing processes for automotive applications. *MRS Bull.* **2015**, *40*, 1045–1054. [CrossRef]
2. Hovorun, T.P.; Berladir, K.V.; Pererva, V.I.; Rudenko, S.G.; Martynov, A.I. Modern materials for automotive industry. *J. Eng. Sci.* **2017**, *4*, f8–f18. [CrossRef]
3. Oliveira, J.P.; Ponder, K.; Brizes, E.; Abke, T.; Edwards, P.; Ramirez, A.J. Combining resistance spot welding and friction element welding for dissimilar joining of aluminum to high strength steels. *J. Mater. Process. Technol.* **2019**, *273*, 116192. [CrossRef]
4. Yang, J.; Oliveira, J.P.; Li, Y.; Tan, C.; Gao, C.; Zhao, Y.; Yu, Z. Laser techniques for dissimilar joining of aluminum alloys to steels: A critical review. *J. Mater. Process. Technol.* **2022**, *301*, 117443. [CrossRef]
5. Cole, G.S.; Sherman, A.M. Light weight materials for automotive applications. *Mater. Charact.* **1995**, *35*, 3–9. [CrossRef]
6. Khorasani, M.; Ghasemi, A.; Leary, M.; Sharabian, E.; Cordova, L.; Gibson, I.; Downing, D.; Bateman, S.; Brandt, M.; Rolfe, B. The effect of absorption ratio on meltpool features in laser-based powder bed fusion of IN718. *Opt. Laser Technol.* **2022**, *153*, 108263. [CrossRef]
7. Linares, J.; Chaves-Jacob, J.; Lopez, Q.; Sprauel, J.-M. Fatigue life optimization for 17-4Ph steel produced by selective laser melting. *Rapid Prototyp. J.* **2022**, *28*, 1182–1192. [CrossRef]
8. Giganto, S.; Martínez-Pellitero, S.; Cuesta, E.; Zapico, P.; Barreiro, J. Proposal of design rules for improving the accuracy of selective laser melting (SLM) manufacturing using benchmarks parts. *Rapid Prototyp. J.* **2022**, *28*, 1129–1143. [CrossRef]
9. Khan, H.M.; Waqar, S.; Koç, E. Evolution of temperature and residual stress behavior in selective laser melting of 316L stainless steel across a cooling channel. *Rapid Prototyp. J.* **2022**, *28*, 1272–1283. [CrossRef]
10. Paul, H.; Chulist, R.; Mania, I. Structural Properties of Interfacial Layers in Tantalum to Stainless Steel Clad with Copper Interlayer Produced by Explosive Welding. *Metals* **2020**, *10*, 969. [CrossRef]
11. Kang, M.; Zhou, L.; Deng, Y.; Luo, Y.; He, M.; Zhang, N.; Huang, Z.; Dong, L. Microstructure and Mechanical Properties of 4343/3003/6111/3003 Four-Layer Al Clad Sheets Subjected to Different Conditions. *Metals* **2022**, *12*, 777. [CrossRef]
12. Murzin, S.P.; Palkowski, H.; Melnikov, A.A.; Blokhin, M.V. Laser Welding of Metal-Polymer-Metal Sandwich Panels. *Metals* **2022**, *12*, 256. [CrossRef]
13. Xu, J.; Fu, J.; Li, S.; Xu, G.; Li, Y.; Wang, Z. Effect of annealing and cold rolling on interface microstructure and properties of Ti/Al/Cu clad sheet fabricated by horizontal twin-roll casting. *J. Mater. Res. Technol.* **2022**, *16*, 530–543. [CrossRef]
14. Zhao, H.; Zhao, C.; Yang, Y.; Wang, Y.; Sheng, L.; Li, Y.; Huo, M.; Zhang, K.; Xing, L.; Zhang, G. Study on the Microstructure and Mechanical Properties of a Ti/Mg Alloy Clad Plate Produced by Explosive Welding. *Metals* **2022**, *12*, 399. [CrossRef]
15. Song, J.; She, J.; Chen, D.; Pan, F. Latest research advances on magnesium and magnesium alloys worldwide. *J. Magnes. Alloy.* **2020**, *8*, 1–41. [CrossRef]
16. Zhang, N.; Wang, W.; Cao, X.; Wu, J. The effect of annealing on the interface microstructure and mechanical characteristics of AZ31B/AA6061 composite plates fabricated by explosive welding. *Mater. Des.* **2015**, *65*, 1100–1109. [CrossRef]
17. Wang, J.; Li, Y.; Liu, P.; Geng, H. Microstructure and XRD analysis in the interface zone of Mg/Al diffusion bonding. *J. Mater. Process. Technol.* **2008**, *205*, 146–150. [CrossRef]
18. Jin, H.; Javaid, A. A new cladding technology to bond aluminium on magnesium. *Mater. Sci. Technol.* **2020**, *36*, 1037–1043. [CrossRef]
19. Matsumoto, H.; Watanabe, S.; Hanada, S. Fabrication of pure Al/Mg–Li alloy clad plate and its mechanical properties. *J. Mater. Process. Technol.* **2005**, *169*, 9–15. [CrossRef]
20. Kim, J.-S.; Lee, K.S.; Kwon, Y.N.; Lee, B.-J.; Chang, Y.W.; Lee, S. Improvement of interfacial bonding strength in roll-bonded Mg/Al clad sheets through annealing and secondary rolling process. *Mater. Sci. Eng. A* **2015**, *628*, 1–10. [CrossRef]
21. Zhang, J.; Liang, W.; Li, H. Effect of thickness of interfacial intermetallic compound layers on the interfacial bond strength and the uniaxial tensile behaviour of 5052 Al/AZ31B Mg/5052 Al clad sheets. *RSC Adv.* **2015**, *5*, 104954–104959. [CrossRef]
22. Wang, P.; Chen, Z.; Hu, C.; Li, B.; Mo, T.; Liu, Q. Effects of annealing on the interfacial structures and mechanical properties of hot roll bonded Al/Mg clad sheets. *Mater. Sci. Eng. A* **2020**, *792*, 139673. [CrossRef]
23. Li, S.; Liu, X.; Jia, Y.; Han, J.; Wang, T. Interface Characteristics and Bonding Performance of the Corrugated Mg/Al Clad Plate. *Materials* **2021**, *14*, 4412. [CrossRef]
24. Cao, X.; Xu, C.; Li, Y.; Cao, X.; Peng, R.; Fang, J. Effect of secondary rolling on the interfacial bonding strength and mechanical properties of Al/Mg/Al clad plates. *Philos. Mag. Lett.* **2022**, *102*, 200–208. [CrossRef]
25. Zeng, X.; Wang, Y.; Li, X.; Li, X.; Zhao, T. Effect of inert gas-shielding on the interface and mechanical properties of Mg/Al explosive welding composite plate. *J. Manuf. Process.* **2019**, *45*, 166–175. [CrossRef]
26. Inao, D.; Mori, A.; Tanaka, S.; Hokamoto, K. Explosive Welding of Thin Aluminum Plate onto Magnesium Alloy Plate Using a Gelatin Layer as a Pressure-Transmitting Medium. *Metals* **2020**, *10*, 106. [CrossRef]
27. Rouzbeh, A.; Sedighi, M.; Hashemi, R. Comparison between Explosive Welding and Roll-Bonding Processes of AA1050/Mg AZ31B Bilayer Composite Sheets Considering Microstructure and Mechanical Properties. *J. Mater. Eng. Perform.* **2020**, *29*, 6322–6332. [CrossRef]
28. Wang, Q.; Li, X.; Shi, B.; Wu, Y. Experimental and Numerical Studies on Preparation of Thin AZ31B/AA5052 Composite Plates Using Improved Explosive Welding Technique. *Metals* **2020**, *10*, 1023. [CrossRef]

29. Atifeh, S.M.; Rouzbeh, A.; Hashemi, R.; Sedighi, M. Effect of annealing on formability and mechanical properties of AA1050/Mg-AZ31B bilayer sheets fabricated by explosive welding method. *Int. J. Adv. Manuf. Technol.* **2022**, *118*, 775–784. [CrossRef]
30. Bae, J.H.; Prasada Rao, A.K.; Kim, K.H.; Kim, N.J. Cladding of Mg alloy with Al by twin-roll casting. *Scr. Mater.* **2011**, *64*, 836–839. [CrossRef]
31. Park, J.; Song, H.; Kim, J.-S.; Sohn, S.S.; Lee, S. Three-Ply Al/Mg/Al Clad Sheets Fabricated by Twin-Roll Casting and Post-treatments (Homogenization, Warm Rolling, and Annealing). *Metall. Mater. Trans. A* **2017**, *48*, 57–62. [CrossRef]
32. Haga, T.; Nakamura, R.; Kumai, S.; Watari, H. A vertical type twin roll caster for an aluminium alloy clad strip. *Arch. Mater. Sci. Eng.* **2013**, *62*, 36–44.
33. Haga, T. Twin Roll Caster for Clad Strip. *Metals* **2021**, *11*, 776. [CrossRef]
34. Feng, G.Y.; Watari, H.; Suzuki, M.; Haga, T.; Shimizu, T. Novel Direct Cladding of Magnesium and Aluminum Alloys Using a Horizontal Twin Roll Caster. *Key Eng. Mater.* **2021**, *880*, 17–22. [CrossRef]
35. Haga, T. High Speed Roll Caster for Aluminum Alloy. *Metals* **2021**, *11*, 520. [CrossRef]
36. Liu, N.; Liu, C.; Liang, C.; Zhang, Y. Influence of Ni Interlayer on Microstructure and Mechanical Properties of Mg/Al Bimetallic Castings. *Metall. Mater. Trans. A* **2018**, *49*, 3556–3564. [CrossRef]
37. Li, G.; Jiang, W.; Guan, F.; Zhu, J.; Zhang, Z.; Fan, Z. Microstructure, mechanical properties and corrosion resistance of A356 aluminum/AZ91D magnesium bimetal prepared by a compound casting combined with a novel Ni-Cu composite interlayer. *J. Mater. Process. Technol.* **2021**, *288*, 116874. [CrossRef]
38. Shu, J.; Yamaguchi, T.; Hara, Y. Influence of a Ni Foil Interlayer on Interface Properties of Mg-Clad Al Materials by Vacuum Roll Bonding. *Mater. Trans.* **2020**, *61*, 1020–1025. [CrossRef]
39. Li, S.; Zheng, Z.; Chang, L.; Guo, D.; Yu, J.; Cui, M. A two-step bonding process for preparing 6061/AZ31 bimetal assisted with liquid molten zinc interlayer: The process and microstructure. *J. Adhes. Sci. Technol.* **2021**, 1–23. [CrossRef]
40. Dong, S.; Lin, S.; Zhu, H.; Wang, C.; Cao, Z. Effect of Ni interlayer on microstructure and mechanical properties of Al/Mg dissimilar friction stir welding joints. *Sci. Technol. Weld. Join.* **2022**, *27*, 103–113. [CrossRef]

Article

Mechanical and Tribological Behavior of Gravity and Squeeze Cast Novel Al-Si Alloy

Vadlamudi Srinivasa Chandra [1,2], Koorella S. V. B. R. Krishna [2], Manickam Ravi [3], Katakam Sivaprasad [1,*], Subramaniam Dhanasekaran [2] and Konda Gokuldoss Prashanth [4,5,6,*]

1. Advanced Materials Processing Laboratory, Department of Metallurgical and Materials Engineering, National Institute of Technology, Tiruchirapalli 620015, India; Srinivasachandra.V@ashokleyland.com
2. Ashok Leyland Limited, Technical Center, Chennai 600103, India; Koorella.krishna@ashokleyland.com (K.S.V.B.R.K.); s.dhanasekaran@ashokleyland.com (S.D.)
3. National Institute for Interdisciplinary Science and Technology, Thiruvananthapuram 695019, India; ravi@niist.res.in
4. Department of Mechanical and Industrial Engineering, Tallinn University of Technology, Ehitajate Tee 5, 19086 Tallinn, Estonia
5. Erich Schmid Institute of Materials Science, Austrian Academy of Sciences, Jahnstrasse 12, 8700 Leoben, Austria
6. Center for Biomaterials, Cellular, and Molecular Theranostics, Vellore Institute of Technology, Vellore 632014, India
* Correspondence: ksp@nitt.edu (K.S.); kgprashanth@gmail.com (K.G.P.)

Citation: Chandra, V.S.; Krishna, K.S.V.B.R.; Ravi, M.; Sivaprasad, K.; Dhanasekaran, S.; Prashanth, K.G. Mechanical and Tribological Behavior of Gravity and Squeeze Cast Novel Al-Si Alloy. *Metals* 2022, *12*, 194. https://doi.org/10.3390/met12020194

Academic Editor: Wenming Jiang

Received: 3 January 2022
Accepted: 19 January 2022
Published: 21 January 2022

Publisher's Note: MDPI stays neutral with regard to jurisdictional claims in published maps and institutional affiliations.

Copyright: © 2022 by the authors. Licensee MDPI, Basel, Switzerland. This article is an open access article distributed under the terms and conditions of the Creative Commons Attribution (CC BY) license (https://creativecommons.org/licenses/by/4.0/).

Abstract: The automotive industry traditionally reduces weight primarily by value engineering and thickness optimization. However, both of these strategies have reached their limits. A 6% reduction in automotive truck mass results in a 13% improvement in freight mass. Aluminum alloys have lower weight, relatively high specific strength, and good corrosion resistance. Therefore, the present manuscript involves manufacturing Al-based alloy by squeeze casting. The effect of applied pressure during the squeeze cast and gravity cast of a novel Al-Si alloy on microstructural evolution, and mechanical and wear behavior was investigated. The results demonstrated that squeeze casting of the novel Al-Si alloy at high-pressure exhibits superior mechanical properties and enhanced wear resistance in comparison to the gravity die-cast (GDC) counterpart. Squeeze casting of this alloy, at high pressure, yields fine dendrites and reduced dendritic arm spacing, resulting in grain refinement. The finer dendrites and reduced dendritic arm spacing in high-pressure squeeze cast alloy than in the GDC alloy were due to enhanced cooling rates observed during the solidification process, as well as the applied squeeze pressure breaks the initial dendrites that started growing during the solidification process. Reduced casting defects in the high-pressure squeeze cast alloy led to a reduced coefficient of friction, resulting in improved wear resistance even at higher loads and higher operating temperatures. Our results demonstrated that squeeze casting of the novel Al-Si alloy at high-pressure exhibits a 47% increase in tensile strength, 33% increase in hardness, 10% reduction in coefficient of friction, and 15% reduction in wear loss compared to the GDC counterpart.

Keywords: squeeze casting; novel Al-Si alloy; wear analysis; microstructure; mechanical properties; pin on disc wear testing

1. Introduction

According to the International Energy Association report of 2019, the transportation industry is the second largest contributor to global CO_2 emissions (at 27%) [1]. The automotive industry globally is striving to reduce CO_2 emissions by light-weighting, improving the efficiency of internal combustion engines, usage of alternate fuels, etc. A 10% reduction in mass results in a 6% improvement in fuel efficiency, by which CO_2 emission will be reduced significantly over the lifetime of the vehicle [1].

Aluminum, being light in comparison to steel, is an excellent choice of material for weight reduction in automobile and aerospace sectors (where it is available in both sheet and cast forms) [2–5]. The aluminum industry offers a wide range of aluminum alloys with various combinations of strength, ductility, wear, and corrosion resistance [6–8]. Many elemental combinations are used to alloy with aluminum as solute [9,10]. Small quantities of elements such as silicon, manganese, iron, chromium, molybdenum, etc. are added in aluminum to enhance its mechanical and physical properties and to improve some of the specific properties required in strategic applications [11–13]. In almost all production industries and in day-to-day life, Al-Si cast alloys play a vital role due to their ease in castability, corrosion resistance, and high mechanical properties [14–16]. Al-Si alloys are widely used in various industries due to their excellent mechanical properties, improved wear and thermal behavior, supreme corrosion resistance, and excellent castability [8,17–19]. The addition of Cu and Fe to this alloy further enhances the mechanical and wear behavior of the material without heat treatment [20–22]. The addition of Copper imparts strength and hardness to the casting [20,22]. The properties achieved by the addition of Fe are comparable to those alloys with various heat-treated and aged Aluminum alloys. Optimized Fe alloying aids for the possibility to reduce the aging time without the addition of Mg, resulting in significant cost saving which is the need of the hour in any industry [23,24].

Iron tends to form intermetallic with other alloying elements resulting in strengthening of the alloy with enhanced wear and thermal behavior [24]. The intermetallics formed are usually hard and brittle with a superior high-temperature behavior. Thus, novelty in conventional Al-Si casting alloys, by the addition of Cu and Fe during the casting process, can result in enhanced performance of the cast product [20–23]. Further, the addition of Fe improves fluidity, a vital requirement to produce a sound casting. Taylor et al. suggested having a critical percentage of Fe based on the silicon percentage in the alloy. If Fe exceeds the critical percentage, it would influence loss of ductility due to shrinkage porosity [23,25]. Enhanced fluidity provides an opportunity for the production of thin-walled castings [26]. Intricate shapes with a near-net finish are possible due to the improved fluidity. 4XXX series wrought alloy has a UTS of ~134 MPa and YS of ~64 MPa and novel Al-Si alloy is expected to have the UTS of ~385 MPa and YS of ~240 MPa [18,19,27–29]. There is a significant enhancement of mechanical properties by the addition of an optimum percentage of Fe, without any thermal treatment leading to huge cost savings [24]. Hence, such novel Al-based alloys can be utilized in a variety of applications, due to their superior mechanical and thermal properties. In addition, it can find a prominent place in automotive industry, where light-weighting at a lower cost is a beneficial advantage. Some of the remarkable advantages of die-casting over conventional sand casting are an increase in productivity, dimensional accuracy of as-cast components, and better mechanical properties as a result of improved microstructural features [30].

Squeeze casting is both economical and has the potential to create cast components with minimal defects, often achieving near-net-shaped components [31–33]. In addition, die-casting reduces the metal wastages which arise due to the use of feeders and risers as in conventional sand casting [34]. Squeeze casting is a combination of the casting and forging processes where the solidification of molten metal takes place under pressure, thereby reducing the casting defects created due to gas entrapments as well as increasing the ductility of the resultant component. The squeeze casting process parameters play an important role in determining the microstructure of the cast components. The process parameters such as squeeze pressure, squeeze pressure duration, pouring temperature, and die temperatures have overall control on the microstructure [35]. The squeezing pressure increases the heat transfer rate in between the mold interfaces that enhance the surface finish and also help to create a uniform microstructure from surface to core [36]. For any new alloys or modified alloys, process parameters have to be optimized for better microstructure and mechanical properties [37]. In the squeeze casting process, the desirable mechanical and microstructure features are based on the combination of mold casting and die forging due to the fact that the molten metal is solidified under hydrostatic pressure. This would have better control over mechanical properties as we all as lead to a homogenous microstructure [38]. In addition, the squeeze-casting process creates a very fine microstructure and also eliminates the gas and shrinkage porosities [39]. Reports on the Fe-addition to Al-Si alloy show that for the Al-Si alloy with Fe, the mechanical properties depend not only on chemical composition but also on the microstructural morphology of the phases (such as the Al-rich alpha phase and eutectic Si phase) [40].

Since the addition of alloys elements to Al lead to the formation of coarse microstructure in the as-cast condition and to obtain finer microstructure suitable strategies need to be followed [41–45]. Fine microstructure can be the addition of grain refiners [44–46], severe plastic deformation [47,48], cryomilling [49], high pressure solidification [50,51], laser processing [52,53], etc. Accordingly, the present investigation aims to study utilize the low-cost fabrication technique (squeeze pressure casting) and explore the influence of applied squeeze pressure on the microstructure during the solidification of a molten Al-Si-Cu-Fe alloy. The influence of squeeze pressure on the mechanical and wear behavior of the alloy is investigated in detail.

2. Experimental Procedure

Samples considered in the present study are processed through gravity die casting (GDC), low-pressure squeeze casting (LPSC) at 5 MPa, and high-pressure squeeze cast (HPSC) at 12 MPa respectively using a cylindrical mold made up of H13 steel. The chemical composition of the alloy in various cast conditions are determined using an optical emission spark (OES) spectrometer and are listed in Table 1.

Table 1. Chemical composition of novel alloy designed.

Element/Weight %	Al	Si	Cu	Mg	Fe	Zn	Mn	Ni	Cr
GDC	91.02	5.41	2.97	0.373	0.135	0.019	0.01	0.005	0.001
LPSC	91.09	5.41	2.91	0.372	0.135	0.016	0.01	0.006	0.001
HPSC	90.99	5.41	2.98	0.391	0.142	0.017	0.01	0.005	0.001

Microstructural analysis of the samples (under various casting conditions) was observed using an optical microscope (LEICA DMLM, Mumbai, India; 50× to 1000× range). The hardness measurements were carried out using a Zwick Roell Vickers microhardness tester (from Zwick, Gurugrum, India) at a test load of 0.1 kgf with a dwell time of 10 s. Tensile testing was performed using a Tinius Olsen H25KL tabletop tensile testing unit (from Tinius Olsen, Noida, India) with a strain rate of 5×10^{-4}/s as per ASTM-E08-2016 standard using a sub-sized specimen [54]. Wear testing was carried out using a pin on disc wear testing machine (Ducom, Bangalore, India) based on the ASTM G99-05 standard [55]. Wear tests were carried out with a sliding velocity set to 0.314 m/s and measured for the sliding distance of 1000 m [56–58]. The sliding disc diameter is 30 mm, the speed of the machine is held at 200 rpm and the test time is considered to be around 3185 s. The machine disc is made up of EN31 material with a roughness of 10 μm and hardness ~60 HRC. The wear testing was carried out with different test variables to understand the behavior of Al-Si alloys as a function of changing parameters. Three different loads were applied (20 N, 40 N, and 60 N) at a higher operating temperature of 200 °C, refer to Table 2. The schematic of the wear testing unit is illustrated in Figure 1. The wear testing machine consists of a specimen in the form of a pin and it is tested against a disc made of EN31 material according to the ASTM G99-05 standards. In addition, the load is applied through the loading panel, and the entire equipment is operated using a computer-based controller. All of the parameters including depth, force, temperature, speed of the disc, time, etc. can be controlled using the controller in an acute fashion. The surface morphological features of all of the tensile fractured samples and worn-out surfaces from wear tests were studied using an FEI Quanta 200 Scanning electron microscope (SEM) (FEI, Bangalore, India).

Table 2. Al-Si alloys wear test input parameters.

Experiment Number	Casting Route	Applied Load (N)	Temperature (°C)
1-A1	GDC	20	200
1-A2	GDC	60	200
1-A3	GDC	40	200
1-B1	LPSC	20	200
1-B2	LPSC	60	200
1-B3	LPSC	40	200
1-C1	HPSC	20	200
1-C2	HPSC	60	200
1-C3	HPSC	40	200

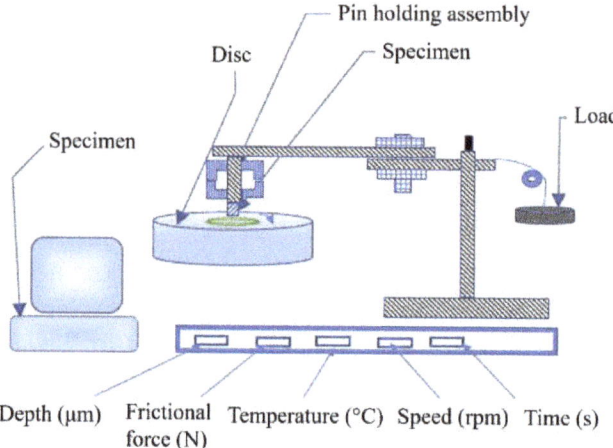

Figure 1. Schematic representation of the pin-on-disc wear testing unit.

3. Results and Discussion

3.1. Microstructure

The microstructure of GDC, LPSC, and HPSC samples are shown in Figure 2 using optical (Figure 2a–c) and scanning electron microscopy (Figure 2d–f). The microstructure of the GDC specimen has coarse dendrites as shown in Figure 2a. The microstructure gets refined with the application of pressure. The LPSC and HPSC samples have shown the presence of small dendrites, which are also deformed. They are not in a continuous state like the GDC samples. An increase in the squeeze-pressure increases the cooling rate, resulting in higher nucleation and finer dendritic size with large dendrite spacing. Similar observations were made by Amar et al. [59], where the 2017A alloy was squeeze cast using GDC and at high pressures. Moreover, Amar et al. have shown that with the application of pressure, a refined and homogeneous microstructure was observed, which is in agreement with the present results. The heat inside the mold and pressure have a significant effect on the size of the dendrites, dendritic morphology, and the distribution of microstructural constituents. Increasing the squeeze casting pressure refined all microstructural features (including the size of the microstructural features and arm spacing of dendrites) and modified the morphology of Al-Si eutectic phases. Further, dendrites were small and almost spherical in shape in squeeze cast conditions. In GDC alloys, the dendrites were observed to have an elongated plate-like morphology (Figure 2d), whereas, in the other two alloys (Figure 2e,f), cast microstructures consist of needle-like morphologies. In all of the samples, the Al-Si-Fe regions are constrained within the inter-dendritic regions due to kinetic differences between the phases. These phases were formed as curved crystals and in some regions, it exhibits plate-like morphology joined along with irregular, curved surfaces.

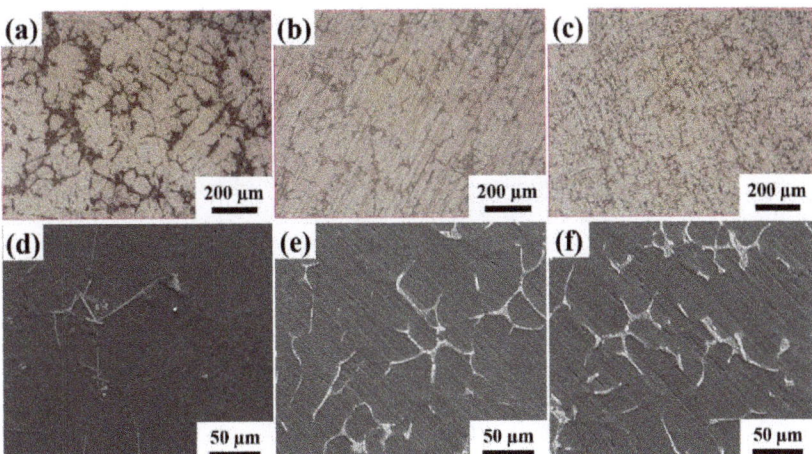

Figure 2. (**a**–**c**) Optical micrographs and (**d**–**f**) scanning electron microscopy images of the cast samples fabricated by (**a**,**d**) GDC route, (**b**,**e**) LPSC route, and (**c**,**f**) HPSC route, respectively.

3.2. Hardness

The hardness analysis was carried out to study the variation of hardness along the cast cross-section from the surface to the middle of the cast sample in all three-process conditions, *viz.*, GDC, LPSC, and HPSC, respectively. The results shown in Figure 3 indicate that the squeeze-cast sample with higher pressure exhibits a higher hardness. Lin et al. studies on the Al-based alloys showed a hardness of 75 HV and 85 HV for GDC and high-pressure squeeze cast materials [38]. Similarly, Thirumal et al. [39] studies on AA6061 alloy castings as a function of different squeeze-cast pressures show an increase in the hardness of the alloy with an increase in the pressure. The results from Lin et al. and Thirumal et al. are in agreement and are similar to the results from the present study. In addition, there is significant variation in hardness values as observed from the surface to core, indicating the absence of porosity and other casting defects.

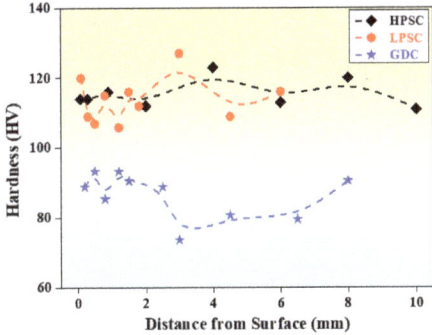

Figure 3. Microhardness survey taken for the cast samples in all three conditions (gravity die casting (GDC), low-pressure squeeze-casting (LPSC), and high-pressure squeeze-casting (HPSC) taken from surface to center of the casting).

The hardness observations from the surface to the core also indicate that the cast structure is homogenous and uniform. On the other hand, the LPSC sample shows similar hardness values to the high-pressure squeeze-cast sample along the surface. However, the hardness values show some fluctuations when measured from the surface to the core, unlike the high-pressure squeeze-cast samples. This corroborates the presence of defects (such as porosity) in these LPSC samples. Similarly, the GDC sample shows inferior hardness when compared to the squeeze-case samples due to reduced cooling rates. In addition, the hardness fluctuates between 95 HV to 75 HV as we move from the surface to the core showing the presence of defects/imperfections in these samples. Based on the hardness survey and microstructural correlation, it is evident that squeeze pressure is one of the most significant process parameters for achieving higher material properties with uniform distribution in the squeeze-casting process. This is in good agreement with the discussion carried out by Azhagan et al. [39] and Mohamed et al. [59]. In addition, the hardness of the alloy increases with the application of pressure. This enhanced behavior in HPSC and LPSC alloys in comparison with GDC alloy was due to improvements in heat transfer rates during solidification due to the applied pressure, resulting in refinement of microstructure and the improved contact area between the die and molten metal surface [50,51,60,61].

3.3. Tensile Properties

Tensile properties of the investigated GDC and other two-squeeze cast samples are shown in Figure 4. The HPSC sample has shown a tensile strength of ~540 MPa against LPSC at ~382 MPa and GDC at ~367 MPa. On comparing GDC and squeeze cast alloys, the mechanical properties are superior for the LPSC and HPSC alloys. The results explain that the samples fabricated by the squeeze-casting process exhibit higher yield and tensile strength as compared to samples fabricated by the GDC process. In the squeeze-cast samples, the tensile and yield strength of the alloy increases with increasing pressure. A decrease in the grain size with an increase in squeeze casting pressure results in an increased grain boundaries volume. The increased grain boundary volume increases the resistance to dislocation movement, resulting in enhanced strength properties [62–64]. As pressure was held on molten metal during the squeeze casting process until the end of the solidification process, the rate of heat transfer was increased and macro and microporosity had been eliminated in comparison to the GDC process, resulting in enhanced mechanical properties. The elongation observed for the cast samples (GDC, to be almost similar with LPSC, and HPSC) is similar within the experimental conditions.

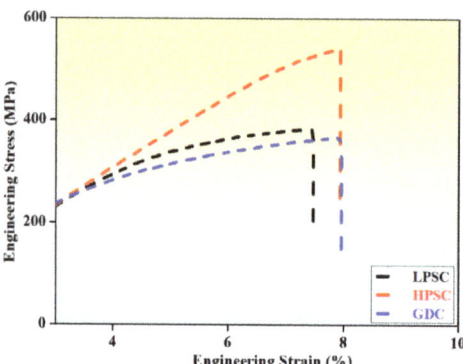

Figure 4. Engineering stress-strain curves of the cast samples fabricated under different conditions (gravity die-casting, low-pressure squeeze casting, and high-pressure squeeze casting).

3.4. Wear Behavior

The wear behavior of all three samples fabricated by the three casting routes (namely GDC, LPSC, and HPSC) were studied to understand their tribological behavior. The wear test results in terms of coefficient of friction (COF) and wear loss are shown in Figure 5. The COF increases with an increase in the working temperature. However, with the application of pressure at the same condition, the COF decreases in general (Figure 5a). The results suggest that HPSC samples show better wear resistance compared to LPSC and GDC samples at every given load and temperature combination (Figure 5b). Such improved tribological performance of the HPSC samples is attributed to the reduction in porosity and shrinkage defects. A higher wear rate is observed for the non-pressurized cast samples due to its high coefficient of friction, which is the result of poor surface quality along with the presence of porosities and shrinkage defects, whereas the coefficient of friction is less in pressurized cast samples, thereby increasing its tribological response. Samples fabricated by squeeze-casting process demonstrated lesser wear rate in comparison to GDC process. Squeeze pressure maintains the molten metal closer to the wall surfaces of the die, which in turn gives a higher cooling rate at the surface. Higher cooling rate results in a more refined dendritic structure, resulting in a smoother surface. Finer microstructures offer improved hardness, which in turn offer higher wear resistance [65–67]. It may be observed that in general, the wear resistance of the HPSC decreases with increasing load and/temperature combination due to accelerated conditions (which is as expected). Ashiri et al. [36] have shown similar wear properties on the Al–Si–Mg–Ni–Cu alloy fabricated by GDC and pressure squeeze cast samples. They have demonstrated that both wear rate and COF decrease with an increase in the pressure at a given load. The wear loss increases with an increase in the applied load. In addition, the COF of pressure squeeze-cast materials is lower than the GDC counterpart, and the results are in agreement with the present study.

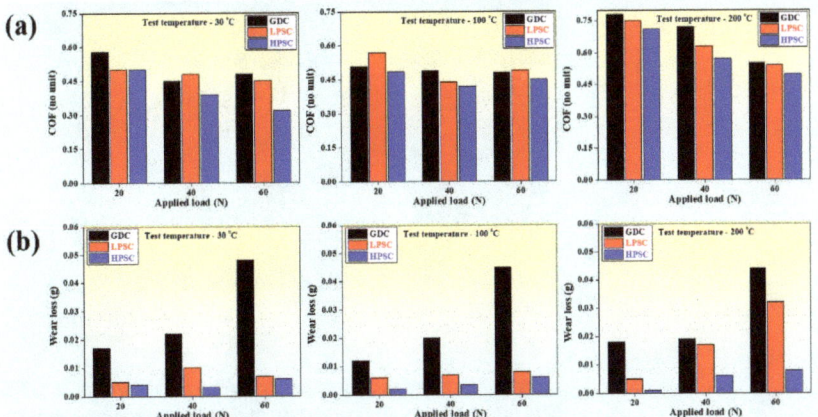

Figure 5. Tribological performance including (**a**) coefficient of friction and (**b**) wear loss comparison between GDC, LPSC, and HPSC samples as a function of load and operating temperature.

3.5. SEM Surface Analysis of the Worn out Samples

Fracture analysis conducted on worn-out samples by using SEM micrographs is shown in Figure 6. The wear surface of Al-based cast samples through the GDC route (Figure 6(a1,a2)) shows the presence of excessive material loss due to digging and penetration (deeper ploughing grooves [68,69]) of the pin at higher loads applied at elevated temperature. On the other hand, the LPSC samples wear surface shows minor digging and smearing observed due to frequent rubbing of the pin (Figure 6(b1,b2)). In addition, delamination and micro cracking (Figure 6(a1,a2)) may be observed in the samples produced through the GDC route due to its lower hardness compared to LPSC and HPSC samples. However, deep ploughing grooves and considerable delamination were not observed in the samples fabricated through LPSC and HPSC. The HPSC samples wear surface shows minimal rubbing/wear pattern at higher loads applied at elevated temperature. This difference in the wear rate of HPSC samples (Figure 6(c1,c2)) is due to the molten metal being solidified under high pressure, which reduces gas entrapments and shrinkages or gas porosity thereby improving its tribological properties, in addition to the microstructural refinement [70–72]. The present results are very similar to the studies conducted by Ashiri et al. [36], where deeper ploughing grooves are observed for the GDC samples compared with the pressure squeeze-cast samples leading to severe damage in the

GDC samples. Hence, the present results demonstrate the role of pressure during the casting/solidification process and its influence in refining the microstructure and improving their mechanical and tribological performance.

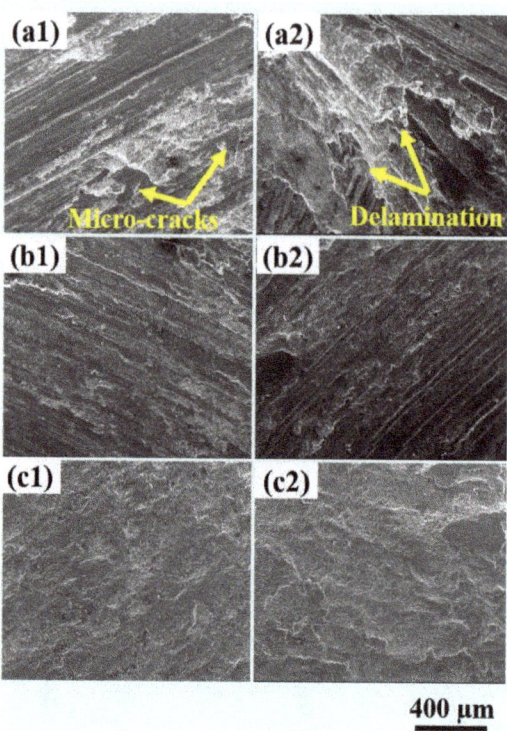

Figure 6. Worn out surface of the cast alloys in GDC, LPSC, an HPSC at two different load conditions 20 N (**a1,b1,c1**) and 60 N (**a2,b2,c2**) at a higher operating temperature of 200 °C.

4. Conclusions

In this study, a varying cast pressure was applied to squeeze cast novel Al-Si alloys, and their effect on microstructure, mechanical properties, and wear behavior at higher operating temperatures were investigated and compared with GDC counterparts. The investigation reveals better mechanical behavior for squeeze cast Al-Si alloys compared to their GDC counterpart. In all cases, the HPSC alloy shows better mechanical behavior (hardness (115 HV) and tensile strength (540 MPa)) as compared to its GDC counterpart (hardness (86 HV), and tensile strength (367 MPa)). The microstructural study reveals reduced grain size, increased grain boundaries volume, reduction in dendrite arm spacing, small and rounded Si eutectic phases, and dendrites in the HPSC alloy. Wear behavior was studied at different loads 20 N and 60 N for the samples fabricated by GDC, LPSC, and HPSC routes. The results reveal the remarkable differences in wear rate, even at higher operating temperature and load for HPSC samples (wear loss 0.09 g), compared to the GDC (wear loss 0.045 g) and LPSC samples (wear loss 0.033 g) under nominal pressure due to minimal casting defects and lesser coefficient of friction. Under the same test condition, the HPSC samples show the least COF of 0.481 as against the GDC (COF—0.534) and LPSC (COF—0.516) samples. This study helps in determining the high-temperature wear-resistant behavior of a novel Al-Si HPSC alloy, making it suitable for critical industrial applications.

Author Contributions: V.S.C. has conducted the literature survey, development, testing and validation of the alloy. K.S.V.B.R.K. helped supporting in literature survey and manuscript draft preparation. M.R. supported in developing the alloy. K.S. contributed in conceptualization, methodology, critical analysis of data, reviewing and editing the manuscript. S.D. supported in formal analysis, data curation and resource allocation. K.G.P. supported in resources, review and editing the manuscript. All authors have read and agreed to the published version of the manuscript.

Funding: This research received no external funding.

Institutional Review Board Statement: Not applicable.

Informed Consent Statement: Not applicable.

Data Availability Statement: The results are a part of an ongoing study and the data will be made available on reasonable requests.

Acknowledgments: The authors would like to acknowledge the support rendered by Ramya Gopalakrishnan and Divya Sekar of M/s Ashok Leyland Ltd.—MED during testing of the samples.

Conflicts of Interest: The authors declare no conflict of interest.

References

1. Brooker, A.D.; Ward, J.; Wang, L. Lightweighting impacts on fuel economy, cost, and component losses. *SAE Tech. Pap.* **2013**, 2. [CrossRef]
2. Cann, J.L.; De Luca, A.; Dunand, D.C.; Dye, D.; Miracle, D.B.; Oh, H.S.; Olivetti, E.A.; Pollock, T.M.; Poole, W.J.; Yang, R.; et al. Sustainability through alloy design: Challenges and opportunities. *Prog. Mater. Sci.* **2021**, *117*, 100722. [CrossRef]
3. Imran, M.; Khan, A.R.A. Characterization of Al-7075 metal matrix composites: A review. *J. Mater. Res. Technol.* **2019**, *8*, 3347–3356. [CrossRef]
4. Vashisht, S.; Rakshit, D. Recent advances and sustainable solutions in automobile air conditioning systems. *J. Clean. Prod.* **2021**, *329*, 129754. [CrossRef]
5. Milman, Y.V. High-Strength Aluminum Alloys. *Met. Mater. High Struct. Effic.* **2004**, 139–150. [CrossRef]
6. Inoue, A. Amorphous, nanoquasicrystalline and nanocrystalline alloys in Al-based systems. *Prog. Mater. Sci.* **1998**, *43*, 365–520. [CrossRef]
7. Lin, T.C.; Cao, C.; Sokoluk, M.; Jiang, L.; Wang, X.; Schoenung, J.M.; Lavernia, E.J.; Li, X. Aluminum with dispersed nanoparticles by laser additive manufacturing. *Nat. Commun.* **2019**, *10*, 1–9. [CrossRef] [PubMed]
8. Wang, Z.; Ummethala, R.; Singh, N.; Tang, S.; Suryanarayana, C.; Eckert, J.; Prashanth, K.G. Selective laser melting of aluminum and its alloys. *Materials* **2020**, *13*, 4564. [CrossRef] [PubMed]
9. Sankaran, K.K.; Mishra, R.S. Aluminum Alloys. *Metall. Des. Alloy. Hierarchical Microstruct.* **2017**, 57–176. [CrossRef]
10. Kim, S.Y.; Lee, G.Y.; Park, G.H.; Kim, H.A.; Lee, A.Y.; Scudino, S.; Prashanth, K.G.; Kim, D.H.; Eckert, J.; Lee, M.H. High strength nanostructured Al-based alloys through optimized processing of rapidly quenched amorphous precursors. *Sci. Rep.* **2018**, *8*, 1–12. [CrossRef]
11. Gao, T.; Hu, K.; Wang, L.; Zhang, B.; Liu, X. Morphological evolution and strengthening behavior of α-Al(Fe,Mn)Si in Al–6Si–2Fe–xMn alloys. *Results Phys.* **2017**, *7*, 1051–1054. [CrossRef]
12. Fabrizi, A.; Ferraro, S.; Timelli, G. The Influence of Fe, Mn and Cr Additions on the Formation of Iron-Rich Intermetallic Phases in an Al-Si Die-Casting Alloy. In *Shape Casting: 5th International Symposium 2014*; Wiley: Hoboken, NJ, USA, 2014; pp. 277–284. [CrossRef]
13. Wang, K.; Tang, P.; Huang, Y.; Zhao, Y.; Li, W.; Tian, J. Characterization of microstructures and tensile properties of recycled Al-Si-Cu-Fe-Mn alloys with individual and combined addition of titanium and cerium. *Scanning* **2018**, *2018*, 1–14. [CrossRef]
14. Prashanth, K.G.; Scudino, S.; Klauss, H.J.; Surreddi, K.B.; Löber, L.; Wang, Z.; Chaubey, A.K.; Kühn, U.; Eckert, J. Microstructure and mechanical properties of Al-12Si produced by selective laser melting: Effect of heat treatment. *Mater. Sci. Eng. A* **2014**, *590*, 153–160. [CrossRef]
15. Ma, P.; Jia, Y.; Prashanth, K.G.; Yu, Z.; Li, C.; Zhao, J.; Yang, S.; Huang, L. Effect of Si content on the microstructure and properties of Al-Si alloys fabricated using hot extrusion. *J. Mater. Res.* **2017**, *32*, 2210–2217. [CrossRef]
16. Ma, P.; Prashanth, K.; Scudino, S.; Jia, Y.; Wang, H.; Zou, C.; Wei, Z.; Eckert, J. Influence of Annealing on Mechanical Properties of Al-20Si Processed by Selective Laser Melting. *Metals* **2014**, *4*, 28–36. [CrossRef]
17. Prashanth, K.G.; Damodaram, R.; Scudino, S.; Wang, Z.; Prasad Rao, K.; Eckert, J. Friction welding of Al-12Si parts produced by selective laser melting. *Mater. Des.* **2014**, *57*, 632–637. [CrossRef]
18. Gokuldoss Prashanth, K.; Scudino, S.; Eckert, J. Tensile Properties of Al-12Si Fabricated via Selective Laser Melting (SLM) at Different Temperatures. *Technologies* **2016**, *4*, 38. [CrossRef]
19. Prashanth, K.G.; Scudino, S.; Chaubey, A.K.; Löber, L.; Wang, P.; Attar, H.; Schimansky, F.P.; Pyczak, F.; Eckert, J. Processing of Al-12Si-TNM composites by selective laser melting and evaluation of compressive and wear properties. *J. Mater. Res.* **2016**, *31*, 55–65. [CrossRef]

20. Zeren, M.; Karakulak, E.; GümÜ, S. Influence of Cu addition on microstructure and hardness of near-eutectic Al-Si-xCu-alloys. *Trans. Nonferrous Met. Soc. China* **2011**, *21*, 1698–1702. [CrossRef]
21. Ahn, S.S.; Pathan, S.; Koo, J.M.; Baeg, C.H.; Jeong, C.U.; Son, H.T.; Kim, Y.H.; Lee, K.H.; Hong, S.J. Enhancement of the Mechanical Properties in Al–Si–Cu–Fe–Mg Alloys with Various Processing Parameters. *Materials* **2018**, *11*, 2150. [CrossRef]
22. Basak, C.B.; Babu, N.H. Influence of Cu on modifying the beta phase and enhancing the mechanical properties of recycled Al-Si-Fe cast alloys. *Sci. Rep.* **2017**, *7*, 1–10. [CrossRef]
23. Taylor, J.A. Iron-Containing Intermetallic Phases in Al-Si Based Casting Alloys. *Procedia Mater. Sci.* **2012**, *1*, 19–33. [CrossRef]
24. Mathew, J.; Remy, G.; Williams, M.A.; Tang, F.; Srirangam, P. Effect of Fe Intermetallics on Microstructure and Properties of Al-7Si Alloys. *JOM* **2019**, *71*, 4362–4369. [CrossRef]
25. Taylor, A.J. The effect of iron in Al-Si casting alloys. In Proceedings of the 35th Australian Foundry Institute National Conference, Adelaide, Australia, 31 October–3 November 2004; pp. 148–157.
26. Górny, M. Fluidity and Temperature Profile of Ductile Iron in Thin Sections. *J. Iron Steel Res. Int.* **2012**, *19*, 52–59. [CrossRef]
27. Prashanth, K.; Scudino, S.; Chatterjee, R.; Salman, O.; Eckert, J. Additive Manufacturing: Reproducibility of Metallic Parts. *Technologies* **2017**, *5*, 8. [CrossRef]
28. Prashanth, K.G.; Scudino, S.; Eckert, J. Defining the tensile properties of Al-12Si parts produced by selective laser melting. *Acta Mater.* **2017**, *126*, 25–35. [CrossRef]
29. Rathod, H.J.; Nagaraju, T.; Prashanth, K.G.; Ramamurty, U. Tribological properties of selective laser melted Al–12Si alloy. *Tribol. Int.* **2019**, *137*, 94–101. [CrossRef]
30. Santosh, M.V.; Suresh, K.R.; Kiran Aithal, S. Mechanical Characterization and Microstructure analysis of Al C355.0 by Sand Casting, Die Casting and Centrifugal Casting Techniques. *Mater. Today Proc.* **2017**, *4*, 10987–10993. [CrossRef]
31. Srivastava, N.; Anas, M. An investigative review of squeeze casting: Processing effects & impact on properties. *Mater. Today Proc.* **2020**, *26*, 1914–1920. [CrossRef]
32. Kwok, T.W.J.; Zhai, W.; Peh, W.Y.; Gupta, M.; Fu, M.W.; Chua, B.W. Squeeze Casting for the Production of Metallic Parts and Structures. *Encycl. Mater. Met. Alloy.* **2022**, 87–99. [CrossRef]
33. Venkatesan, S.; Xavior, M.A. Analysis of Mechanical Properties of Aluminum Alloy Metal Matrix Composite by Squeeze Casting—A Review. *Mater. Today Proc.* **2018**, *5*, 11175–11184. [CrossRef]
34. Weiler, J.P. A review of magnesium die-castings for closure applications. *J. Magnes. Alloy.* **2019**, *7*, 297–304. [CrossRef]
35. Lus, H.M.; Ozer, G.; Guler, K.A. In Situ Composite of (Mg2Si)/Al Fabricated by Squeeze Casting. *TMS Annu. Meet.* **2012**, *1*, 775–781. [CrossRef]
36. Ashiri, R.; Niroumand, B.; Karimzadeh, F. Physical, mechanical and dry sliding wear properties of an Al–Si–Mg–Ni–Cu alloy under different processing conditions. *J. Alloys Compd.* **2014**, *582*, 213–222. [CrossRef]
37. Raji, A.; Khan, R.H. Effects of pouring temperature and squeeze pressure on the properties of AI-8%Si alloy squeeze cast components. In *Proceedings of the Institute of Cast Metals Engineers—67th World Foundry Congress, wfc06: Casting the Future*; Curran Associates Inc.: Red Hook, NY, USA, 2006; Volume 2, pp. 834–843.
38. Lin, B.; Zhang, W.W.; Lou, Z.H.; Zhang, D.T.; Li, Y.Y. Comparative study on microstructures and mechanical properties of the heat-treated Al-5.0Cu-0.6Mn-xFe alloys prepared by gravity die casting and squeeze casting. *Mater. Des.* **2014**, *59*, 10–18. [CrossRef]
39. Thirumal Azhagan, M.; Mohan, B.; Rajadurai, A. Optimization of process parameters to enhance the hardness on squeeze cast aluminium alloy AA6061. *Int. J. Eng. Technol.* **2014**, *6*, 183–189.
40. Brayshaw, W.J.; Roy, M.J.; Sun, T.; Akrivos, V.; Sherry, A.H. Iterative mesh-based hardness mapping. *Sci. Technol. Weld. Join.* **2016**, *22*, 404–411. [CrossRef]
41. Pongen, R.; Birru, A.K.; Parthiban, P. Study of microstructure and mechanical properties of A713 aluminium alloy having an addition of grain refiners Al-3.5 Ti-1.5C and Al-3Cobalt. *Results Phys.* **2019**, *13*, 102105. [CrossRef]
42. Zong, Y.Y.; Chen, L.; Zhao, Z.G.; Shan, D.B. Flow Lines, Microstructure, and Mechanical Properties of Flow Control Formed 4032 Aluminum Alloy. *Mater. Manuf. Processes* **2014**, *29*, 466–471. [CrossRef]
43. Chen, Z.; Yan, K. Grain refinement of commercially pure aluminum with addition of Ti and Zr elements based on crystallography orientation. *Sci. Rep.* **2020**, *10*, 1–8. [CrossRef]
44. Nadendla, H.B.; Nowak, M.; Bolzoni, L. Grain Refiner for Al-Si Alloys. *Miner. Met. Mater. Ser.* **2016**, 1009–1012. [CrossRef]
45. Guan, R.G.; Tie, D. A Review on Grain Refinement of Aluminum Alloys: Progresses, Challenges and Prospects. *Acta Metall. Sin.* **2017**, *30*, 409–432. [CrossRef]
46. Xi, L.; Gu, D.; Guo, S.; Wang, R.; Ding, K.; Prashanth, K.G. *Grain Refinement in Laser Manufactured Al-Based Composites with TiB2 Ceramic*; Elsevier: Amsterdam, The Netherlands, 2020; Volume 9, pp. 2611–2622.
47. Singh, A.; Basha, D.A.; Matsushita, Y.; Tsuchiya, K.; Lu, Z.; Nieh, T.G.; Mukai, T. Domain structure and lattice effects in a severely plastically deformed CoCrFeMnNi high entropy alloy. *J. Alloys Compd.* **2020**, *812*, 152028. [CrossRef]
48. Arzaghi, M.; Fundenberger, J.J.; Toth, L.S.; Arruffat, R.; Faure, L.; Beausir, B.; Sauvage, X. Microstructure, texture and mechanical properties of aluminum processed by high-pressure tube twisting. *Acta Mater.* **2012**, *60*, 4393–4408. [CrossRef]
49. Lavernia, E.J.; Han, B.Q.; Schoenung, J.M. Cryomilled nanostructured materials: Processing and properties. *Mater. Sci. Eng. A* **2008**, *493*, 207–214. [CrossRef]

50. Liu, X.; Ma, P.; Ji, Y.D.; Wei, Z.J.; Suo, C.J.; Ji, P.C.; Shi, X.R.; Yu, Z.S.; Prashanth, K.G. Solidification of Al-xCu alloy under high pressures. *J. Mater. Res. Technol.* **2020**, *9*, 2983–2991. [CrossRef]
51. Ma, P.; Wei, Z.J.; Jia, Y.D.; Yu, Z.S.; Prashanth, K.G.; Yang, S.L.; Li, C.G.; Huang, L.X.; Eckert, J. Mechanism of formation of fibrous eutectic Si and thermal conductivity of SiCp/Al-20Si composites solidified under high pressure. *J. Alloys Compd.* **2017**, *709*, 329–336. [CrossRef]
52. Bayoumy, D.; Schliephake, D.; Dietrich, S.; Wu, X.H.; Zhu, Y.M.; Huang, A.J. Intensive processing optimization for achieving strong and ductile Al-Mn-Mg-Sc-Zr alloy produced by selective laser melting. *Mater. Des.* **2021**, *198*, 109317. [CrossRef]
53. Yap, C.Y.; Chua, C.K.; Dong, Z.L.; Liu, Z.H.; Zhang, D.Q.; Loh, L.E.; Sing, S.L. Review of selective laser melting: Materials and applications. *Appl. Phys. Rev.* **2015**, *2*, 041101. [CrossRef]
54. ASTM E8/E8M-16. *Standard Test Methods for Tension Testing of Metallic Materials*; ASTM International: West Conshohocken, PA, USA, 2016.
55. ASTM G99-05. *Standard Test Method for Wear Testing with a Pin-on-Disk Apparatus*; ASTM International: West Conshohocken, PA, USA, 2014.
56. Khan, S.; Ahmad, Z. Comparative analysis for coefficient of friction of LM 25 alloy and LM 25 granite composite at different sliding speeds and applied pressure. *Int. J. Mech. Prod. Eng. Res. Dev.* **2018**, *8*, 291–300. [CrossRef]
57. Feyzullahoğlu, E.; Şakiroğlu, N. The tribological behaviours of aluminium-based materials under dry sliding. *Ind. Lubr. Tribol.* **2011**, *63*, 350–358. [CrossRef]
58. Shanmughasundaram, P. Investigation on the Wear Behaviour of Eutectic Al-Si Alloy-Al$_2$O$_3$—Graphite Composites Fabricated Through Squeeze Casting. *Mater. Res.* **2014**, *17*, 940–946. [CrossRef]
59. Amar, M.B.; Souissi, S.; Souissi, N.; Bradai, C. Pressure and die temperature effects on microstructure and mechanical properties of squeeze casting 2017A wrought Al alloy. *Int. J. Microstruct. Mater. Prop.* **2012**, *7*, 491–501. [CrossRef]
60. Wei, Z.; Ma, P.; Wang, H.; Zou, C.; Scudino, S.; Song, K.; Prashanth, K.G.; Jiang, W.; Eckert, J. The thermal expansion behaviour of SiCp/Al-20Si composites solidified under high pressures. *Mater. Des.* **2015**, *65*, 387–394. [CrossRef]
61. Ma, P.; Zou, C.M.; Wang, H.W.; Scudino, S.; Fu, B.G.; Wei, Z.J.; Kühn, U.; Eckert, J. Effects of high pressure and SiC content on microstructure and precipitation kinetics of Al-20Si alloy. *J. Alloys Compd.* **2014**, *586*, 639–644. [CrossRef]
62. Maity, T.; Prashanth, K.G.; Balci, Ö.; Kim, J.T.; Schöberl, T.; Wang, Z.; Eckert, J. Influence of severe straining and strain rate on the evolution of dislocation structures during micro-/nanoindentation in high entropy lamellar eutectics. *Int. J. Plast.* **2018**, *109*, 121–136. [CrossRef]
63. Maity, T.; Prashanth, K.G.; Balçi, Ö.; Wang, Z.; Jia, Y.D.; Eckert, J. Plastic deformation mechanisms in severely strained eutectic high entropy composites explained via strain rate sensitivity and activation volume. *Compos. Part B Eng.* **2018**, *150*, 7–13. [CrossRef]
64. Maity, T.; Balcı, Ö.; Gammer, C.; Ivanov, E.; Eckert, J.; Prashanth, K.G. High pressure torsion induced lowering of Young's modulus in high strength TNZT alloy for bio-implant applications. *J. Mech. Behav. Biomed. Mater.* **2020**, *108*, 103839. [CrossRef] [PubMed]
65. Chaubey, A.; Konda Gokuldoss, P.; Wang, Z.; Scudino, S.; Mukhopadhyay, N.; Eckert, J. Effect of Particle Size on Microstructure and Mechanical Properties of Al-Based Composite Reinforced with 10 Vol.% Mechanically Alloyed Mg-7.4%Al Particles. *Technologies* **2016**, *4*, 37. [CrossRef]
66. Mu, Y.; Zhang, L.; Xu, L.; Prashanth, K.; Zhang, N.; Ma, X.; Jia, Y.; Xu, Y.; Jia, Y.; Wang, G. Frictional wear and corrosion behavior of AlCoCrFeNi high-entropy alloy coatings synthesized by atmospheric plasma spraying. *Entropy* **2020**, *22*, 740. [CrossRef]
67. Wang, Z.; Georgarakis, K.; Zhang, W.W.; Prashanth, K.G.; Eckert, J.; Scudino, S. Reciprocating sliding wear behavior of high-strength nanocrystalline Al$_{84}$Ni$_7$Gd$_6$Co$_3$ alloys. *Wear* **2017**, *382–383*, 78–84. [CrossRef]
68. Attar, H.; Prashanth, K.G.; Chaubey, A.K.; Calin, M.; Zhang, L.C.; Scudino, S.; Eckert, J. Comparison of wear properties of commercially pure titanium prepared by selective laser melting and casting processes. *Mater. Lett.* **2015**, *142*, 38–41. [CrossRef]
69. Ehtemam-Haghighi, S.; Prashanth, K.G.; Attar, H.; Chaubey, A.K.; Cao, G.H.; Zhang, L.C. Evaluation of mechanical and wear properties of Ti-xNb-7Fe alloys designed for biomedical applications. *Mater. Des.* **2016**, *111*, 592–599. [CrossRef]
70. Ma, P.; Wei, Z.J.; Jia, Y.D.; Zou, C.M.; Scudino, S.; Prashanth, K.G.; Yu, Z.S.; Yang, S.L.; Li, C.G.; Eckert, J. Effect of high pressure solidification on tensile properties and strengthening mechanisms of Al-20Si. *J. Alloys Compd.* **2016**, *688*, 88–93. [CrossRef]
71. Luo, D.; Zhou, Q.; Ye, W.; Ren, Y.; Greiner, C.; He, Y.; Wang, H. Design and Characterization of Self-Lubricating Refractory High Entropy Alloy-Based Multilayered Films. *ACS Appl. Mater. Interfaces* **2021**, *13*, 55712–55725. [CrossRef]
72. Hua, D.; Xia, Q.; Wang, W.; Zhou, Q.; Li, S.; Qian, D.; Shi, J.; Wang, H. Atomistic insights into the deformation mechanism of a CoCrNi medium entropy alloy under nanoindentation. *Int. J. Plast.* **2021**, *142*, 102997. [CrossRef]

Article

Influence of Bifilm Defects Generated during Mould Filling on the Tensile Properties of Al–Si–Mg Cast Alloys

Mahmoud Ahmed El-Sayed [1], Khamis Essa [2] and Hany Hassanin [3,*]

[1] Department of Industrial and Management Engineering, Arab Academy for Science and Technology and Maritime Transport, P.O. Box 1029, Abu Qir, Alexandria 21599, Egypt; dr.mahmoudelsayed12@gmail.com
[2] School of Engineering, University of Birmingham, Birmingham B15 2TT, UK; k.e.a.essa@bham.ac.uk
[3] School of Engineering, Technology and Design, Canterbury Christ Church University, Canterbury CT1 1QU, UK
* Correspondence: hany.hassanin@canterbury.ac.uk

Abstract: Entrapped double oxide film defects are known to be the most detrimental defects during the casting of aluminium alloys. In addition, hydrogen dissolved in the aluminium melt was suggested to pass into the defects to expand them and cause hydrogen porosity. In this work, the effect of two important casting parameters (the filtration and hydrogen content) on the properties of Al–7 Si–0.3 Mg alloy castings was studied using a full factorial design of experiments approach. Casting properties such as the Weibull modulus and position parameter of the elongation and the tensile strength were considered as response parameters. The results suggested that adopting 10 PPI filters in the gating system resulted in a considerable boost of the Weibull moduli of the tensile strength and elongation due to the enhanced mould filling conditions that minimised the possibility of oxide film entrainment. In addition, the results showed that reducing the hydrogen content in the castings samples from 0.257 to 0.132 cm^3/100 g Al was associated with a noticeable decrease in the size of bifilm defects with a corresponding improvement in the mechanical properties. Such significant effect of the process parameters studied on the casting properties suggests that the more careful and quiescent mould filling practice and the lower the hydrogen level of the casting, the higher the quality and reliability of the castings produced.

Keywords: bifilm; Al–7 Si–0.3 Mg alloy; hydrogen; filtration; Weibull modulus

1. Introduction

Aluminium alloys are the most used materials in the automotive industry due to their excellent properties such as strength, durability, safety and low density, resulting in reduced emissions and an increase in fuel efficiency of the produced vehicles. In addition, aluminium is a fully recyclable material without losing recycled quality. Therefore, there is an accelerated demand for innovations and developments for aluminium alloys, particularly regarding improving the mechanical properties of the cast components [1–6].

Cast aluminium alloys are shown to have low gas absorption, with the exception of hydrogen [7–10]. Oxide film defects (bifilms) are typically created due to surface disturbance of the Al melt during pouring and/or transfer processes. This causes the oxidised surface to be folded upon itself entrapping an air layer within it and then be entrained in the bulk liquid Al [11–14]. Entrainment of bifilm defects is one of the most significant issues in aluminium castings, as they are claimed to deteriorate the tensile and fatigue properties. They are also promoting the creation of other casting defects such as pores and iron intermetallics [15–17].

Research studies showed that the bifilm defects occurrence during the pouring of the melt could be explained through the critical ingate velocity concept. As the melt enters the mould cavity with a speed more than a critical value (about 0.5 m/s for most aluminium alloys), the flow front becomes unstable and allows the creation of surface oxide foldable layers [18–22]. Literature has also confirmed that it is challenging to produce reliable casting and avoid oxide film entrainment using top-pouring methods. Therefore, bottom-pouring gating techniques are preferred because they avoid instability of melt flow behaviour during mould filling by controlling the ingate-velocity requirements for sound castings [23–26].

During solidification of the cast, the solubility of the hydrogen in Al melt considerably drops causing the former to be rejected by the growing dendrites. Concurrently, entrained bifilms, initially being compacted due to bulk turbulence, start to unfurl due to the negative pressure ascending from the shrinkage of the cast. This behaviour causes hydrogen to diffuse into the developed bifilm and inflate them into pores [27]. The results recently supported these findings by El-Sayed, Chen and Griffiths that demonstrated a harmful influence of hydrogen on the mechanical properties of an Al casting [28–31].

Design of experiments (DoE) is a systematic approach used to plan, conduct, and analyse tests to study the effect of different parameters of a given process on the response(s) of that process through performing the minimum number of experiments [32–35]. A two-level full factorial design is one of the most widely used experimental designs in which each of the process parameters is set at two levels. These levels are called 'high' and 'low', 'Good' and 'Bad' or '+1' and '−1'. A factorial design denoted 2^k design is a full factorial design of k parameters—each has two levels, and the design will involve 2^k runs [36].

Previous investigations about casting of light alloys had looked at the effect of different casting parameters on the mechanical properties. However, not many attempts in the literature have been made to identify which of these parameters had a statistically significant effect on the tensile properties of Al–Si–Mg cast alloys. The current research was carried out to cover the research gap and utilizes statistical techniques by means of Full Factorial Design of Experiments (DoE) and Analysis of Variance (ANOVA) to identify the significance of casting process parameters and study their effect on the UTS and % elongation of the castings produced. In addition, the application of DoE not only allowed to statistically assess the significance of the studied parameters, it also provided a sort of quantification of the weight of each parameter in impacting the studied process outputs through the calculation of the standardised effect of each factor. Finally, the use of DoE allowed also to assess the interaction effect between the studied parameters, which is very difficult, if not impossible, to determine otherwise.

In this paper, the effect of the hydrogen content of aluminium casting and the use of filters on the amount and size of bifilm defects, and by implication on the properties of Al–7 Si–0.3 Mg alloy castings was studied. A two-factor DoE was used for the modelling and the analysis of the casting process. The aim of such study is to provide a better understanding of the factors dominating the quality and reproducibility of light metal cast alloys.

2. Experiment

The two-parameter Weibull distribution is an empirical distribution [37] introduced by Weibull in 1951, and the distribution function is expressed as

$$P = 1 - \exp\{-[x/x_0]^m\} \quad (1)$$

where:

P = the cumulative fraction of failures in the mechanical property, e.g., a tensile test;
x = variable being measured, e.g., tensile strength;
x_0 = position parameter or characteristic value at which 63% of the samples failed;
m = Weibull modulus.

Taking the logarithm of Equation (1) twice yields a linear equation:

$$\ln[-\ln(1-P)] = m\ln(x) - m\ln(x_0) \quad (2)$$

with a slope of "m" and an intercept of "−m ln(x_0)". When "ln [−ln(1−P)]" is plotted versus "ln(x)", Weibull probability plot is obtained and therefore the values of "m" and "x_0" could be determined [20].

Literature suggests that Weibull distribution could better explain the failure of materials under a mechanical loading than a normal distribution [37,38]. A greater Weibull modulus and position parameters mean that the samples have fewer defects, which indicate higher and more reproducible properties. This study employed a two-parameter Weibull distribution to quantify the variability of the ultimate tensile strength and the % elongation of Al-Si-Mg cast alloys.

In this study, castings from Al–7 wt.%Si–0.3 wt.%Mg alloy were produced using gravity casting technique. Hydrogen contents and filtration were the two factors of the sand casting process that were considered for the experimentation. In addition, four process outputs or study responses were considered in this study. They were the cast Weibull modulus and position parameter of the UTS (denoted m_{UTS} and $x_{0)UTS}$, respectively), and the Weibull modulus and position parameter of the elongation (denoted m_{ELONG} and $x_{0)ELONG}$, respectively). Each process parameter was varied over two levels: "−1" and "+1". The experiment, therefore, contained four combinations of hydrogen contents and filtration. The design matrix of the experimental work is shown in Table 1.

Table 1. Experimental plan.

Factor	Coded Symbol	Experiment			
		1	2	3	4
Hydrogen content of the casting	A	−1	+1	−1	+1
Level	-	(High)	(Low)	(High)	(Low)
Filtration	B	−1	−1	+1	+1
Level	-	(Unfiltered)	(Unfiltered)	(Filtered)	(Filtered)

A two-level full factorial study was applied to explore the effects of hydrogen contents and filtration and their interaction using Design-Expert Software Version 7.0.0 (Stat-Ease Inc., Minneapolis, MN, USA). Figure 1a shows a sketch of the pattern geometry used to produce the resin-bonded sand moulds. The mould consists of ten tensile test bars of a length of 100 mm and a diameter of 11 mm. The runner used in this mould had a thickness of 25 mm. Two moulds were cast to produce 20 tensile test bars for each of the four experiments listed in Table 1. In each experiment, about 6 kg of Al–7 Si–0.3 Mg alloy were melted in an induction furnace (Inductotherm, Droitwich, United Kingdom). The melt was kept at 800 °C under a partial vacuum atmosphere of 0.2 bar for 2 h before pouring to allow the expansion of the charge oxide films and consequent floating to the melt surface, which helps to remove them as suggested in the literature [29]. The melt was then poured into the moulds in such a way to promote the creation of fresh bifilms.

In order to evaluate the effect of the hydrogen content in the casting samples, the experiments were grouped into two categories—the first category includes samples with high hydrogen content (Experiments 1, 3), and the second category includes samples low hydrogen content (Experiments 2, 4). For Experiments 1 and 3 the molten aluminium was poured into the moulds that had been prepared one day before the experiment. As for Experiments 2 and 4, and in order to ensure obtaining castings with low hydrogen content, degassing was carried out using AlSCAN equipment (ABB Measurement & Analytics, Pennsylvania, USA) for 30 min before pouring. In addition, and to eliminate hydrogen picking up from the mould walls, the moulds were kept under a reduced pressure of 0.5 bar for 14 days before the experiment in order to allow the resin solvent to evaporate completely from the sand moulds [28]. For Experiments 3 and 4, two ceramic filters (Fesoco, Birmingham, United Kingdom) of 10 pores per linear inch (PPI) and dimensions of 50 × 50 × 20 mm, were placed in the filter prints at the locations shown in Figure 1a. This would allow better control of the melt flow inside the mould, aiming to reduce the possibility of oxide film entrainment. After solidification of the cast, samples were cut from the runner bar, and hydrogen measurement was analysed using LECO™ hydrogen analyser (LECO, St. Joseph, MI, USA) for solid-state hydrogen measurement of the castings from different experiments.

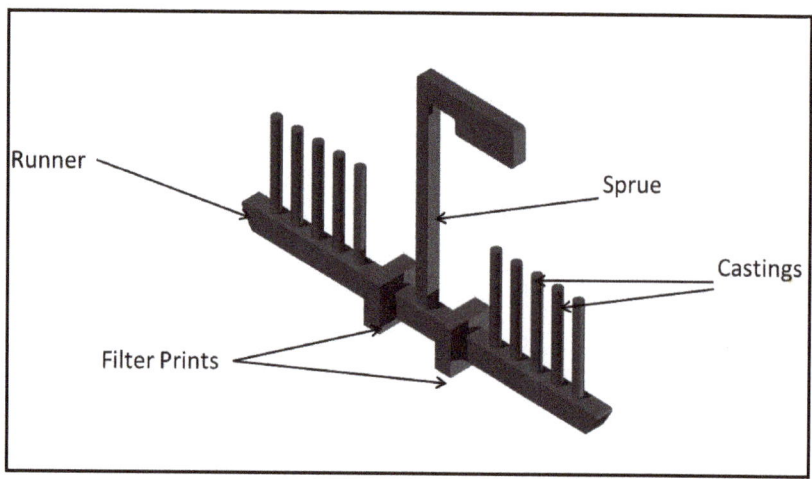

(a)

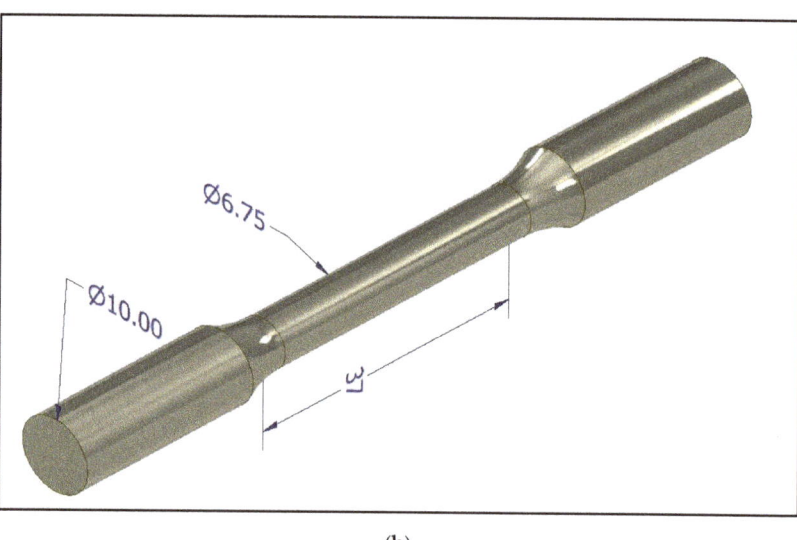

(b)

Figure 1. A schematic diagram of (**a**) the pattern used to create the mould, (**b**) a tensile test sample (dimensions in mm).

Tensile test samples were prepared by machining the solidified castings using a turning machine. Figure 1b shows a schematic of the tensile test samples. Testing was performed with a WDW-100E universal testing machine (Time Group Inc., Beijing, China) with a strain rate of 1 mm min^{-1}. The ultimate tensile strength (UTS) and % elongation results were assessed using a two-parameter Weibull distribution to evaluate the effect of the casting parameters on the scatter of the casting tensile properties. Finally, the broken tensile test samples were examined using a Philips XL-30 scanning electron microscope (SEM) (SEMTech Solutions, Inc., North Billerica, MA, USA) equipped with an energy dispersive X-ray analyser (EDS) for the evidence of bifilm. For each of Experiments 1 and 4, three specimens were selected for investigation that showed the lowest UTS, as they were expected to include more oxide film defects.

3. Results and Discussion

As discussed, aluminium alloy melt was held under vacuum to eliminate the effect of previously introduced oxides in the raw material and ensure that the castings' variability is only due to the changed casting process parameters [27,36]. These process parameters or casting conditions are the amount of hydrogen in the solidified casting and the filtration.

Table 2 lists the casting conditions of the experiments performed and the corresponding Weibull analysis results for different properties.

Table 2. Position parameter and Weibull modulus of cast specimens produced under different conditions.

Exp. No.	Hydrogen Level (cm^3/100 g Al)	Filtration	UTS (MPa)		% Elongation	
			Position Parameter	Weibull Modulus	Position Parameter	Weibull Modulus
1	0.257	Unfiltered	81	4.15	4.08	2.17
2	0.132	Unfiltered	106	8.95	4.92	6.14
3	0.257	Filtered	105	7.78	4.72	4.88
4	0.132	Filtered	148	21.67	8.2	10.87

The results showed a significant effect of the degassing treatment as well as the holding of the sand mould under reduced pressure, for a given time before pouring in, on the casting hydrogen content. The average hydrogen content of the samples cut from the solidified undegassed castings (Experiments 1 and 3) and degassed castings (Experiments 2 and 4) were 0.257 and 0.132 cm^3/100 g Al, respectively. The noticeable reduction in the hydrogen level in Experiments 2 and 4 is reasoned to the use of a degassing treatment that was able to decrease the hydrogen content in the melt before pouring, as well as the vacuum treatment of the moulds before the use that seemed to minimise the amount of hydrogen picked by the poured melt from mould walls [39].

Figure 2 shows the Weibull distribution results of the tensile samples. Corresponding plots for the % elongation are also presented in Figure 3. As shown in the two figures, the data representing both properties are linearly distributed, as suggested by correlation coefficients. In both figures the slope of the data trend lines (that represent the Weibull moduli) of both the UTS and % elongation in Experiment 4, where degassing and ceramic filters were employed, were the largest among all castings.

By applying the DoE approach to investigate the effect of the process parameters, Figure 4a–c shows the effect of hydrogen content, filtration and the interaction between the two parameters, respectively, on m_{UTS}. Corresponding plots related to m_{ELONG} are presented in Figure 5a–c, respectively. Note that in both Figures 4b and 5b the points representing the responses are connected using dotted lines, not solid lines, to indicate a categoric factor.

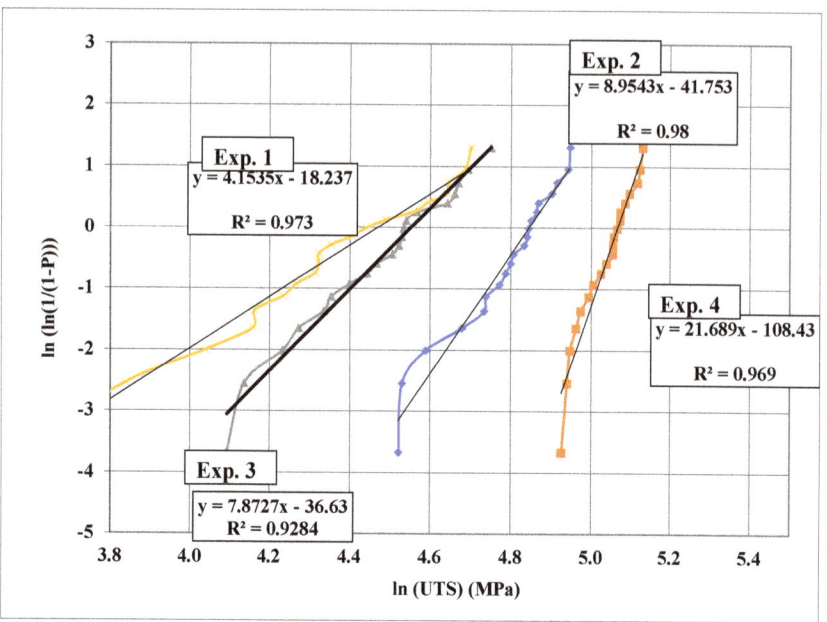

Figure 2. Weibull distribution of UTS of Al–7 Si–0.3 Mg alloy for different experiments itemised in Table 1.

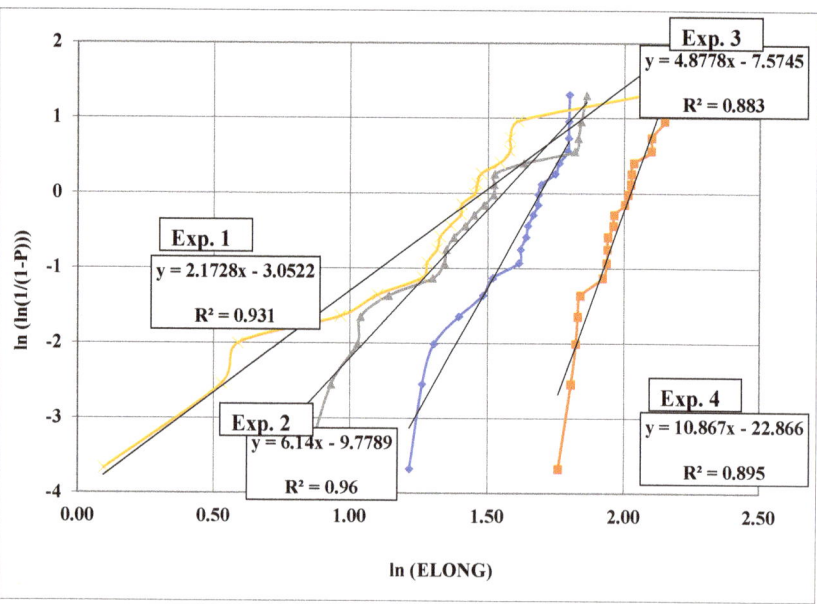

Figure 3. Weibull distribution of % elongation of Al–7 Si–0.3 Mg alloy for different experiments itemised in Table 1.

The two figures indicate that both moduli had been increased consistently with either the decrease of hydrogen content and/or the use of filters. The value of m_{UTS}, at a hydrogen level of 0.257 cm^3/100 g Al and without the use of filters, was about 4 (Experiment 1). Decreasing the hydrogen level to 0.132 cm^3/100 g Al increased the modulus to about 9, while the use of 10 PPI filters increased the modulus to 8. However, the application of degassing and mould treatment (that tended to decrease hydrogen level) and the implementation of filters resulted in a significant improvement of m_{UTS} to about 22 (Experiment 4). Moreover, m_{ELONG} of about 2 was obtained for the undegassed casting poured in unfiltered moulds. M_{ELONG} values of 6, 5 and 11, respectively, were obtained for castings when hydrogen level was reduced, filters were implemented, and both conducts were adopted. Finally, the results suggest that the hydrogen level and filtration interaction is also significant, especially for m_{UTS}, as shown in Figures 4c and 5c. At lower hydrogen level, the positive effect of filtration on both moduli is clearer. Likewise, the antithesis impact of hydrogen level on the Weibull moduli is more obvious when filtration was adopted.

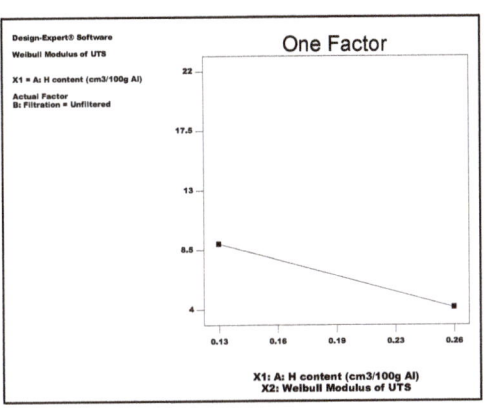

(a)

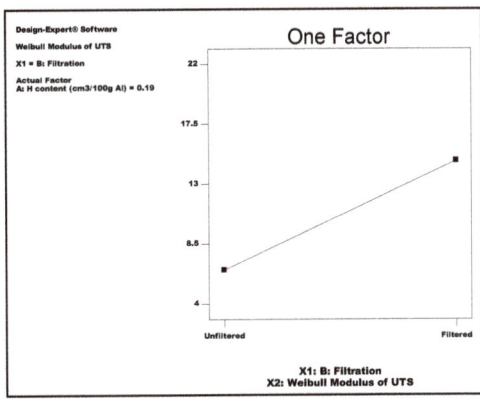

(b)

(c)

Figure 4. Effect of (**a**) hydrogen level, (**b**) use of filters, (**c**) the interaction between hydrogen level and filters on m_{UTS}.

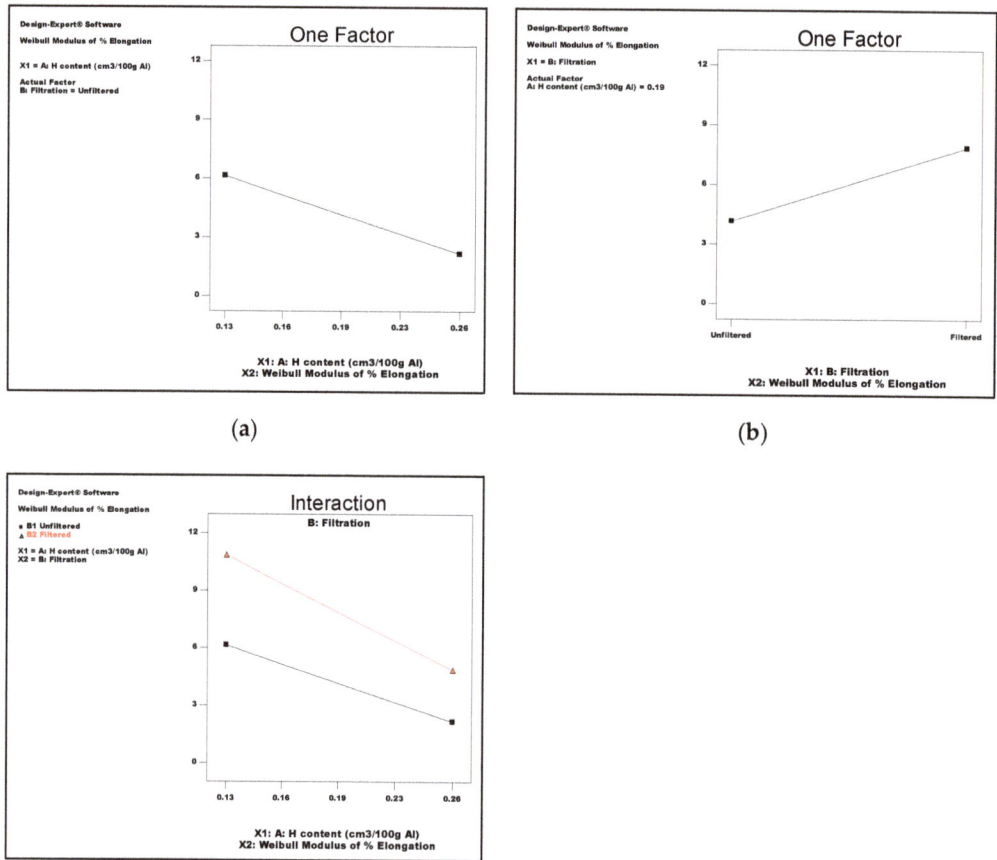

Figure 5. Effect of (**a**) hydrogen level, (**b**) filtration, (**c**) the interaction between hydrogen level and filtration on m_{ELONG}.

The influence of the hydrogen levels, filters, and interaction between them on $x_{0)UTS}$ and $x_{0)ELONG}$ are shown in Figures 6 and 7, respectively. $x_{0)UTS}$ and $x_{0)ELONG}$ exhibited similar trends to those of m_{UTS} and m_{ELONG}. Again, in both Figures 6b and 7b the points representing the responses are connected using dotted lines, not solid lines, to indicate a categoric factor. The parameters were enhanced 81 to 148 MPa for the UTS, and from 4 to 7 for the % elongation upon reducing the hydrogen level and use of filters. Furthermore, the interaction between both factors was also revealed to remarkably influence the position parameters of both tensile properties. Employing ceramic filters enabled a sharper relationship between the hydrogen level and $x_{0)UTS}$ (Figure 6c) and $x_{0)ELONG}$ (Figure 7c), and vice versa.

Generally, it was evident that the implementation of ceramic filters and the application of casting procedures that promoted reducing the hydrogen level of the casting had a significant effect on the enhancement of m_{UTS}, $x_{0)UTS}$ m_{ELONG} and $x_{0)ELONG}$, as could be inferred from Figures 4–7. The Weibull moduli and position parameters were the highest for samples produced in Experiment 4 compared to all castings. This is owed to the use of filters and low hydrogen level castings. This indicates that the casting quality has been improved, and the variability among them has been reduced.

Literature has demonstrated that the use of a poor gating system, with a runner of ≥25 mm height and without the use of filters, was associated with the formation and entrainment of a substantial amount of bifilm defects [28]. They have also advocated that due to the lack of bonding between the inner (dry) sides of a bifilm, the rejected hydrogen during solidification usually penetrates into the defect and easily expands it, like a balloon, creating a hydrogen pore in the final casting [40]. In Experiment 1, the casting experiment was performed without a prior degassing of the aluminium melt to produce a high hydrogen level casting. This resulted in a substantial drop in the tensile strength of the samples (a $x_{0)UTS}$ of 81 MPa and a $x_{0)ELONG}$ of 4%) and also widened their spread (4 and 2, respectively, for m_{UTS} and m_{ELONG}). This could be easily inferred from the results shown in Figures 2–7, as well as in Table 2, which showed that samples from Experiment 1 experienced the worst properties among all experiments.

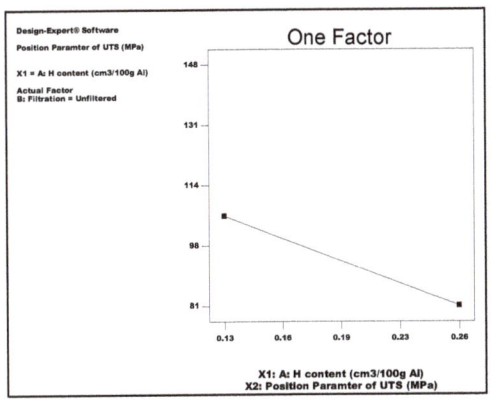

(a)

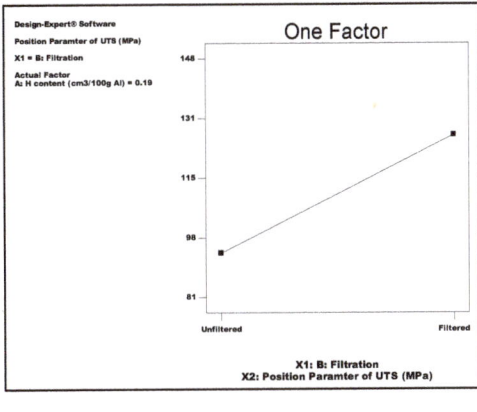

(b)

(c)

Figure 6. Influence of (**a**) hydrogen level, (**b**) filtration (**c**) the interaction between hydrogen level and filtration on $x_{0)UTS}$.

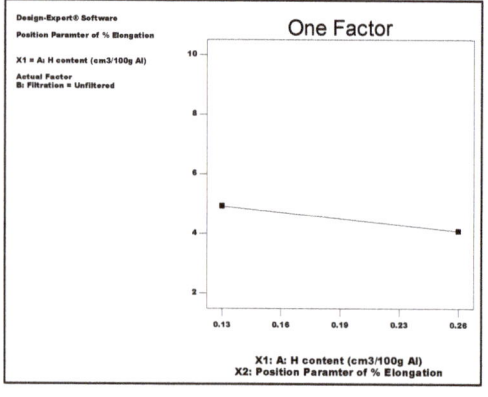
(a)

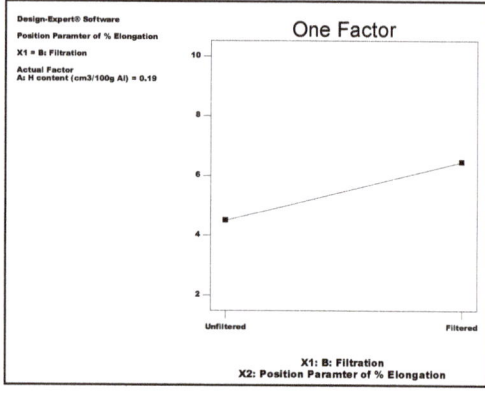
(b)

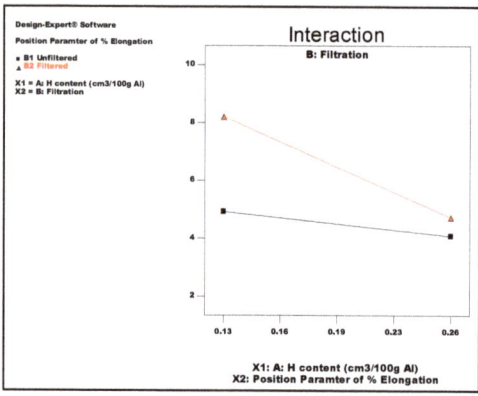
(c)

Figure 7. Graphs of (**a**) hydrogen level, (**b**) filtration, (**c**) the interaction between hydrogen level and filtration against $x_{0)ELONG}$.

Using the experimental data, a factorial analysis using Analysis of Variance (ANOVA) statistical approach was executed to determine the standardised effects of studied parameters (the hydrogen level of the casting and filtration) and their interaction on four responses considered in this study: m_{UTS}, $x_{0)UTS}$, m_{ELONG} and $x_{0)ELONG}$. Table 3 summarises the list of factors and their interaction, as well as the effect of each factor and/or interaction. The effect is the change in the response as the factor changes from the "−1" level to the "+1" level. In other words, the effect of a given factor A is the difference between the mean values of the response at levels "+1" and "−1" of A. A positive value of the effect denotes a lineal influence favouring optimisation, whereas a negative sign signifies a converse repercussion of the parameter on the studied response [41].

It is evident that the H level of the casting had an adverse effect on m_{UTS}, $x_{0)UTS}$, m_{ELONG} and $x_{0)ELONG}$, while filtration showed an advantageous effect on the four outputs evaluated in this study, see Table 3. This could be concluded from the signs of the main effects of the two parameters for different responses. However, and independent of the sign of the effect, ANOVA results had demonstrated that the standardised effects of each of the hydrogen levels and filtration on the four responses were too close. This is a clear indication of the important roles played by both factors in influencing the reproducibility of aluminium castings.

Table 3. Factorial analysis of the properties of the casting.

Term	Standardised Effect			
	Weibull Modulus of UTS	Position Parameter of UTS (MPa)	Weibull Modulus of % Elongation	Position Parameter of % Elongation (%)
A-Hydrogen Level	−9.43	−34	−4.98	−2.16
B-Filtration	8.18	33	3.72	1.96
AB	−4.56	−9	−1.01	−1.32

The DoE results, presented in Table 3, signified that doubling the hydrogen content was associated with a negative effect on m_{UTS} and $x_{0)UTS}$ of about −9 and −34 MPa, respectively, and on m_{ELONG} and $x_{0)ELONG}$ of about −5 and −2%, respectively. On the other hand, filtration was shown to have a favourable effect on m_{UTS} and $x_{0)UTS}$ of about 8 and 33 MPa, respectively, and on m_{ELONG} and $x_{0)ELONG}$ of about 4 and 2%, respectively. This might be due to the role played by the filters in damping the acceleration of the incoming flow of liquid Al during its passage through the runner. This would permit a calmer and smoother flow behaviour of the melt inside the mould, reduce the ingate velocity, minimise oxide film entrainment, and ultimately improve the tensile properties. This confirms the results obtained by Green and Campbell [38,42], who apprised a considerable improvement in the Weibull moduli of the tensile properties of Al–7Si–Mg castings, of about 350%, while applying a turbulent-free filling system that seemed to prevent oxide film entrainment.

Bifilm defects were detected at the specimens' fracture surfaces from the four experiments carried out in this work. Typical examples of such defects from Experiments 1 and 4 are shown in Figures 8 and 9, respectively. Results of the accompanied EDX examination at the suspected oxide films confirmed that spinel films existed at these surfaces. The EDX spectra contained a peak for carbon which was suggested to be caused by the contamination of the atmosphere inside the SEM. The identity of the spinel films was confirmed by EDX analyses that detected relatively large amounts of oxygen at the suspected oxide films (indicated by the high oxygen peak in the EDX spectrum), which had not been detected in other adjacent areas of the fracture surface being examined. It was shown that the approximate average areas of the spinel layers detected on the fracture surfaces of the examined test bars (assuming an elliptical shape for the oxide film) from Experiments 1 and 4 were about 1.4 and 4.5 mm^2, respectively.

In Experiment 1 the bad mould design and the lack of filtration are expected to violate the critical ingate velocity and accordingly plenteous entrainment of oxide films is expected. Additionally, the relatively high hydrogen level of the castings in this experiment (about 0.257 cm^3/100 g Al) was expected to increase the hydrogen ingress into the bifilms and increase their sizes. Therefore, several oxide film defects of comparatively larger sizes were often detected at the fracture surfaces of test bars from this experiment. See Figure 8. In contrast, the area covered with oxide layers detected on the fracture surface of a specimen from Experiment 4 (Figure 9) was considerably smaller than that in the casting from Experiment 1 by about one-third. This is suggested to be a result of the significantly lower hydrogen level of the former experiment due to the application of degassing, as well as the use of filters that seemed to minimise the oxide film entrainment during mould filling.

The substantial reduction in the amount and size of bifilm defects, which is related to the combined effect of the two casting parameters, caused the castings from Experiment 4 to have a noticeable improvement of their m_{UTS} and m_{ELONG} by about 420% and 400%, respectively, compared to those in Experiment 1. Moreover, a less significant increase in $x_{0)UTS}$ and $x_{0)ELONG}$ of about 82% and 101% respectively was obtained due to the reduced hydrogen level and the use of filters. See Table 2.

The connotation of the current findings would be that the implementation of filters could significantly reduce the production of bifilm defects. Moreover, the reduction of casting hydrogen level would minimise the amount of the gas diffuses into the entrained bifilms, decreasing the size of the defects. Thus, if appropriate treatment procedures for both the melt and the sand mould would be considered, this would permit the production of castings with minimum hydrogen level. Additionally, if filtration was applied, this would allow a more quiescent mould filling regime. These considerations would allow a casting producer to reduce both the number and the size of the bifilms in the melt and consequently obtaining an Al cast alloy with improved mechanical properties.

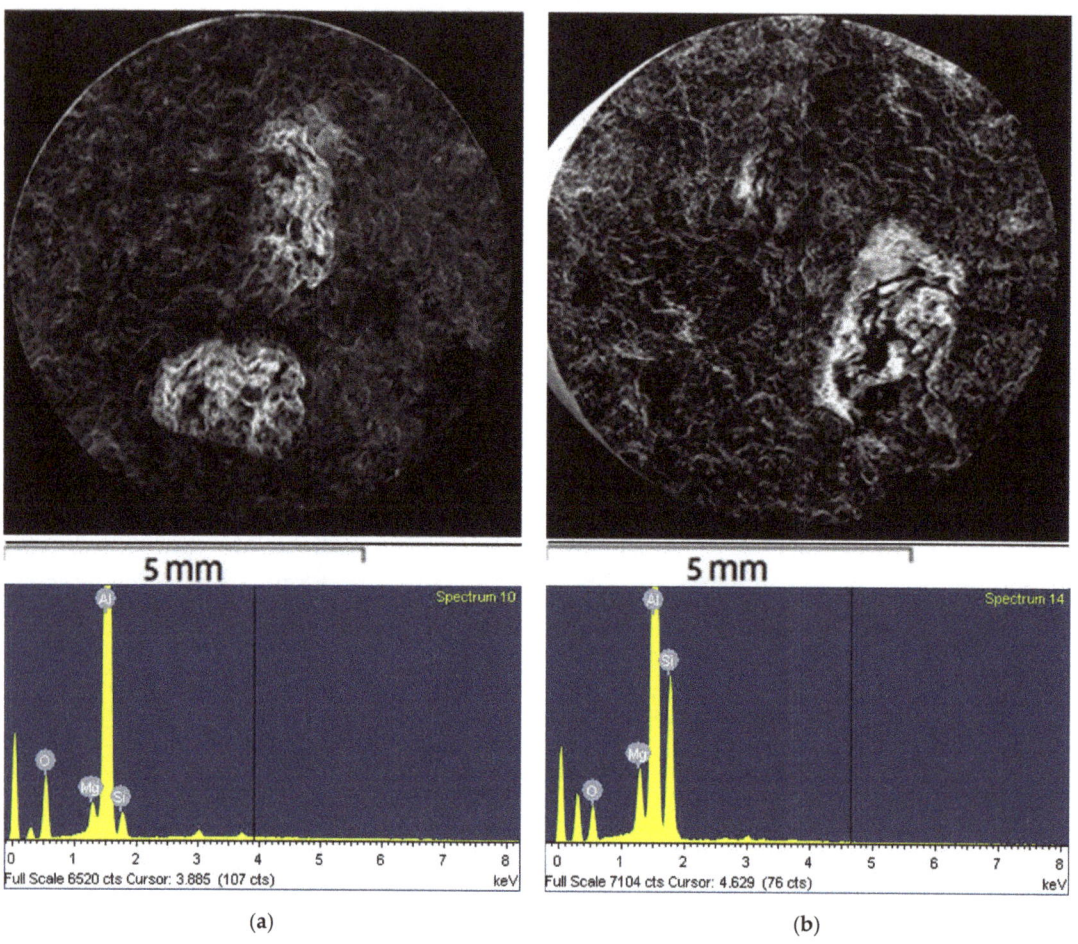

Figure 8. (**a**,**b**) Electron microscopy fractographs and corresponding EDX spectra (at the positions pointed with "X") of oxide films on the fracture surfaces of two tensile-tested bars from Experiment 1.

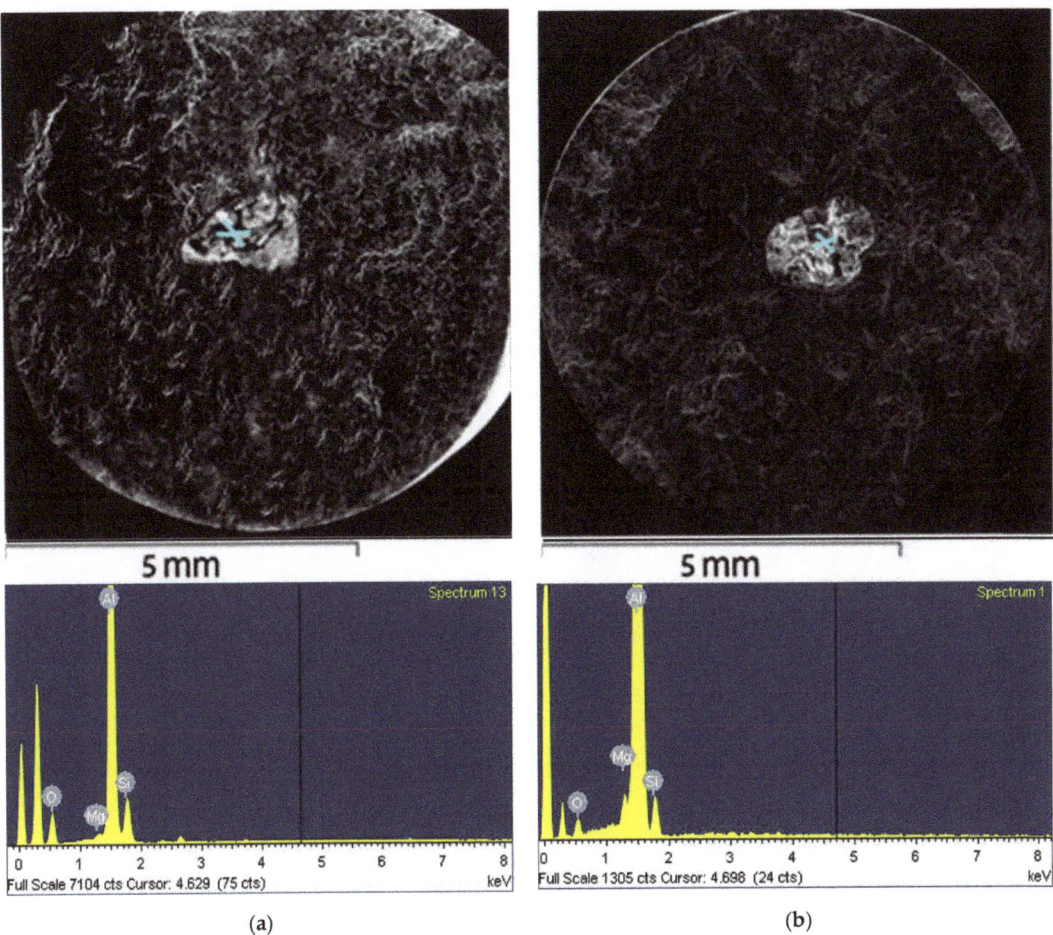

Figure 9. (**a**,**b**) Electron microscopy fractographs and corresponding EDX spectra (at the positions pointed with "X") of oxide films on the fracture surfaces of two tensile-tested bars from Experiment 4.

4. Conclusions

1. The detection of bifilms at fracture surfaces of the majority of tensile samples examined is a suggestion of the deleterious influence of these inclusions on the mechanical properties of Al cast alloys.
2. ANOVA results suggested that reducing the hydrogen content of an Al–7 Si–0.3 Mg cast alloy and the application of 10 PPI filters had remarkable positive standardised effects of about 9 and 4, respectively, on the UTS and % elongation Weibull modulus.
3. Statistical analysis also indicated that both parameters had significant standardised effects on the UTS and % elongation position parameters, of 33 MPa and 2%, respectively.
4. The optimised casting condition involving the implementation of filters and the application of the precautions required to decrease the hydrogen level of the casting resulted in an outstanding improvement of the Weibull moduli of the UTS and % elongation by a factor of about 4.

Author Contributions: Conceptualization, M.A.E.-S., K.E. and H.H.; methodology, M.A.E.-S., K.E. and H.H.; software, M.A.E.-S., K.E. and H.H.; writing—original draft preparation, M.A.E.-S., K.E. and H.H.; writing—review and editing, M.A.E.-S., K.E. and H.H. All authors have read and agreed to the published version of the manuscript.

Funding: This research received no external funding.

Institutional Review Board Statement: Not applicable.

Informed Consent Statement: Not applicable.

Data Availability Statement: The data presented in this study are available on request from the corresponding author.

Conflicts of Interest: The authors declare no conflict of interest.

References

1. Divandari, M.; Campbell, J. A new technique for the study of aluminum oxide films. *Aluminum Trans.* **2000**, *2*, 233–238.
2. Dai, X.; Yang, X.; Campbell, J.; Wood, J. Influence of oxide film defects generated in filling on mechanical strength of aluminium alloy castings. *Mater. Sci. Technol.* **2004**, *20*, 505–513. [CrossRef]
3. Divandari, M.; Campbell, J. Oxide film characteristics of Al–7Si–Mg alloy in dynamic conditions in casting. *Int. J. Cast Met. Res.* **2004**, *17*, 182–187. [CrossRef]
4. Wang, Q.; Crepeau, P.; Davidson, C.; Griffiths, J. Oxide films, pores and the fatigue lives of cast aluminum alloys. *Metall. Mater. Trans. B* **2006**, *37*, 887–895. [CrossRef]
5. Mirak, A.; Divandari, M.; Boutorabi, S.; Campbell, J. Oxide film characteristics of AZ91 magnesium alloy in casting conditions. *Int. J. Cast Met. Res.* **2007**, *20*, 215–220. [CrossRef]
6. Campbell, J. Entrainment defects. *Mater. Sci. Technol.* **2006**, *22*, 127–145. [CrossRef]
7. Erzi, E.; Gürsoy, Ö.; Yüksel, Ç.; Colak, M.; Dispinar, D. Determination of Acceptable Quality Limit for Casting of A356 Aluminium Alloy: Supplier's Quality Index (SQI). *Metals* **2019**, *9*, 957. [CrossRef]
8. Matejka, M.; Bolibruchová, D.; Podprocká, R. The Influence of Returnable Material on Internal Homogeneity of the High-Pressure Die-Cast AlSi$_9$Cu$_3$(Fe) Alloy. *Metals* **2021**, *11*, 1084. [CrossRef]
9. Neuser, M.; Grydin, O.; Andreiev, A.; Schaper, M. Effect of Solidification Rates at Sand Casting on the Mechanical Joinability of a Cast Aluminium Alloy. *Metals* **2021**, *11*, 1304. [CrossRef]
10. Souissi, N.; Souissi, S.; Niniven, C.L.; Amar, M.B.; Bradai, C.; Elhalouani, F. Optimization of Squeeze Casting Parameters for 2017 A Wrought Al Alloy Using Taguchi Method. *Metals* **2014**, *4*, 141–154. [CrossRef]
11. Griffiths, W.; Lai, N.-W. Double oxide film defects in cast magnesium alloy. *Metall. Mater. Trans. A* **2007**, *38*, 190–196. [CrossRef]
12. Raiszadeh, R.; Griffiths, W. The behaviour of double oxide film defects in liquid Al alloys under atmospheric and reduced pressures. *J. Alloys Compd.* **2010**, *491*, 575–580. [CrossRef]
13. El-Sayed, M.A. The behaviour of bifilm defects in cast Al-7Si-Mg alloy. *PLoS ONE* **2016**, *11*, e0160633. [CrossRef]
14. El-Sayed, M.; Salem, H.G.; Kandeil, A.-R.; Griffiths, W.D. A study of the behaviour of double oxide films in Al alloy melts. In *Materials Science Forum*; Trans Tech Publications Ltd.: Zurich, Switzerland, 2013; pp. 260–265.
15. Hamed Basuny, F.; Ghazy, M.; Kandeil, A.-R.; El-Sayed, M.A. Effect of casting conditions on the fracture strength of Al-5 Mg alloy castings. *Adv. Mater. Sci. Eng.* **2016**. [CrossRef]
16. Lai, N.; Griffiths, W.; Campbell, J. Modelling of the potential for oxide film entrainment in light metal alloy castings. *Modeling Cast. Weld. Adv. Solidif. Processes-X* **2003**, 415–422. Available online: https://www.researchgate.net/profile/John-Campbell-38/publication/260629687_Modelling_of_the_potential_for_oxide_film_entrainment_in_light_metal_alloy_castings/links/02e7e5349543c16b2a000000/Modelling-of-the-potential-for-oxide-film-entrainment-in-light-metal-alloy-castings (accessed on 3 December 2021).
17. Raiszadeh, R.; Griffiths, W. The effect of holding liquid aluminum alloys on oxide film content. *Metall. Mater. Trans. B* **2011**, *42*, 133–143. [CrossRef]
18. Reilly, C.; Green, N.; Jolly, M.; Gebelin, J.-C. The modelling of oxide film entrainment in casting systems using computational modelling. *Appl. Math. Modell.* **2013**, *37*, 8451–8466. [CrossRef]
19. Campbell, J. *Castings*; Elsevier: Amsterdam, The Netherlands, 2003.
20. El-Sayed, M.A.M. *Double Oxide Film Defects and Mechanical Properties in Aluminium Alloys*; University of Birmingham: Birmingham, UK, 2012.
21. Griffiths, W.; Caden, A.; El-Sayed, M. The Behaviour of Entrainment Defects in Aluminium Alloy Castings. In Proceedings of the 2013 International Symposium on Liquid Metal Processing and Casting, Austin, TX, USA, 22–25 September 2013; pp. 187–192.
22. Griffiths, W.D.; Caden, A.; El-Sayed, M. An investigation into double oxide film defects in aluminium alloys. In *Materials Science Forum*; Trans Tech Publications Ltd.: Zurich, Switzerland, 2014; pp. 142–147.
23. Bahreinian, F.; Boutorabi, S.M.A.; Campbell, J. Critical gate velocity for magnesium casting alloy (ZK51A). *Int. J. Cast Met. Res.* **2006**, *19*, 45–51. [CrossRef]
24. Cox, M.; Wickins, M.; Kuang, J.P.; Harding, R.A.; Campbell, J. Effect of top and bottom filling on reliability of investment castings in Al, Fe, and Ni based alloys. *Mater. Sci. Technol.* **2000**, *16*, 1445–1452. [CrossRef]
25. Halvaee, A.; Campbell, J. Critical mold entry velocity for aluminum bronze castings. *AFS Trans.* **1997**, *105*, 35–46.
26. Runyoro, J.; Boutorabi, S.M.A.; Campbell, J. Critical gate velocities for film-forming casting alloys: A basic for process specification. *AFS Trans.* **1992**, *100*, 225–234.

27. Dispinar, D.; Campbell, J. Critical assessment of reduced pressure test. Part 1: Porosity phenomena. *Int. J. Cast Met. Res.* **2004**, *17*, 280–286. [CrossRef]
28. El-Sayed, M.; Griffiths, W. Hydrogen, bifilms and mechanical properties of Al castings. *Int. J. Cast Met. Res.* **2014**, *27*, 282–287. [CrossRef]
29. El-Sayed, M.; Hassanin, H.; Essa, K. Effect of casting practice on the reliability of Al cast alloys. *Int. J. Cast Met. Res.* **2016**, *29*, 350–354. [CrossRef]
30. Chen, Q.; Griffiths, W. The investigation of the floatation of double oxide film defect in liquid aluminium alloys by a four-point bend test. *Int. J. Cast Met. Res.* **2019**, *32*, 221–228. [CrossRef]
31. Chen, Q.; Griffiths, W. Modification of Double Oxide Film Defects with the Addition of Mo to An Al-Si-Mg Alloy. *Metall. Mater. Trans. B.* **2021**, *52*, 502–516. [CrossRef]
32. Laakso, P.; Riipinen, T.; Laukkanen, A.; Andersson, T.; Jokinen, A.; Revuelta, A.; Ruusuvuori, K. Optimization and simulation of SLM process for high density H13 tool steel parts. *Phys. Procedia* **2016**, *83*, 26–35.
33. Leary, M.; Mazur, M.; Elambasseril, J.; McMillan, M.; Chirent, T.; Sun, Y.; Qian, M.; Easton, M.; Brandt, M. Selective laser melting (SLM) of AlSi12Mg lattice structures. *Mater. Des.* **2016**, *98*, 344–357. [CrossRef]
34. Pawlak, A.; Rosienkiewicz, M.; Chlebus, E. Design of experiments approach in AZ31 powder selective laser melting process optimization. *Arch. Civ. Mech. Eng.* **2017**, *17*, 9–18. [CrossRef]
35. El-Sayed, M.A.; Essa, K.; Ghazy, M.; Hassanin, H. Design optimization of additively manufactured titanium lattice structures for biomedical implants. *Int. J. Adv. Manuf. Technol.* **2020**, *110*, 1–12. [CrossRef]
36. Croarkin, C.; Tobias, P.; Zey, C. *Engineering Statistics Handbook*; The Institute Gaithersburg: Gaithersburg, MD, USA, 2001.
37. Weibull, W. A statistical distribution function of wide applicability. *J. Appl. Mech.* **1951**, *18*, 293–297. [CrossRef]
38. Green, N.R.; Campbell, J. Statistical distributions of fracture strengths of cast Al-7Si-Mg alloy. *Mater. Sci. Eng. A* **1993**, *173*, 261–266. [CrossRef]
39. Winardi, L.; Griffin, R.D.; Littleton, H.E.; Griffin, J.A. Variables Affecting Gas Evolution Rates and Volumes from Cores in Contact with Molten Metal. *AFS Trans.* **2008**, *116*, 505.
40. Farhoodi, B.; Raiszadeh, R.; Ghanaatian, M.-H. Role of double oxide film defects in the formation of gas porosity in commercial purity and Sr-containing Al alloys. *J. Mater. Sci. Technol.* **2014**, *30*, 154–162. [CrossRef]
41. Nayak, A.K.; Pal, D.; Santra, K. Ispaghula mucilage-gellan mucoadhesive beads of metformin HCl: Development by response surface methodology. *Carbohydr. Polym.* **2014**, *107*, 41–50. [CrossRef] [PubMed]
42. Green, N.; Campbell, J. Influence of oxide film filling defects on the strength of Al-7Si-Mg alloy castings. *AFS Trans.* **1994**, *102*, 341–347.

Article

Complex Structure Modification and Improvement of Properties of Aluminium Casting Alloys with Various Silicon Content

Anastasiya D. Shlyaptseva [1], Igor A. Petrov [1], Alexandr P. Ryakhovsky [1], Elena V. Medvedeva [2] and Victor V. Tcherdyntsev [3,*]

1. Moscow Aviation Institute, Orshanskaya 3, 125993 Moscow, Russia; ShlyaptsevaAD@mai.ru (A.D.S.); petrovia2@mai.ru (I.A.P.); ryahovskiyap@mai.ru (A.P.R.)
2. Institute of Electrophysics, Ural Branch, Russian Academy of Science, Amudsena str., 106, 620016 Yekaterinburg, Russia; lena@iep.uran.ru
3. Institute of New Materials and Nanotechnology, National University of Science and Technology "MISIS", Leninskii Prosp, 4, 119049 Moscow, Russia
* Correspondence: vvch@misis.ru; Tel.: +7-9104002369

Abstract: The possibility of using complex structure modification for aluminium casting alloys' mechanical properties improvement was studied. The fluxes widely used in the industry are mainly intended for the modification of a single structural component of Al–Si alloys, which does not allow unifying of the modification process in a production environment. Thus, a new modifying flux that has a complex effect on the structure of Al–Si alloys has been developed. It consists of the following components: TiO_2, containing a primary α-Al grain size modifier; BaF_2 containing a eutectic silicon modifier; KF used to transform titanium and barium into the melt. The effect of the complex titanium dioxide-based modifier on the macro-, microstructure and the mechanical properties of industrial aluminium–silicon casting alloys containing 5%, 6%, 9%, 11% and 17% Si by weight was studied. It was found that the tensile strength (σ_B) of Al–Si alloys exceeds the similar characteristics for the alloys modified using the standard sodium-containing flux to 32%, and the relative elongation (δ) increases to 54%. The alloys' mechanical properties improvement was shown to be the result of the flux component's complex effect on the macro- and microstructure. The effect includes the simultaneous reduction in secondary dendritic arm spacing due to titanium, the refinement and decreasing size of silicon particles in the eutectic with barium and potassium, and the modifying of the primary silicon. The reliability of the studies was confirmed using up-to-date test systems, a significant amount of experimental data and the repeatability of the results for a large number of samples in the identical initial state.

Keywords: aluminium casting alloys; complex structure modification; titanium; barium; titanium dioxide; mechanical properties; microstructure; macrostructure

1. Introduction

Aluminium–silicon casting alloys are in demand due to the most favourable combination of casting, mechanical properties and a number of special operational properties [1]. Currently, there are several directions for the improvement of the alloys' properties, but melt modification does not become irrelevant, thus allowing achievement of the required level of the mechanical properties of alloys.

A large number of research works cover the issue of Al–Si alloy modifications [2–5], but there are no unified and reliable methods of Al–Si alloy modification with the various silicon rates so far. Most modifiers do not fully meet the production requirements. The fluxes widely used in the industry are mainly intended for the modification of a single structural component of Al–Si alloys, which does not allow unifying of the modification process in the production environment.

Sodium-based modifiers are widely used [6] due to the availability of sodium salts and a good modifying ability of sodium as a modifier of the Al–Si eutectic. However, the scope of application of such modifiers is limited to alloys with the eutectic as the main structural component. Moreover, a significant disadvantage of alloy modification with sodium is the short modification effect and the increased tendency to form gas porosity [6].

In recent years, researchers have paid particular attention to the issues of the modification of Al–Si alloys using the complex impact on their structure [7–13]. The use of modifiers that give an impact on various structural components of the alloys is more effective than modification of a single phase. As a result, the scope of application of the modifier is expanded. However, such complex compositions often contain expensive substances, and this disadvantage limits the scope of their application.

Therefore to reduce the cost of such compositions, it is necessary to find new, available complex modifiers, develop production technologies and use such modifiers for the aluminium melt.

This paper describes the changes in the mechanical properties, macro- and microstructure of various alloys as a result of melt treatment with a complex modifying flux consisting of 19–29 wt.% TiO_2 + 32–40 wt.% BaF_2 + 34–42 wt.% KF mixture [14].

To ensure eutectic modification (α-Al + Si), the flux contains barium, a surface-active element for eutectic silicon, which is injected using fluoride. When barium is injected into the melt, the particles of eutectic silicon are refined and rounded. The main advantage of barium use is the long-term preservation of the modifying ability [15–19]. Barium fluoride is an available substance with a relatively low cost.

To modify the primary α-Al grain size, we propose to use titanium, the most efficient grain modifier in aluminium alloys, which refines the α-Al dendrites in Al–Si alloys [20–24]. Titanium is introduced using its oxide compounds instead of traditional ligatures and salts. Titanium dioxide is the most available and inexpensive compound among titanium-containing substances.

To modify Al–Si alloys, titanium dioxide should be reduced to titanium by melting with the subsequent formation of additional $TiAl_3$ crystallization centres. Therefore, the flux consists of potassium fluoride, which improves the wettability of titanium dioxide with aluminium and dissolves the oxide, thereby ensuring the aluminothermic reduction of the oxide to metal [25–28]. Potassium fluoride has a relatively low cost and increases the probability of barium transition from fluoride to the melt due to the formation of the low-melting eutectic [29].

The paper describes the study of the complex flux versatility, i.e., the use of this modifier for aluminium casting alloys with the various silicon content.

2. Materials and Methods

The following widely-used alloys were selected for the study:

Hypoeutectic alloy with copper of Al–Si–Cu system (Alloy 1); hypoeutectic alloys with magnesium of Al–Si–Mg system (Alloy 2) and (Alloy 3); eutectic binary alloy of Al–Si system (Alloy 4); hypereutectic piston alloy of Al–Si–Cu–Mg–Ni system (Alloy 5). The chemical composition of the alloys is listed in Table 1.

Table 1. Elemental composition of the study alloys according to spectral analysis data (average values).

Alloy			Elements (wt. %)								
System	N°	Si	Fe	Cu	Mg	Mn	Ni	Zn	Ti	Al
Al-5wt.% Si-Cu	Alloy 1	5.45	0.64	1.21	0.35	0.19	0.02	0.22	0.05	base
Al-6 wt.% Si-Mg	Alloy 2	5.94	0.37	-	0.19	0.006	-	0.01	0.011	base
Al-9 wt.% Si-Mg	Alloy 3	9.28	0.37	0.09	0.31	0.25	0.04	0.02	0.026	base
12 wt.% Si	Alloy 4	11.53	0.36	0.002	0.0006	0.0026	-	0.01	0.009	base
Al-17 wt.% Si-Cu-Mg-Ni	Alloy 5	17.02	0.36	1.03	0.88	0.009	1.35	0.012	0.007	base

Experimental alloy meltings were carried out in a muffle electric resistance furnace. The weight of a single melt was 900 g. The melt was pre-degassed by blowing with an inert gas (argon).

The melt surface was covered with an even layer of the flux under study at a temperature of 770–790 °C. After holding for 8–10 min at a given temperature, the flux was thoroughly mixed deep into the melt for 3–5 min. Then, the melt was held for 15–20 min, and the slag was removed from the melt surface. The temperature of the melt was brought to 710 °C, then, the melt was poured into the prepared sandy-clay mould.

The melt was treated with the standard sodium-containing flux (25% NaF + 62.5% NaCl + 12.5% KCl [30]) at a temperature of 730–750 °C. The melt surface was covered with an even layer of the flux in the amount of 1.5% by the melt weight. After holding for 10 min at a given temperature, the flux was thoroughly mixed deep into the melt. Then, the slag was removed from the melt surface, and the melt was held.

Alloys 1 and 2, after modification, were subjected to heat treatment in accordance with ASTM B917/B917M—12 in the T6 mode, alloy 3 was subjected to heat treatment in the T62 mode.

The mechanical properties of the alloys (ultimate strength (tensile strength) σ_B (MPa), relative elongation δ (%)) were determined in accordance with ASTM B557M—15 using Instron 5982 testing machine.

In each experiment, 4 samples were tested. Each experiment was repeated three times.

Microstructural studies were conducted using Imager.Z2m AXIO universal research microscope (Carl Zeiss, Microscopy GmbH, Göttingen, Germany).

For etching the macrostructure of alloys 2–5, the Hume-Rothery reagent (15 g of $CuCl_2$, 100 mL of H_2O) was used, for alloy 1, the Keller's reagent (2.5% HNO_3, 1.5% HCl, 0.5% HF) was used [31].

Quantitative analysis of the microstructure (the average area of eutectic silicon, the average size of primary silicon and α-Al dendrites) was carried out using specialized ImageExpert Pro 3.7 software, version 3.7.5.0, NEXSYS, Moscow, Russia using three images for each sample. The photos were processed according to the following operations: photo resizing, scale selection, binarization, determination of the object of study by colour, etc. For images processing, ImageExpert Pro 3.7 software uses built-in algorithms (techniques corresponding to the international standard ASTM E112-10).

To measure the size of an α-Al dendrite, secondary dendrite arm spacing (SDAS) was determined. SDAS was evaluated as reported in [32] by measuring thirty dendrites for each sample, employing three images at 50× magnification.

The primary Si particle size was evaluated as the maximum Feret's diameter.

The chemical (elemental) composition of the prototypes was studied using CCD-based Q4 TASMAN-170 spark optical emission spectrometer. The control of Q4 TASMAN-170 spectrometer is carried out from a desktop computer using special QMatrix software, version 3.8.1, Bruker Quantron GmbH, Kalkar, Germany.

Before the start of the chemical (elemental) composition analysis, Q4 TASMAN-170 spectrometer was set up using special calibration samples. The samples for the chemical (elemental) composition analysis were made from the sink head of the casting and had the size of 25 × 25 × 10 mm.

3. Results and Discussion

The results of mechanical tests of various silumins treated with the developed flux and the standard sodium-containing flux are given in Table 2. The experiments showed that the treatment with the developed complex flux improves the mechanical properties (tensile strength σ_B and relative elongation δ) of eutectic, hypoeutectic and hypereutectic Al–Si alloys, compared to the similar parameters of the alloys treated with the conventional industrial flux.

Table 2. Mechanical properties of Al–Si alloys, the average area of eutectic Si and the average size of primary Si, SDAS, depending on the type of treatment (sand casting).

Alloy	Properties	Unmodified	Standard Flux	Complex Flux
Alloy 1 (T6)	σ_B, MPa	316 ± 6	317 ± 5	367 ± 7
	δ, %	0.67 ± 0.1	1.25 ± 0.2	1.92 ± 0.15
	$S_{Si\,eut}$, μm^2	8.84 ± 0.25	6.75 ± 0.41	5.42 ± 0.5
	SDAS, μm	35.71 ± 4.15	35.59 ± 5.18	25.47 ± 2.14
Alloy 2 (T6)	σ_B, MPa	239 ± 10	235 ± 6	262 ± 4
	δ, %	0.98 ± 0.25	3.72 ± 0.3	4.89 ± 0.3
	$S_{Si\,eut}$, μm^2	19.48 ± 2.2	2.21 ± 0.17	1.99 ± 0.2
	SDAS, μm	37.69 ± 2.15	35.23 ± 2.18	30.33 ± 2.08
Alloy 3 (T62)	σ_B, MPa	255 ± 10	245 ± 10	323 ± 6
	δ, %	1.05 ± 0.2	2.9 ± 0.15	3.60 ± 0.15
	$S_{Si\,eut}$, μm^2	12.55 ± 1.61	3.91 ± 0.28	2.68 ± 0.37
	SDAS, μm	28.35 ± 3.1	30.98 ± 2.9	21.72 ± 2.07
Alloy 4	σ_B, MPa	140 ± 6	160 ± 9	175 ± 9
	δ, %	2.27 ± 0.2	8.05 ± 0.7	12.2 ± 0.8
	$S_{Si\,eut}$, μm^2	53.71 ± 6.0	0.56 ± 0.05	0.48 ± 0.09
	SDAS, μm	37.98 ± 5.03	36.01 ± 4.5	27.6 ± 2.2
Alloy 5	σ_B, MPa	150 ± 4	-	182 ± 4
	δ, %	0.38 ± 0.05	-	0.74 ± 0.15
	$S_{Si\,eut}$, μm^2	30.34 ± 4.01	-	21.70 ± 3.89
	Si^I, μm	92 ± 12	-	46 ± 9

The significant improvement in the alloys' mechanical properties upon modification with an experimental flux is caused by the complex impact on the alloy structure.

Measurements of SDAS and the average area of eutectic Si were carried out for all samples to observe the difference between the modified and unmodified specimens. As Table 2 demonstrates, the average area of eutectic Si and the SDAS decreased in all alloys after the addition of complex flux.

The microstructure of the eutectic unmodified alloy Al-12 wt% Si (alloy 4) features a coarse eutectic (α-Al + Si) in the form of large needles and plates ($S_{Si\,eut}$ = 53.71 μm^2) with large α-Al dendrites (SDAS = 37.98 μm) (Figure 1a).

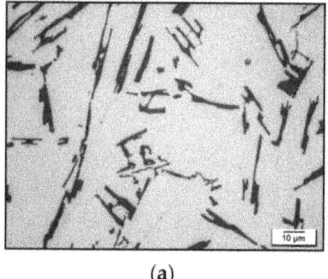 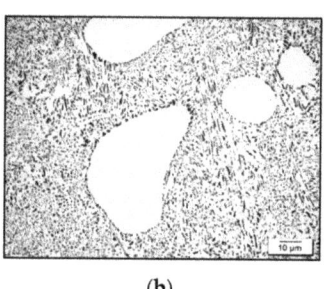 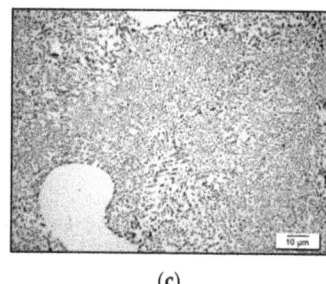

(a)   (b)   (c)

Figure 1. Microstructure of eutectic alloy 4 depending on treatment: (**a**) untreated; (**b**) standard flux; (**c**) complex flux.

Treatment of alloy 4 with such fluxes results in the refinement of eutectic silicon, and the particles of eutectic silicon become globular (Figure 1b,c). The average area of eutectic silicon particles in a modified alloy is two orders of magnitude smaller than the area of eutectic silicon in an unmodified alloy (Table 2). However, treatment with a complex flux (Figure 1c) leads to a greater fragmentation of the eutectic compared to treatment with a standard flux (Figure 1b). The average particle area of eutectic silicon decreases from 0.56 µm^2 for standard flux to 0.48 µm^2 for complex flux. The developed flux also affects the α-Al dendrites, SDAS in alloy 4 was reduced by 27% (Table 2). Titanium, as a component of the flux, facilitates refinement of the alloy macrograin, as clearly seen in the photographs of the macrostructure (Figure 2c). The standard flux does not change the macrograin size (Figure 2b) compared to the unmodified alloy (Figure 2a). Due to grain refinement, the mechanical properties of the alloy are also improved. Compared to the unmodified alloy, the relative elongation increases 5.4 times (12.2%) and the tensile strength increases by 25% (175 MPa). The resulting values are lower in the case of treatment with a standard flux, δ, by 52%; σ$_B$ by 9.4%.

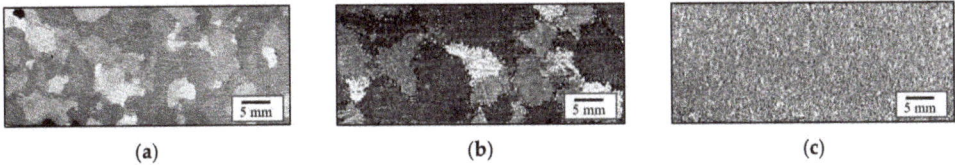

Figure 2. Macrostructure of eutectic alloy 4 depending on treatment: (**a**) untreated; (**b**) standard flux; (**c**) complex flux.

The main structural components of hypoeutectic Al–Si casting alloys are α-Al dendrites and aluminium–silicon eutectic (Figure 3a,b).

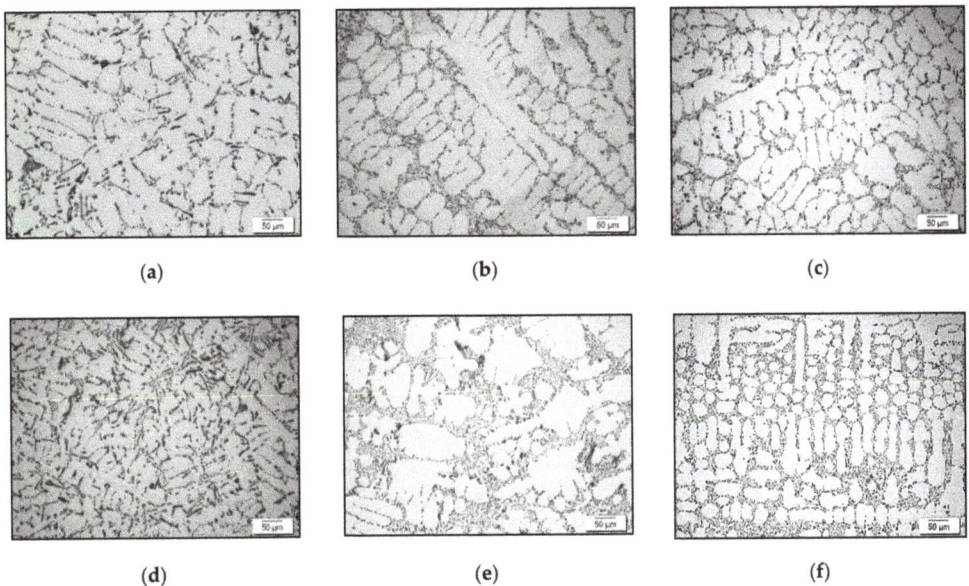

Figure 3. Microstructure of hypoeutectic alloys: (**a**) Alloy 2, untreated; (**b**) Alloy 2, standard flux; (**c**) Alloy 2, complex flux; (**d**) Alloy 3, untreated; (**e**) Alloy 3, standard flux; (**f**) Alloy 3, complex flux.

Modification with the complex flux significantly affects all structural components of hypoeutectic silumins (Figure 3c,f). Compared to the alloy treated with the standard flux

(Figures 3b,e and 4b,e), the complex flux reduces SDAS, which is most expressed in Al-9 wt.% Si-Mg alloy (Alloy 3) (Figure 3e), as well as refines alloy macrograins (Figure 4c,f). In detail, SDAS in alloy 2 was reduced by 19% and in alloy 3 by 23%. This effect on the structure is exerted by titanium transformed from the dioxide into the melt and contained in the modified alloys Al-7 wt.% Si-Mg (Alloy 2) and Al-9 wt.% Si-Mg (Alloy 3) in the amount of 0.119 wt.% and 0.126 wt.%, respectively. Moreover, the complex flux refines eutectic silicon more intensely than sodium-containing flux. The size of the eutectic area was strongly reduced after complex flux addition: in alloy 2 it was reduced by 89% and in alloy 3 by 78% compared to the unmodified alloy. The effect on the eutectic in the complex flux is exerted by two surface-active elements—barium and potassium.

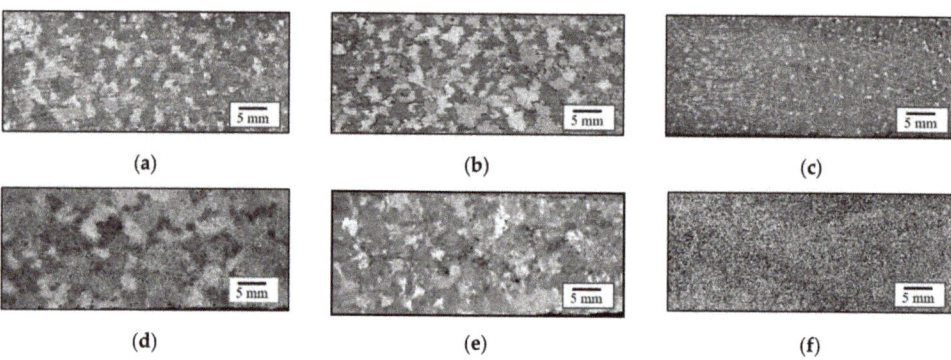

Figure 4. Macrostructure of hypoeutectic alloys: (**a**) Alloy 2, untreated; (**b**) Alloy 2, standard flux; (**c**) Alloy 2, complex flux; (**d**) Alloy 3, untreated; (**e**) Alloy 3, standard flux; (**f**) Alloy 3, complex flux.

The mechanical properties of hypoeutectic alloys were determined after heat treatment in the T6 mode (Alloy 2) and the T62 mode (Alloy 3) (Table 2). The greatest improvement in properties is achieved when using the complex flux. Compared to unmodified alloy 2 and 3, δ increases 5 times and 3.4 times, respectively, σ_B increases by 7% and 27%, respectively. δ of alloys 2 and 3 treated with a complex flux is higher than upon treatment with a standard flux by 31% and 24%, respectively; σ_B is higher by 11% and 32%, respectively.

Al-5 wt.% Si-Cu alloy (Alloy 1) has a wider crystallization range compared to the previously considered silumins and has a relatively high strength—more than 300 MPa (Table 2). However, the study alloy has low values of the relative elongation—0.67%.

The treatment of alloy 1 with a complex modifying flux based on titanium dioxide makes it possible to increase the relative elongation of the alloy by 2.9 times, and the strength by 16%. The resulting values of the mechanical properties of the experimental alloy are also higher than upon treatment with the standard flux: the relative elongation increases by 54%, and the strength increases by 16%.

The significant improvement of the alloy properties upon modification with an experimental flux is a consequence of the complex impact upon the structure of the copper Al–Si alloy. The main structural components in the alloy are aluminium dendrites and aluminium–silicon eutectic (Figure 5a). Figure 5 shows the structure of the alloy heat treated according to the T6 mode. The result of heat treatment is not only the complete dissolution of copper and magnesium in aluminium but also the partial refinement and spheroidization of eutectic silicon [32].

Complete refinement and spheroidization of eutectic silicon is achieved as a result of modification. The average area of eutectic Si shrank by 39% after complex flux addition and by 23% after standard flux addition. When treated with the standard flux, the eutectic is modified with sodium (Figure 5b), when modified with the experimental flux—with barium, successfully transferred into the melt in the amount of 0.019 wt.% (Figure 5c), and potassium transferred into the melt in amount of 0.002 wt.%. However, in contrast to the standard flux, the developed modifier also refines aluminium dendrites (Figure 5c) with the volume fraction exceeding 70% [33]. SDAS in alloy 1 was reduced by 29%. The macrograin is refined (Figure 6c). Such an effect on the structure is exerted by titanium transformed from the dioxide into the alloy and contained in the amount of 0.186% of the alloy weight.

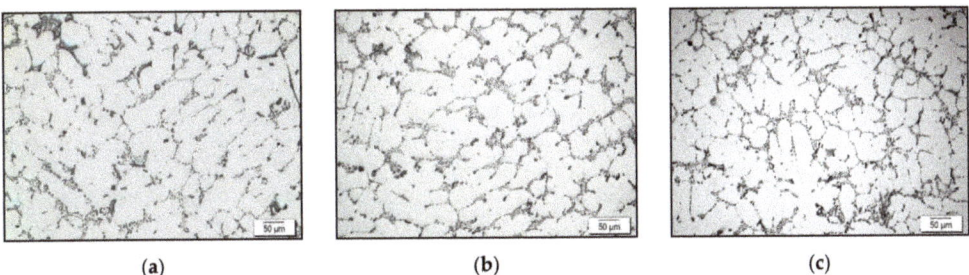

(a)   (b)   (c)

Figure 5. Microstructure of hypoeutectic alloy with copper, Alloy 1: (**a**) untreated; (**b**) standard flux; (**c**) complex flux.

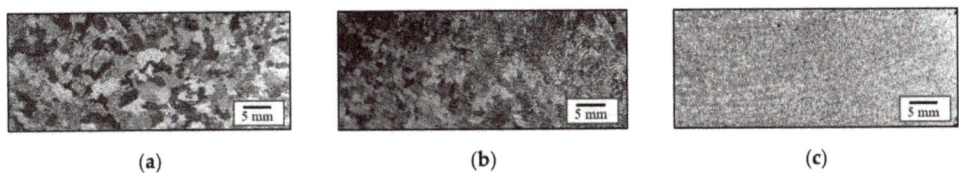

(a)   (b)   (c)

Figure 6. Macrostructure of hypoeutectic alloy with copper, Alloy 1: (**a**) untreated; (**b**) standard flux; (**c**) complex flux.

Al-17wt.% Si-Cu-Mg-Ni alloy (Alloy 5) is assigned to piston alloys and has high hardness, even at elevated temperatures. However, it has low plasticity due to the high volume fraction of excess phases. Modification can improve the mechanical properties of the alloy. The studies of the influence of the developed modifier on the mechanical properties of the hypereutectic alloy showed that the alloy treatment with a complex flux increases the alloy strength by 1.2 times (182 MPa), and the relative elongation by about 2 times (0.74%).

Figure 7a shows the structure of the hypereutectic alloy. The casting alloy structure has primary crystals of the silicon phase in the shape of polyhedrons. Aluminium–silicon eutectic has a coarse structure. Modification with a complex flux based on titanium dioxide reduces the size of primary silicon by 2 times to 46 µm, contributes to the refinement of the eutectic (Figure 7b), and refines macrograins (Figure 8b), while the standard flux does not affect the properties and structure of alloy 5.

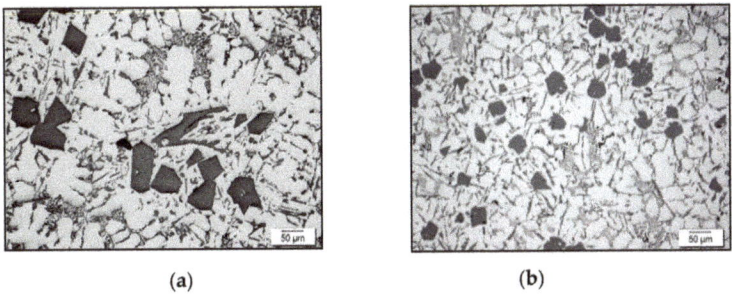

(a)   (b)

Figure 7. Microstructure of hypereutectic alloy 5: (**a**) untreated; (**b**) complex flux.

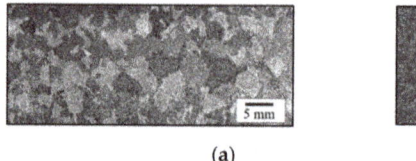
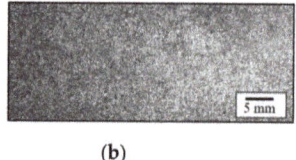

(a) (b)

Figure 8. Macrostructure of hypereutectic alloy 5: (**a**) untreated; (**b**) complex flux.

It should be noted that the eutectic phase's size is different among the unmodified alloys: 8–19 µm^2 for alloys with less silicon (alloys 1, 2 and 3) to 53 µm^2 for eutectic alloy 4, but such a difference is reduced after modification with flux. Unmodified alloy 4, having a lower content of alloying elements, has larger eutectic particle sizes. This behaviour is due to the large presence of alloying elements in alloys 1, 2 and 3 favouring nucleation of intermetallics into the eutectic area, thus limiting the growth of silicon [34].

The positive effect of the modifier on various structural components of Al–Si alloys is stipulated by the modifying impact of titanium and barium. Spectral analysis data (Figure 9) confirm that Ti and Ba are successfully transformed into the melt in the amount sufficient for modification and within permissible concentration (Figure 9). The concentration of modifying elements in alloys is provided in optimal amounts: Ti—0.1–0.2 wt% and Ba—0.010–0.020 wt%.

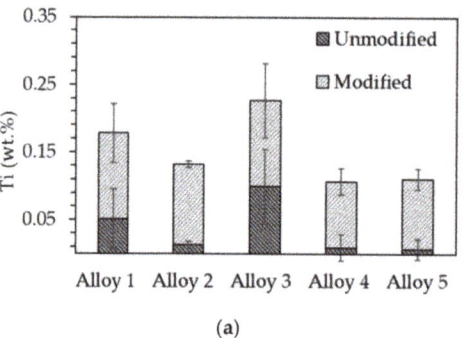
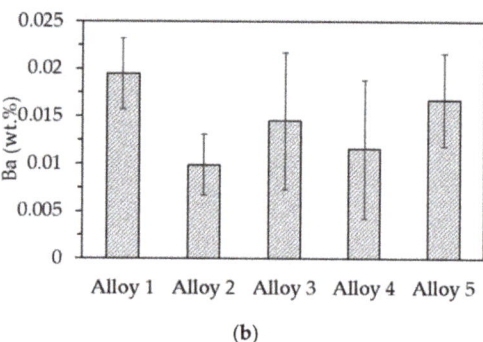

(a) (b)

Figure 9. Mass concentration of modifying elements in Al–Si alloys according to spectral analysis data: (**a**) Ti, (**b**) Ba.

The long-term retention of the modification effect is critical for the casting of aluminium alloys. The comparative study of the effect of melt holding up to 5 h on the structure and mechanical properties of Al-12 wt% Si binary alloy (alloy 4) treated with a complex flux was carried out.

The melt was held at a temperature of 720–740 °C. The results were compared with the time of action of the sodium salt flux (standard flux).

The results of mechanical tests are given in Figure 10. According to the data obtained, the time of action of the complex flux during casting is 5 h, while the standard flux retains its modifying effect only for 30 min.

The structure of the alloy held up to 2 h is completely modified and represents a eutectic with the refined and spheroidized silicon (Figure 11). The average area of eutectic silicon increases from 0.48 µm^2 for no exposure to 2.1 µm^2 for 5 h exposure. Nevertheless, after 5 h of exposure, silicon is significantly refined compared to an unmodified alloy. The extension of the holding time up to 5 h results in the growth of the particles of eutectic silicon, which is caused by the partial burn-off of the modifying elements (Figure 11).

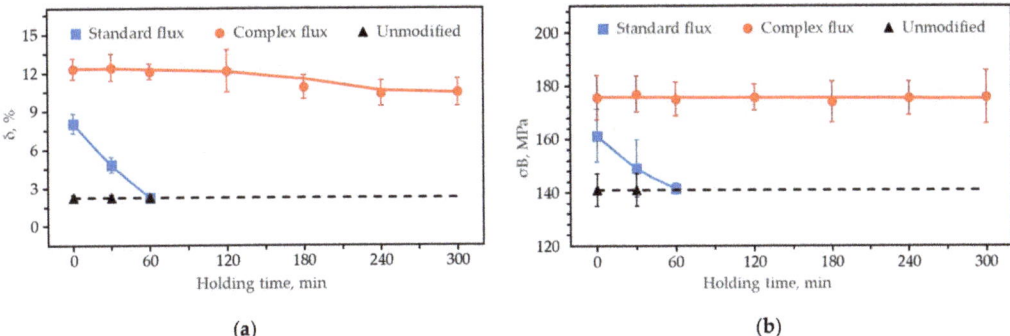

Figure 10. Influence of the holding time of the melt of the binary eutectic alloy Al-12% wt. Si (alloy 4) on the relative elongation (**a**), ultimate strength (**b**).

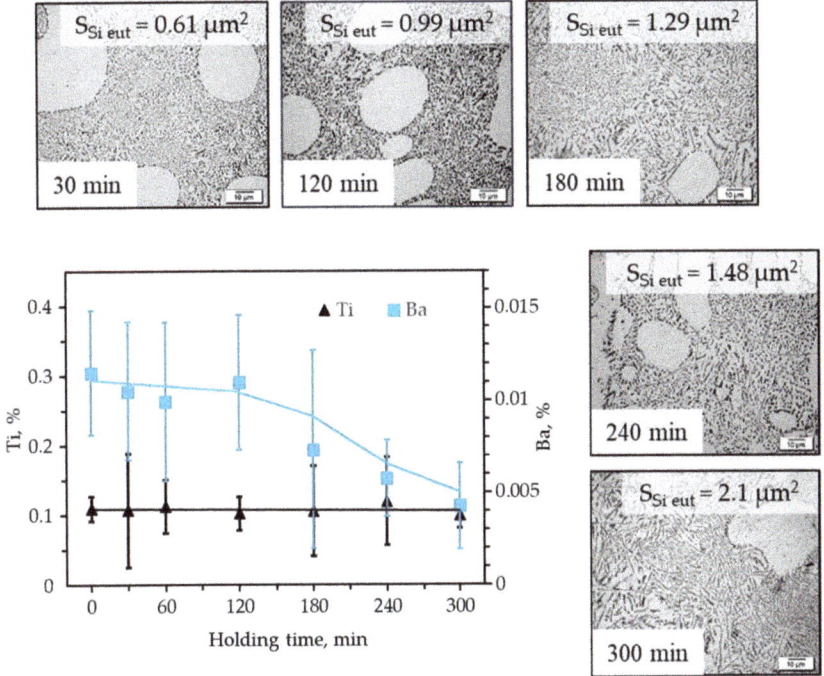

Figure 11. Microstructure and concentration (wt.%) of modifying elements in Al-12% wtSi alloy treated with a complex flux.

If the titanium concentration exceeds 0.2 wt%, rough needles of the TiAl$_3$ intermetallic phase are formed in the alloy, and the mechanical properties decrease [25]. The value of less than 0.05 wt.% Ti is insufficient to modify the grain and α-Al dendrites. The increased barium content may also worsen characteristics due to the formation and growth of the intermetallic phase based on barium, for example, Al$_2$Si$_2$Ba [17], and an insufficient amount may lead to incomplete modification of eutectic silicon.

4. Conclusions

The developed complex flux consists of available substances that are widely produced by the industry and has a significant effect on the mechanical properties and structure of hypoeutectic, eutectic, hypereutectic aluminium–silicon alloys.

1. Compared to the modifier that is widely used in the industry, modification with a complex flux increases the strength of hypoeutectic and eutectic silumins by 10–32% and plasticity by 24–54% (depending on the alloy). Moreover, the developed flux reduces the size of primary silicon, which results in the increase in the properties of the hypereutectic alloy Al-17 wt% Si-Cu-Mg-Ni.
2. The improvement in the mechanical properties of alloys results from the complex effect of the flux components on the structure. The microstructural analysis established that all the main structural components of silumins, i.e., α-Al dendrites, aluminium-silicon eutectic and primary silicon, are refined in the alloys modified with the complex flux. The SDAS and average area of eutectic silicon decreases in all the alloys as a consequence of addition of complex flux.
3. Another advantage of the complex flux compared to the standard sodium-containing flux is the long-term retention of the modifying effect.
4. The study results showed that the use of the complex modifier based on titanium dioxide is a promising direction for the improvement of the structure of aluminium casting alloys and increasing their mechanical properties.
5. Further studies are planned in relation to the combined effect of modifying elements of various types (Ti, Ba, K) on the crystallization parameters and the size, shape and composition of the structural components of Al–Si casting alloys.

Author Contributions: Supervision, A.P.R.; writing—original draft preparation, A.D.S. and I.A.P.; writing—review and editing, A.D.S., I.A.P. and V.V.T.; investigations, A.D.S., I.A.P., E.V.M. and A.P.R. All authors have read and agreed to the published version of the manuscript.

Funding: The research was carried out within the state assignment (No. FUER-2019-0004).

Institutional Review Board Statement: Not applicable.

Informed Consent Statement: Not applicable.

Data Availability Statement: Not applicable.

Conflicts of Interest: The authors declare no conflict of interest.

References

1. Davis, J.R. (Ed.) *Aluminium and Aluminium Alloys*; ASM International: Materials Park, OH, USA, 1993; p. 627.
2. Hegde, S.; Prabhu, K.N. Modification of eutectic silicon in Al–Si alloys. *J. Mater. Sci.* **2008**, *43*, 3009–3027. [CrossRef]
3. Mazahery, A.; Shabani, M.O. Modification mechanism and microstructural characteristics of eutectic Si in casting Al-Si alloys: A review on experimental and numerical studies. *JOM* **2014**, *66*, 726–738. [CrossRef]
4. Moniri, S.; Shahani, A.J. Chemical modification of degenerate eutectics: A review of recent advances and current issues. *J. Mater. Res.* **2019**, *34*, 20–34. [CrossRef]
5. Gursoy, O.; Timelli, G. Lanthanides: A focused review of eutectic modification in hypoeutectic Al–Si alloys. *J. Mater. Res. Technol.* **2020**, *9*, 8652–8665. [CrossRef]
6. Lu, L.; Nogita, K.; Dahle, A.K. Combining Sr and Na additions in hypoeutectic Al–Si foundry alloys. *Mater. Sci. Eng. A* **2005**, *399*, 244–253. [CrossRef]
7. Wang, J.; Liu, Y.; Zhu, J.; Liu, Y.; Su, X. Effect of Complex Modification on Microstructure and Mechanical Properties of Hypoeutectic Al–Si. *Metallogr. Microstruct. Anal.* **2019**, *8*, 833–839. [CrossRef]
8. Nikitin, K.V. *Modification and Complex Treatment of Silumins: A Textbook*, 2nd ed.; Samara State Technical University: Samara, Russia, 2016; 92p.
9. Shlyaptseva, A.D.; Petrov, I.A.; Ryakhovskii, A.P. Prospects of using titanium dioxide as a component of modifying composition for aluminium casting alloys. *Mater. Sci. Forum* **2019**, *946*, 636–643. [CrossRef]
10. Volochko, A.T. Modification of eutectic and primary silicon particles in silumins. Development Prospects. *Cast. Metall.* **2015**, *4*, 38–44.
11. Wu, D.Y.; Kang, J.; Feng, Z.H.; Su, R.; Liu, C.H.; Li, T.; Wang, L.S. Utilizing a novel modifier to realize multi-refinement and optimized heat treatment of A356 alloy. *J. Alloys Compd.* **2019**, *791*, 628–640. [CrossRef]
12. Zhao, H.L.; Bai, H.L.; Wang, J.; Guan, S.K. Preparation of Al–Ti–C–Sr master alloys and their refining efficiency on A356 alloy. *Mater. Character.* **2009**, *60*, 377–383. [CrossRef]
13. Qiu, C.; Miao, S.; Li, X.; Xia, X.; Ding, J.; Wang, Y.; Zhao, W. Synergistic effect of Sr and La on the microstructure and mechanical properties of A356.2 alloy. *Mater. Des.* **2017**, *114*, 563–571. [CrossRef]

14. Shlyaptseva, A.D.; Petrov, I.A.; Ryakhovsky, A.P.; Moiseev, V.S.; Bobryshev, B.L.; Azizov, T.N. Method for Modifying Aluminum-Silicon Alloys. Patent RU 2,743,945,C1, 1 March 2021.
15. Knuutinen, A.; Nogita, K.; McDonald, S.D.; Dahle, A.K. Modification of Al–Si alloys with Ba, Ca, Y and Yb. *J. Light Met.* **2001**, *1*, 229–240. [CrossRef]
16. Sha, X.; Chen, X.; Ning, H.; Xiao, L.; Yin, D.; Mao, L.; Zheng, J.; Zhou, H. Modification of Eutectic Si in Al-Si-(Ba) Alloy by Inducing a Novel 9R Structure in Twins. *Materials* **2018**, *11*, 1151. [CrossRef]
17. Rao, J.; Zhang, J.; Liu, R.; Zheng, J.; Yin, D. Modification of eutectic Si and the microstructure in an Al-7Si alloy with barium addition. *Mater. Sci. Eng. A* **2018**, *728*, 72–79. [CrossRef]
18. Zhang, X.H.; Su, G.C.; Ju, C.W.; Wang, W.C.; Yan, W.L. Effect of modification treatment on the microstructure and mechanical properties of Al–0.35%Mg–7.0%Si cast alloy. *Mater. Des.* **2010**, *31*, 4408–4413. [CrossRef]
19. Nogita, K.; Knuutinen, A.; McDonald, S.D.; Dahle, A.K. Mechanisms of eutectic solidification in Al-Si alloys modified with Ba, Ca, Y and Yb. *J. Light Met.* **2001**, *1*, 219–228. [CrossRef]
20. Samuel, E.; Golbahar, B.; Samuel, A.M.; Doty, H.W.; Valtierra, S.; Samuel, F.H. Effect of grain refiner on the tensile and impact properties of Al–Si–Mg cast alloys. *Mater. Des.* **2014**, *56*, 468–479. [CrossRef]
21. Lipinski, T. *Influence of Ti and Melt Number on Microstructure and Mechanical Properties of Al-Si Alloy on Agriculture Machine Parts*; Engineering for Rural Development: Jelgava, Latvia, 2018; pp. 1431–1436. [CrossRef]
22. Wang, S.; Liu, Y.; Peng, H.; Lu, X.; Wang, J.; Su, X. Microstructure and mechanical properties of Al-12.6Si eutectic alloy modified with Al-5Ti master alloy. *Adv. Eng. Mater.* **2017**, *19*, 1700495. [CrossRef]
23. Wu, Y.; Zhang, J.; Liao, H.; Li, G.; Wu, Y. Development of high performance near eutectic Al–Si–Mg alloy profile by micro alloying with Ti. *J. Alloys Compd.* **2016**, *660*, 141–147. [CrossRef]
24. Li, P.; Liu, S.; Zhang, L.; Liu, X. Grain refinement of A356 alloy by Al–Ti–B–C master alloy and its effect on mechanical properties. *Mater. Des.* **2013**, *47*, 522–528. [CrossRef]
25. Napalkov, V.I.; Makhov, S.V.; Popov, D.A. Production of additions for aluminum alloys. *Met. Sci. Heat Treat.* **2012**, *53*, 478–483. [CrossRef]
26. Makhov, S.V.; Kozlovskiy, G.A.; Moskvitin, V.I. The concepts of the process of aluminothermic obtaining of Al-Ti Master alloy from TiO2 dissolved in chloride-fluoride melt. *Tsvetnye Met.* **2015**, *11*, 34–38. [CrossRef]
27. Maeda, M.; Yahata, T.; Mitugi, K.; Ikeda, T. Aluminothermic Reduction of Titanium Oxide Materials Transactions. *JIM* **1993**, *34*, 599–603.
28. Shlyaptseva, A.D.; Petrov, I.A.; Ryakhovsky, A.P.; Moiseev, V.S. Investigation of aluminothermic reduction of titanium dioxide in the aluminium melt. *Found. Russ.* **2019**, *11*, 8–19.
29. Hatem, G. Calculation of phase diagrams for the binary systems BaF2-KF and KF-ZrF4 and the ternary system BaF2-KF-ZrF4. *Thermochim. Acta* **1995**, *260*, 17–28. [CrossRef]
30. Frolov, K.V. Blank production technology. In *Mechanical Engineering*; Masinostroenie: Moscow, Russia, 2004; Available online: https://bookree.org/reader?file=476656& (accessed on 15 October 2021).
31. William, F.; Gale Terry, C. (Eds.) Smithells Metals Reference Book, 8th ed.Totemeier Imprint: Oxford, UK, 2003; p. 2080.
32. Belov, N.A. *Phase Composition of Industrial and Promising Aluminium Alloys: Monograph*; MISiS Publishing House: Moscow, Russia, 2010; 511p.
33. Vandersluis, E.; Ravindran, C. Comparison of Measurement Methods for Secondary Dendrite Arm Spacing. *Metallogr. Microstruct. Anal.* **2017**, *6*, 89–94. [CrossRef]
34. Fracchia, E.; Gobber, F.S.; Rosso, M. Effect of Alloying Elements on the Sr Modification of Al-Si Cast Alloys. *Metals* **2021**, *11*, 342. [CrossRef]

Review

Recent Advances in the Grain Refinement Effects of Zr on Mg Alloys: A Review

Ming Sun [1], Depeng Yang [1], Yu Zhang [2,*], Lin Mao [3], Xikuo Li [1] and Song Pang [4]

1. School of Materials and Chemistry, University of Shanghai for Science and Technology, Shanghai 200093, China
2. College of Materials Science and Engineering, Chongqing University, Chongqing 400044, China
3. Shanghai Institute for Minimally Invasive Therapy, School of Health Science and Engineering, University of Shanghai for Science and Technology, Shanghai 200093, China
4. Shanghai Metal Materials Near-Net-Shape Engineering Research Center, Shanghai Spaceflight Precision Machinery Institute, Shanghai 201600, China
* Correspondence: yu.zhang@cqu.edu.cn

Abstract: As the lightest structural materials, Mg alloys show great effectiveness at energy saving and emission reduction when applied in the automotive and aerospace fields. In particular, Zr-bearing Mg alloys (non-Al containing) exhibit high strengths and elevated-temperature usage values. Zr is the most powerful grain refiner, and it provides fine grain sizes, uniformities in microstructural and mechanical properties and processing formability for Mg alloys. Due to the importance of Zr alloying, this review paper systematically summarizes the latest research progress in the grain refinement effects of Zr on Mg alloys. The main points are reviewed, including the alloying process of Zr, the grain refinement mechanism of Zr, factors affecting the grain refinement effects of Zr, and methods improving grain refinement efficiency of Zr. This paper provides a comprehensive understating of grain refinement effects of Zr on Mg alloys for the researchers and engineers.

Keywords: magnesium alloy; Mg-Zr master alloy; Zr refiner; grain refinement; casting

1. Introduction

Due to low densities and high specific strengths, Mg alloys show great potential in aerospace and automobile areas because the application of Mg alloys to structural components can effectively reduce weight, fuel consumption and CO_2 emissions [1,2]. However, the mechanical properties of Mg alloys are relatively low, and require further improvement by means of various strengthening methods. During the casting process of Mg alloys, grain refinement is one of the most important steps for improving castability, microstructure uniformity, mechanical properties, and post-formability [3,4]. In particular, the low ductility of Mg alloys due to the hexagonal close-packed (HCP) crystal structure can be effectively improved by grain refinement.

On the basis of differences in the grain refinement method employed, Mg alloys can be roughly divided into two groups, i.e., Al-bearing and Al-free Mg alloys. For the Al-bearing Mg alloys, there are still no ideal grain refiners [4]. Conversely, for Al-free Mg alloys, Zr has become the most effective grain refiner since it was found in the 1940s by Sauerwald [5]. Driven by the needs of aircraft components, Zr-containing Mg alloys with higher strengths and elevated temperature resistances have been developed [6,7] that take advantage of grain refinement, the precipitation strengthening effect, and other strengthening factors [8]. For example, WE43 alloy (Mg-4% Y-3% RE-0.5% Zr, RE = rare earth) shows high strength and long-term thermal stability at temperatures as high as 250 °C [6], and Mg-8Gd-4Y-0.8Zr alloy exhibits a better creep resistance than WE54 at an applied stress of 100 MPa and temperatures from 250 °C to 300 °C [7], which can be applied in military and aerospace areas [9]. Thus, it is believed that without the grain refiner Zr, the development of high-strength Mg alloys remains a challenge.

To date, there have been some high-impact review papers on the grain refinement of light alloys [1,3,4,10,11], on the basis of which some basic aspects of the effect of grain refinement with Zr on Mg alloys can be found. However, most of the information from the references is so fragmented that readers must browse through more papers to acquire sufficient knowledge. In other words, there is still a lack of a systematic review on the effect of grain refinement with Zr on Mg alloys. Considering the importance of Zr refinement, this paper comprehensively summarizes the recent advances in effect of grain refinement with Zr on Mg alloys. The principal text includes the alloying process of Zr, the mechanism of grain refinement with Zr, factors influencing the grain refinement behaviors of Zr, methods to improve grain refinement efficiency of Zr, and prospective research on Zr grain refinement. This review will provide researchers and engineers with a full understanding of Zr grain refinement in Mg alloys.

2. Alloying Process of Zr

2.1. The Methods of Zr addition

The direct addition of pure Zr metal into Mg melt is very difficult, because Zr has a much higher melting point (~1852 °C) than Mg (~650 °C) and a stable oxide film that normally presents on Zr particle surfaces [9,12]. Before the popular use of the Mg-Zr master alloy, many approaches were tried, such as alloying with pure Zr powder, ZrO_2, or Zr-rich salt mixtures [6,13–15]. Table 1 shows some examples of early patents claiming methods for alloying Zr in Mg alloy melt [16–19]. For example, Sauerwald et al. tried using the Zr powder as a source [16]. However, Zr powder is prone to ignite, and the oxide film inhibits the diffusion of Zr in the Mg melt. In addition, the direct addition of Zr-containing chloride or fluoride salt mixtures into Mg alloy melt has also been tried, which showed that the residuals $MgCl_2$ or MgF_2 are difficult to remove from the melt [6]. In a word, early methods were not satisfactory due to their complicated processes and the contamination of the melt. Recent research has attempted the traditional Zr-rich salts method again, through the addition of K_2ZrF_6-NaCl-KCl [20,21], $KZrF_5$-LiCl-KCl-CaF_2 [22] or $ZrCl_4$ [23] salts to Mg-RE alloys, further confirming that good grain refinement effects can be achieved. Nevertheless, the salt inclusions are really difficult to remove.

Table 1. Early patents of Zr alloying methods in Mg alloy.

Ref.	Composition or Route	Aim
[16]	Pure Zr powders + Mg melt at 700 °C → solidified → annealed at about 600 °C for 4~5 days → Mg-Zr master alloy	Preparing Mg-Zr master alloy
[17]	$ZrCl_4$ (15~60%) + KCl ($\geq$10%) + $BaCl_2$ ($\geq$30%)	Preparing Zr-rich salts mixture used in Mg melt
[18]	$ZrCl_4$ (5~25%) + ZrO_2 ($\leq$25%) + KF + $BaCl_2$. ZrO_2 as inspissation agent	
[19]	$ZrCl_4$ (33~45%) + KCl (33~45%) + $MnCl_2$ (12~33%) + $BaCl_2$ ($\leq$20%)	

In addition to the direct use of Zr-rich salt mixtures, earlier researchers have also tried to prepare a Mg-Zr master alloy. For example, Robert [24] tried three different routes to produce Mg-Zr master alloys: (1) Distillation of Mg out of low-Zr Mg-Zr alloy to obtain a high-Zr Mg-Zr alloy. Mg could either be sublimated at pressures of mercury below about 3 mm, or boiled at higher pressures. However, the alloys showed serious compositional segregation. (2) Fusing compacts of ZrH_2 powder and Mg shavings. The reason ZrH_2 was used is that ZrH_2 is not pyrophoric, unlike pure Zr powder. A compact of ZrH_2 powder and Mg shavings was formed using a hydraulic press. Then, the compact was heated to a temperature above 900 °C for a couple of hours in an electric vacuum furnace in order to remove the H_2 and form Mg-Zr alloy. However, several trials only demonstrated limited success. (3) Melting compacts of Zr and Mg shavings in a sealed steel "bomb". The steel bomb was heated in an electric muffle furnace at a temperature below 500 °C for a couple of hours. After being taken out of the furnace, it was shaken vigorously and then quenched in water. The results showed that the Mg-Zr alloy was of sufficient quality. However, due to their complex processes and low production capacity, these methods have been abandoned.

In terms of the development of a production method for Mg-Zr master alloys, the chemical reduction of Zr-rich fluoride or chloride salt with molten Mg has become a suitable route [14]. During the production process of the Mg-Zr master alloy, the Zr particles were formed in-situ within the Mg matrix, and thus the particle surfaces were not contaminated with O_2 or other impurities. This kind of Zr particle is clean, active, and quickly diluted in Mg melt, helping to easily achieve the saturation of Zr content [6]. Therefore, since about 1960, only Mg-Zr master alloy has been widely used as a satisfactory Zr refiner in Mg alloys. A famous commercial Mg-Zr product named Zirmax, developed by Magnesium Elektron in the UK, has become one of the popular products [6]. This product contains nominally more than 30% Zr in a relatively homogeneous distribution of Zr particles.

2.2. The Features of the Mg-Zr master Alloy

Generally, Mg-Zr master alloy contains a large number of Zr particles, the sizes of which range from the sub-micron level to hundreds of microns. The backscattered electron (BSE) scanning electron microscopy (SEM) images in Figure 1 show the typical microstructural features of different Mg-Zr products available on the market [25–27]. It can be seen that the Zr sizes exhibit a log-normal distribution. However, the number density of each size range has great differences, depending on the preparation parameters.

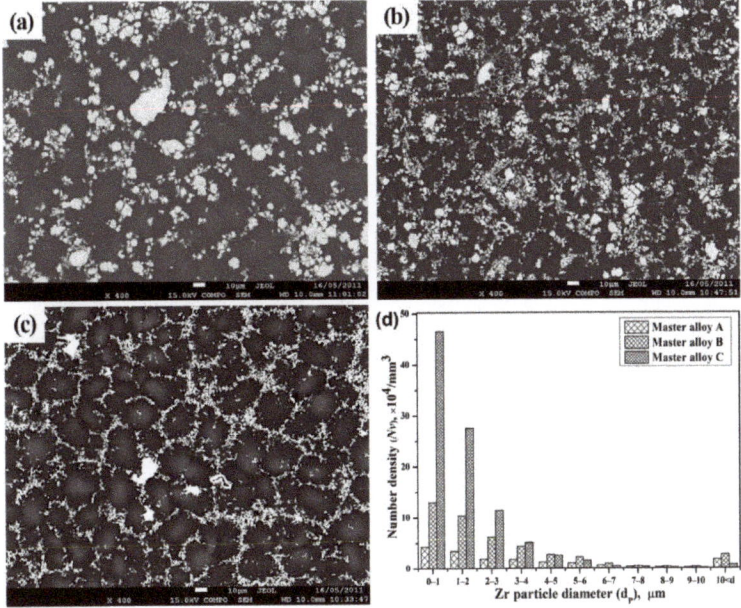

Figure 1. SEM backscattered electron (BSE) images of different Mg-Zr master alloys [25]: (**a**) Mg-30 wt.% Zr; (**b**) Zirmax® Mg-33% Zr, (**c**) AM-CAST® Mg-25 wt.% Zr; and (**d**) Zr particle size distributions [26]. Reprinted with permission from ref. [26]. Copyright 2013 Wiley.

To ensure the high absorptivity of Zr in Mg melt, some aspects should be considered, such as excessive addition of Zr, high alloying temperature, and adequate stirring. In particular, high temperature and adequate stirring accelerate the dissolution of Zr [28]. With the aid of stirring, the dissolution of Zr can be completed within a few minutes in the temperature range 730~780 °C [29]. However, if the temperature is above 780 °C, the difficulties in melt protection obviously increase, since Mg melt is prone to oxidation and burning, although protective gas (e.g., CO_2 + SF_6) can be used. Normally, in industrial practice, temperatures below 780 °C are regarded as suitable for Zr alloying.

2.3. Settling Behavior of Zr Particles during Alloying

The settling of Zr particles, forming sediment, is the most serious problem during Zr alloying process. This is due to their having a higher density (~6.52 g/cm³) than liquid Mg (~1.7 g/cm³), leading to the spontaneous settling of Zr particles. The bigger the Zr particle size is within the Mg-Zr master alloy, the faster the rate of settling will be. The settling of Zr reduces the absorption of Zr, causes waste of Zr, and forms a non-uniform microstructure from the top to the bottom of the ingot [3,30]. To compensate for the loss of Zr, the addition of Zr is always as high as 1~3 wt.% in practical operations, which is 2~4 times the nominal Zr composition of Mg alloys [29,31]. Therefore, the expensive Zr waste increases the cost of the Mg alloy.

Figure 2a,b show the experiment and simulation results, respectively, of Zr settling reported by Qian [30]. The settling behavior of Zr particles follows the Stokes law:

$$S \approx \frac{g(\rho_{Zr} - \rho_{Mg})d_P^2}{18\eta_{Mg}}t \qquad (1)$$

where S is settling distance, g is gravitational acceleration, ρ_{Zr} is the density of Zr, ρ_{Mg} is the density of Mg melt, η_{Mg} is the viscosity of Mg melt, d_P is the diameter of Zr particle, and t is the holding time. It can be seen that larger Zr particles settle more quickly than smaller ones; S increases with prolonging time; a higher melt temperature favors quicker settling due to the lower viscosity of Mg melt. These results can guide the industrial operations of the Zr alloying process in order to avoid severe loss of Zr. Generally, the melt should be cast after settling for 15~30 min, in order to prevent continuous settling and the fading of the grain refinement. Moreover, after setting for more than 60 min, the melt can be re-stirred to recover the settled Zr particles [29,32], which makes the casting process more complicated.

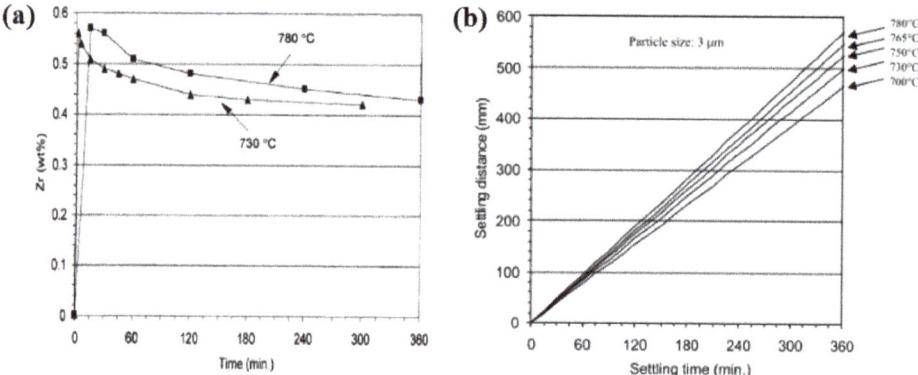

Figure 2. (a) Experimental results for settling; (b) predicted settling results of 3 μm Zr particles in pure Mg melt as a function of time and temperature [30]. Reprinted with permission from ref. [30]. Copyright 2001 Elsevier.

Recently, to avoid the settling problem and reduce costs, the in-situ Al_2RE nucleating particles were trialed to replace Zr [33–35]. The grain refinement effect of Al_2RE is comparable to that of Zr. However, the aging precipitation ability is reduced due to the consumption of RE element following the addition of Al, and the formation of needle-like Al_xRE_y phases is detrimental to the mechanical properties.

3. Grain Refinement Mechanisms of Zr

It is generally accepted that the grain refinement mechanism of Zr is dictated by both soluble and insoluble Zr. With the help of quantitative analysis of Zr_S and Zr_T, the grain refinement behaviors of Zr can be well estimated [36]. On the one hand, when the Zr content reaches the peritectic point, the peritectic reaction is believed to be the core mechanism of grain refinement [3,4,31,37,38]. On the other hand, a grain refinement effect can also be observed when the Zr content is far below the peritectic point, which is thought to be due to the constitutional supercooling (CS) effect generated by soluble Zr [3,4,31,37,38]. This section will briefly discuss these two aspects.

3.1. Nucleation Effect
3.1.1. Peritectic Reaction

According to Mg-Zr binary phase diagram, Zr and Mg do not form compounds, and there is a peritectic reaction at 653.5 °C. The peritectic reaction leads to Mg nucleation at the primary Zr-rich sites, which plays a very important role in the grain refinement process of Mg by Zr [10,13,39]. The peritectic point of Zr composition was previously found to be 0.6% Zr. However, re-assessment shows that it is substantially lower, at about 0.45% [4]. Thus, it is possible to use a smaller amount of Zr, thus reducing the cost of grain refinement [40].

Table 2 presents a comparison between the crystal structures of Mg and Zr, indicating that both α-Zr and α-Mg have HCP crystal structures, and their lattice parameters are quite similar. Thus, Zr has outstanding potency for acting as a nucleation substrate for Mg phase [41]. The perfect nucleation ability of Zr can be further verified by low undercooling ($\Delta T_n \approx 0.15$ °C [42]) equal to that for Al grains nucleating on TiB_2 substrates (~0.2 °C [43]).

Table 2. The crystal structure and lattice parameters of Mg and Zr [41].

Phase	Crystal Structure	Lattice Parameters
α-Mg	HCP	α = 0.320 nm, c = 0.520 nm
α-Zr	HCP	α = 0.323 nm, c = 0.514 nm

Figure 3 shows an example demonstrating that 0.5% Zr content can strongly refine the grains of Mg alloy, where the grain size of WE54 Mg alloy (Mg-5Y-4HRE-0.5Zr, HRE = heavy rare earth elements) was refined from 295 μm to 83 μm [44] under the conditions of pouring at about 780 °C into a steel mold preheated to 300 °C. This content of Zr is basically in accordance with the peritectic point in the Mg-Zr phase diagram. The simulation work performed by Zhao et al. [45] further showed that the growth velocity of dendrite tip of Mg-4Y alloy can be reduced to be one-sixth (1/6) using 0.5Zr, forming a fine grain size. However, this perspective still requires further evidence.

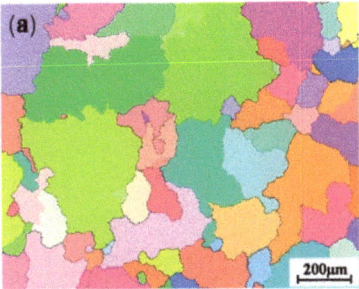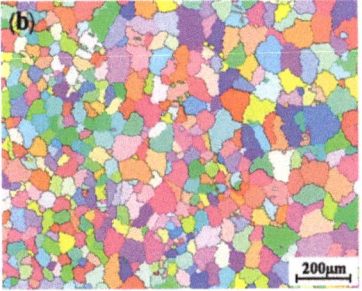

Figure 3. An example showing the powerful effect of grain refinement with Zr on Mg alloy. Electron backscattered diffraction technology (EBSD) image of: (**a**) Zr-free WE54 alloy; (**b**) Zr-containing WE54 alloy. Poured at 780 °C into a steel mold (300 °C). Reprinted with permission from ref. [44]. Copyright 2013 Elsevier.

In early work, it was thought that only the dissolved Zr was effective for grain refinement, because saturated dissolved Zr content meets the standards of a peritectic composition. However, later work verified that Mg grains can nucleate not only onto Zr particles precipitating from the melt during cooling, but also onto undissolved Zr particles through heterogeneous nucleation [4,31,38,46]. The products of peritectic solidification in the microstructure of Zr-containing Mg alloy are Zr-rich halos or cores, present in most Mg grains, as shown in Figure 4 [26,37]. The Zr-rich halo shows either dendritic or nearly spherical growth, depending on the soluble Zr content. A higher level of soluble Zr (close to the solubility) usually leads to spherical halo, while a lower level of soluble Zr leads to a dendritic halo [38].

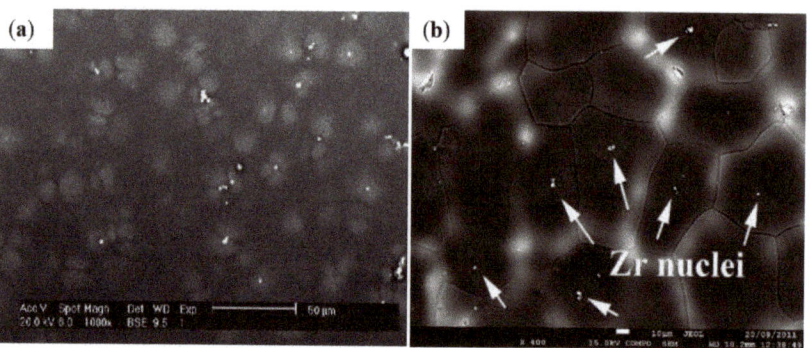

Figure 4. SEM BSE image of Zr-rich cores in: (**a**) Mg-0.56Zr alloy [37]; Reprinted with permission from ref. [37]. Copyright 2002 Elsevier. (**b**) Mg-5Gd-1.5Y-0.55Zr alloy [26]; Reprinted with permission from ref. [44]. Copyright 2013 Wiley.

3.1.2. HRTEM Observations of Zr Nucleus

To facilitate the understanding of the grain refinement mechanism, the orientation relationships (ORs) between Mg grains and Zr nuclei were observed via high-resolution transmission electron microscopy (HRTEM). Table 3 summarizes some of the reported coherent or semi-coherent ORs in grain-refined Mg alloys [23,47–51]. All the ORs have low misfits, which are responsible for the good nucleating potency of Zr. For example, the misfit of OR $[\bar{1}2\bar{1}3]_{Mg} \parallel [\bar{1}2\bar{1}3]_{Zr} + (1\bar{2}12)_{Mg} \overset{\wedge}{1°} (1\bar{2}12)_{Zr}$, OR $[2\bar{1}\bar{1}0]_{Mg} \parallel [2\bar{1}\bar{1}0]_{Zr} + (0001)_{Mg} \parallel (0001)_{Zr}$ and OR $[01\bar{1}1]_{Mg} \parallel [01\bar{1}1]_{Zr} + (\bar{1}011)_{Mg} \parallel (\bar{1}011)_{Zr}$ is only 0.13% [47], 0.9% [49] and 0.41% [50], respectively. Figure 5 shows a typical example of OR $[2\bar{1}\bar{1}0]_{Mg} \parallel [2\bar{1}\bar{1}0]_{Zr} + (01\bar{1}1)_{Mg} \parallel (01\bar{1}1)_{Zr}$ observed in the sand-cast Mg-8Gd-3Y-0.82Zr alloy [23].

Table 3. Reported ORs between Mg and Zr in grain-refined Mg alloys [1].

ORs	Alloys	Processing	Ref.
$[2\bar{1}\bar{1}0]_{Mg} \parallel [2\bar{1}\bar{1}0]_{Zr} + (01\bar{1}\bar{1})_{Mg} \parallel (01\bar{1}\bar{1})_{Zr}$	Mg-7.43Gd-2.74Y-0.82Zr	Sand-cast	[23]
$[\bar{1}2\bar{1}3]_{Mg} \parallel [\bar{1}2\bar{1}3]_{Zr} + (11\bar{2}2)_{Mg} \overset{\wedge}{1°} (11\bar{2}2)_{Zr}$	Mg-0.5Zr	Gravity cast	[47]
$[\bar{1}2\bar{1}3]_{Mg} \parallel [\bar{1}2\bar{1}3]_{Zr} + (10\bar{1}0)_{Mg} \parallel (10\bar{1}0)_{Zr}$	Mg-1.0Zr	Gravity cast	[47]
$[2\bar{1}\bar{1}0]_{Mg} \parallel [2\bar{1}\bar{1}0]_{Zr} + (0001)_{Mg} \parallel (0001)_{Zr}$	Mg-0.5Zr	Gravity cast	[48]
	Mg-1.0Zr	IMS + Gravity cast	[49]
$[01\bar{1}1]_{Mg} \parallel [01\bar{1}1]_{Zr} + (\bar{1}011)_{Mg} \parallel (\bar{1}011)_{Zr}$	Mg-1.0Zr	HPDC	[50]
$[2\bar{4}2\bar{3}]_{Mg} \parallel [2\bar{4}2\bar{3}]_{Zr} + (10\bar{1}0)_{Mg} \parallel (10\bar{1}0)_{Zr}$	Mg-0.1Zr	HPDC	[51]
	Mg-0.52Zr	Gravity cast	[27]

[1] In Table 3, ∧—tilted directions; ∥—parallel directions; IMS—intensive melt shearing; HPDC—high-pressure die casting.

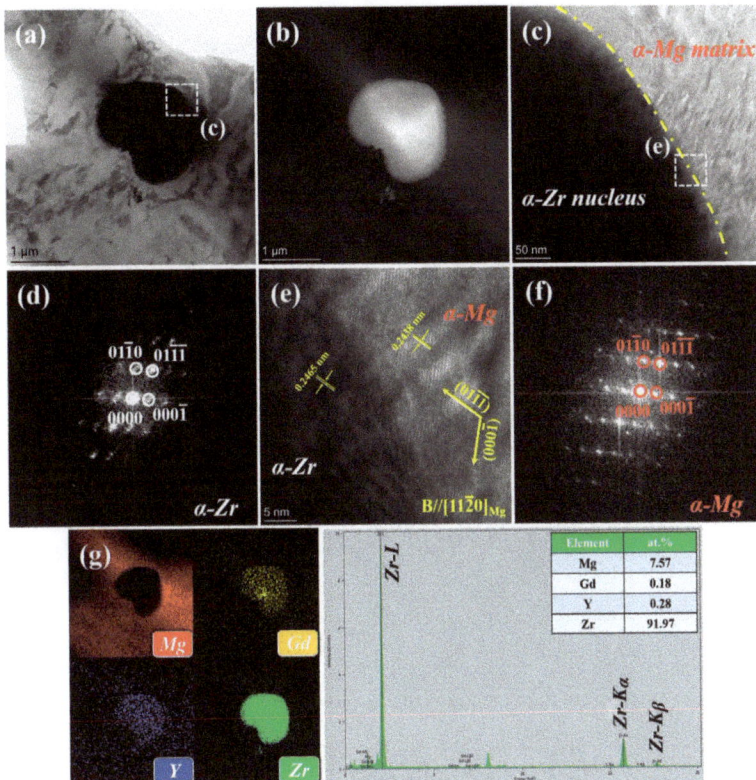

Figure 5. TEM showing OR between Mg matrix and Zr nucleus in Mg-8Gd-3Y-0.82Zr alloy: (**a**) bright-field image; (**b**) dark-field image; (**c**) enlarged view of the selected area in (**a**); (**e**) HRTEM image of the interface between Mg matrix and Zr nucleus (beam $\parallel [2\bar{1}\bar{1}0]_{Zr}$); (**d,f**) fast Fourier transform (FFT) spectrum of α-Zr and α-Mg, respectively; (**g**) EDS analysis of Zr nucleus. Reprinted with permission from ref. [23]. Copyright 2021 Elsevier.

With the analysis of HRTEM and the selected area diffraction pattern (SADP), Saha concluded that Mg grains only nucleate on the "faceted" crystal planes of Zr particles [47,48], e.g., the basal planes {0001} or the prismatic planes {10$\bar{1}$0}. However, contradicting Saha, Peng et al. [49] found that the interface between the Zr nucleus and the Mg matrix was "curved" rather than "faceted". The curved interface, in combination with coherent ORs, provides perfect lattice matching along various directions and across various planes, benefitting the nucleation of Mg grains on the Zr particle surface.

Although various ORs have been reported, as shown in Table 3, Yang et al. [51] suggested that very strict ORs are not required for Zr nuclei, because Zr particles may be wetted by Mg melt on all exposed crystal planes. This viewpoint was validated by HRTEM observations in Ref. [49], as shown in Figure 6 [51]. Perfect coherent ORs can be observed, among which even the slightly higher index OR $[\bar{2}4\bar{2}3]_{Mg} \parallel [\bar{2}4\bar{2}3]_{Zr}$ + $(10\bar{1}0)_{Mg} \parallel (10\bar{1}0)_{Zr}$ is present, which is similar to that described in Ref. [27]. Thus, α-Mg grains can grow epitaxially on any suitable planes of Zr nuclei. Recently, in-situ neutron diffraction observations have provided further evidence of the grain refinement mechanism of Zr, showing that with the addition of Zr, all of the diffraction intensities of the $(10\bar{1}0)$, (0002) and $(10\bar{1}1)$ planes of Mg-5Zn-0.7Zr alloy increase at similar rates during the early stages of solidification, leading to the formation of a uniform grain structure [52].

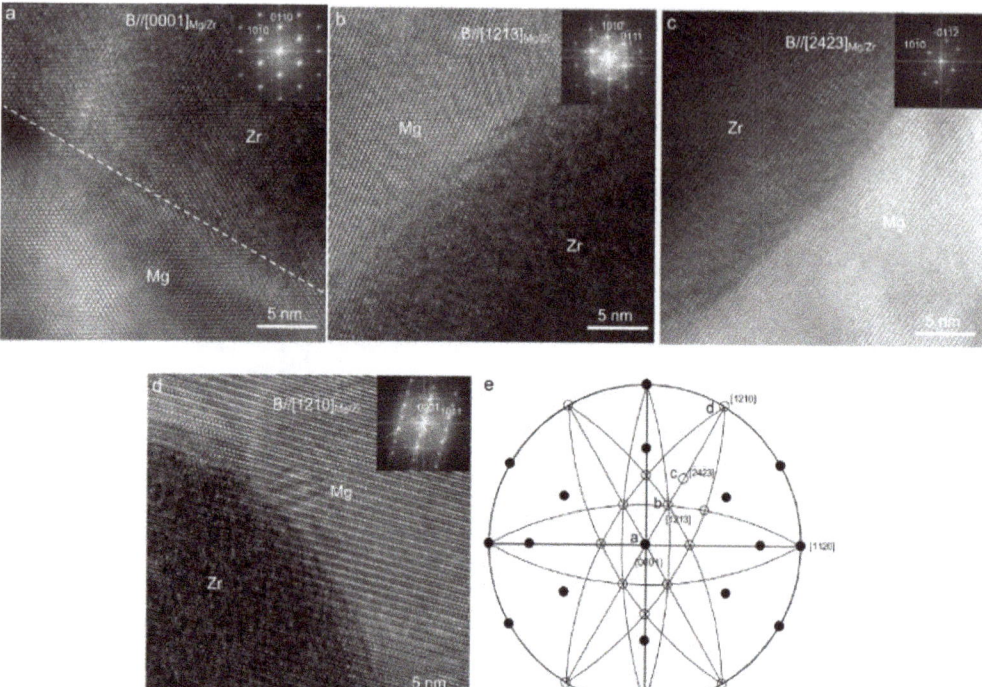

Figure 6. HRTEM images showing: (**a**) faceted interface; (**b**–**d**) curved interfaces between α-Mg grains and Zr nuclei in Mg-0.1Zr alloy along four different low-index zone axes. (**e**) Stereographic projection of *HCP* structure showing the various ORs (**a**–**d**) within a 90° range [51]. Reprinted with permission from ref. [51]. Copyright 2015 Elsevier.

3.2. Constitutional Supercooling (CS) Effect

Zr can affect both the nucleation and the growth of the dendritic phase in Mg alloy depending on its status in the melt [53]. When the addition of Zr content is higher, grain refinement can be achieved due to the large number of Zr nucleating particles. When the addition of Zr is low, an obvious grain refinement effect can still be observed. This is mainly due to the strong CS effect of solute [4,10,26,43,54–56], since Zr is also a solute element in the Mg-Zr system. The growth restriction factor (*GRF*, *Q*) is defined as follows to reflect the CS effect [4,43,54]:

$$Q = m(k-1) \qquad (2)$$

where m is the liquidus gradient and k is the partition coefficient. The larger the Q value is (unit K or °C), the stronger the grain refinement effect will be. On the basis of phase diagram analysis, the Q value of Zr was calculated to be 38.29, which is much higher than that of most of the alloying elements, such as Al (4.32), Y (1.70), Zn (5.31), etc. [4,11], indicating that soluble Zr produces a strong grain refinement effect [1,4]. Peng et al. [57] proved that the grain refinement of Mg-9Gd-3Y-0.25Zr alloy (low Zr) was mainly contributed by the *GRF* effect, while the grain refinement of Mg-9Gd-3Y-0.51Zr alloy (high Zr) was mainly contributed by the heterogeneous nucleation of Zr particles. The CS effect of Zr was recently verified by Zhang et al. [57]. As shown in Figure 7, remarkable grain refinement (from 74 μm to 3.5 μm) occurred in the melt pool of the laser-surface-remelted (LSR) Mg-3Nd-1Gd-0.5Zr alloy, which was mostly due to the high CS effect of soluble Zr caused by LSR.

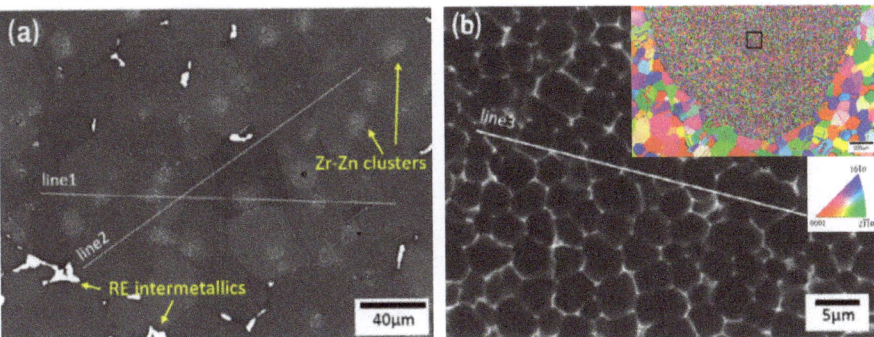

Figure 7. SEM BSE images of the laser-remelted Mg-3Nd-1Gd-0.5Zr alloy showing the microstructure of: (**a**) the substrate; (**b**) the melt pool, the cross-section EBSD analysis of which is included as an inset in the upper-right corner [58]. Reprinted with permission from ref. [58]. Copyright 2020 Elsevier.

In summary, the nucleation effect combined with the CS effect leads to a high nucleation rate and low undercooling, resulting in the powerful effect of grain refinement with Zr. Such effects are not only of benefit to achieving equiaxed grains during the casting process, they also facilitate the formation of slurries with fine and spherical primary particles during the semi-solid forming process [59,60].

4. Factors Influencing the Grain Refinement Behaviors of Zr

This section briefly summarizes the effects that influence the Zr alloying efficiency during the melting process.

4.1. Effect of Zr Size Distribution in Mg-Zr Master Alloy

In the microstructure of Mg alloy refined by Zr, Qian et al. [30,61] observed that the majority of Zr nuclei had sizes in the range of 1~5 μm. Sun et al. [25–27] further systematically compared the size distribution of different Mg-Zr master alloys (Figure 1) and the subsequent grain refinement efficiency on Mg-Gd-Y alloys (Figure 8). It was found that Mg-Zr master aloys with more Zr particles within the size range 1~5 μm (i.e., master alloy C) exhibited better grain refinement efficiency. This is because both the saturation of soluble Zr content and a greater number of Zr nucleation sites could be achieved when the Mg-Zr master alloy had a finer Zr particle distribution.

4.2. Effect of Cooling Condition

Generally, increasing the cooling rate ($\dot{T}$) results in a decrease in grain size. For example, Liu et al. [62] showed that the grain size of Mg-Zr binary alloys (0.3, 0.7 or 1.2Zr) decreased with increasing cooling rate from sand mold to Cu mold. Sun et al. [63] found that the relationship between the grain size of Mg-Y-Zr alloy (1~9Y, 0.1~0.5Zr) and $1/\sqrt{Q\dot{T}}$ could be linearly fitted at a cooling rate ≤80 °C/s (Figure 9 shows Mg-1Y-Zr alloy as an example of this phenomenon). Yang et al. [51] showed that, compared with traditional gravity casting, HPDC improved the percentage of activated Zr nuclei from 1~2% to 48.78% in the Mg-0.1Zr alloy, and resulting in finer grains.

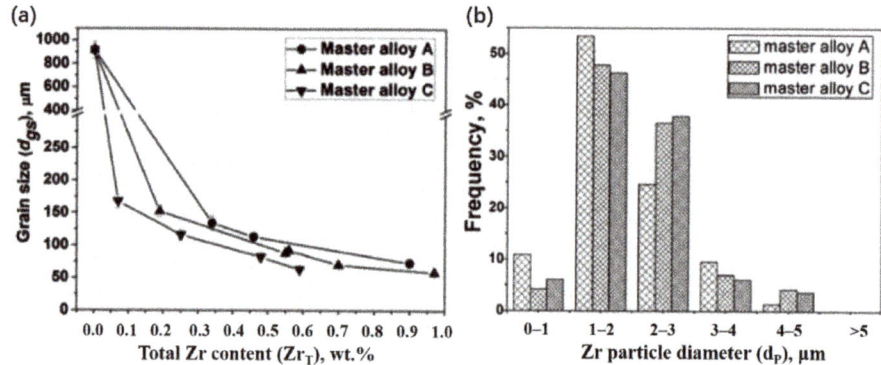

Figure 8. (**a**) The relationship between grain size (d_{gs}) and total Zr content (Zr_T) for Mg-5Gd-1.5Y alloy refined using different Mg-Zr master alloys. (**b**) The distribution of Zr nuclei in Mg-5Gd-1.5Y alloy with Zr_T contents of 0.46, 0.55 and 0.59, inoculated by master alloys A, B and C, respectively [26]. The microstructure and Zr particle size distributions of the Mg-Zr master alloys are shown in Figure 1. Reprinted with permission from ref. [26]. Copyright 2013 Wiley.

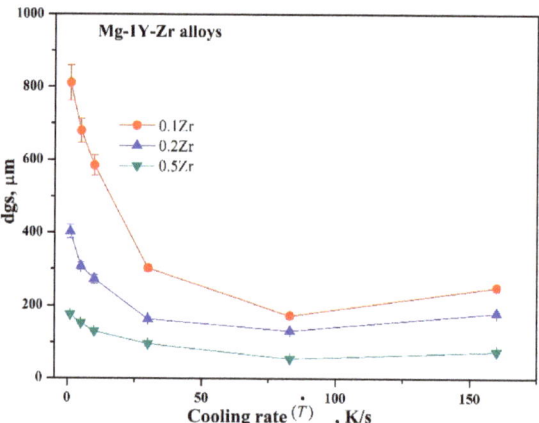

Figure 9. Effect of cooling rate ($\dot{T}$) on the grain size (d_{gs}) of Mg-1Y-Zr alloy [63]. Reprinted with permission from ref. [63]. Copyright 2020 Springer.

Nevertheless, grain coarsening has also been observed under some cases of high cooling rate and steep thermal gradient [58,63]. According to Figure 9, by Sun et al. [63], grain coarsening of the Mg-Y-Zr alloy (1~9Y, 0~0.5Zr) occurred at a cooling rate 160 °C/s (with rapid water-quenching). This was due to the reduction in size of the CS zone, decreasing the likelihood of further nucleation. Yang et al. [50] showed that the grain coarsening effect occurred in HPDC Mg-1Zr alloy, in contrast to HPDC pure Mg. The grain size increased abnormally from 6.7 to 18.9 μm following the addition of 1% Zr. This change was attributed to the competition between the native MgO and the Zr particles. Without the addition of Zr, the high number density of the in-situ MgO particles permitted them to act as nucleation sites, due to a low misfit with Mg (5.46%), leading to a finer grain size. However, with the addition of Zr, the Zr particles acted as nucleation sites first, thereby suppressing the activation of the in-situ MgO particles. More importantly, the number density of the Zr particles was lower than that of the in-situ MgO particles. Therefore, the grain size of Mg-1Zr was coarser than that of pure Mg.

4.3. Effect of Alloying Elements

Contrary effects can be found depending on the elements employed. On the one hand, the elements Al [64,65], Mn, Si, Fe [29,66], Sn, Ni, Co and Sb [29] inhibit the grain refinement ability of Zr, which is referred to as the poisoning effect. This is mainly due to the reactions with Zr forming stable compounds that can no longer act as a nucleant [4,13,29]. Beryllium (Be) also poisons Zr, but via a different mechanism, which is probably the formation of a new coating on the Zr particle surface, thus reducing the potency of the Zr [67]. On the other hand, the elements Ca [68] and Zn [69–71] enhance the grain refinement ability of Zr, mainly via the mechanism increasing the soluble Zr content. This section summarizes the effects of these elements.

4.3.1. Al

Al poisons the effect of grain refinement with Zr, which is the main reason for which Mg-Al alloy cannot be grain-refined using Zr. Tamura et al. [72] found that even the trace addition of 0.015% Zr could severely coarsen the grain size of high-purity Mg-9Al alloy; Kabirian et al. [64] found that the dendrites of AZ91 alloy were greatly coarsened by the addition of 0.2~1.0% Zr; Balasubramani et al. [65] found the grain size of Mg-1Zr was sharply coarsened from 114 ± 38 μm to 1550 ± 38 μm with the addition of 0.2% Al (Figure 10). Two reasons have been proposed. One is that Zr poisons the surface of the native Al-carbide nuclei (such as Al_4C_3 [73]) that are originally present in high-purity Mg-9Al alloy. Another is that the consumption of solute Al by Zr (or solute Zr by Al) reduces the extent of the CS effect. The Al-Zr compounds were identified to be Al_2Zr [64], Al_3Zr_2 [64] or Al_3Zr [65]. Despite the negative effect on grain refinement, Al-Zr compounds can increase the thermal stabilities of grain during heat treatment, refine grains during hot deformation [74], and increase creep resistance [75].

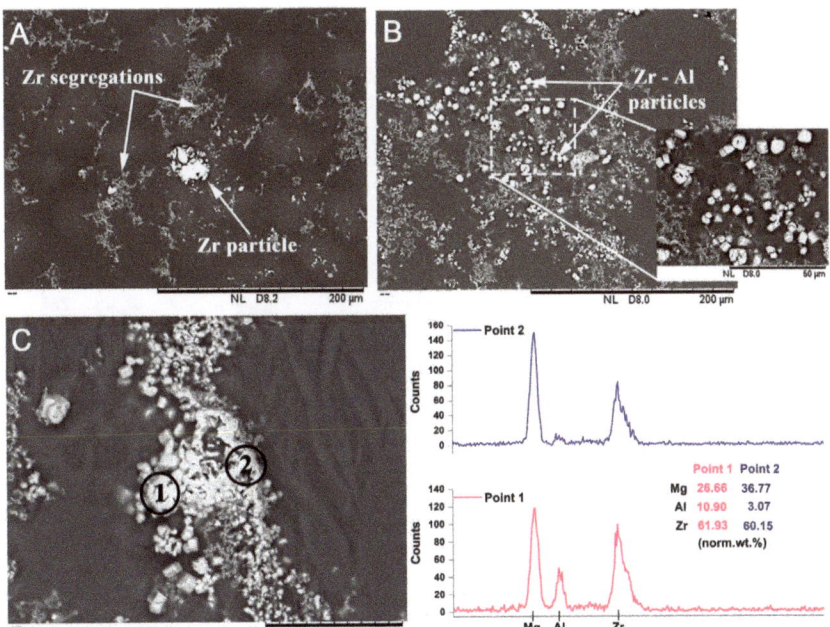

Figure 10. SEM BSE images showing: (**A**) blocky Zr particles and fine Zr particles in Mg-1Zr alloy; (**B**) faceted Zr_xAl_y particles in Mg-1Zr-Al-Be alloy; (**C**) Zr clusters surrounded by fine precipitates of Zr_xAl_y particles in Mg-1Zr-Al-Be alloy, EDS testing point 1 is rich in Al and Zr, whereas point 2 is rich in Zr. The samples were cut from the bottom of an ingot to observe the clusters [65]. Reprinted with permission from ref. [65]. Copyright 2004 Elsevier.

4.3.2. Fe

Fe is one of the most common impurities in Mg alloys, and may result from the raw materials, fluxes, crucible, and stirring tools used. Fe is detrimental to the effect of grain refinement with Zr. This can be understood from two perspectives.

On the one hand, Fe can react with Zr in the Mg melt, forming the compound Fe_2Zr, as verified by Cao et al. [66]. As a result, the Zr content is rapidly consumed, and the grain refinement effect is remarkably reduced. This effect has also been observed during the remelting process of pre-alloyed or commercial ingots of Zr-bearing Mg alloys when using an Fe crucible. An obvious loss of dissolved Zr takes place due to the diffusion of Fe through the Mg melt, finally leading to the addition of extra Zr in order to achieve the required level of grain refinement [76].

On the other hand, trace Fe and Si impurities on tiny Zr particles may alter the surface chemistry of the Zr particles, upsetting the coherent match between the Zr and Mg lattices, as reported Davis et al. [77]. Therefore, the nucleating potency of the Zr particles is decreased, reducing the grain refinement effect.

On the basis of these analyses, it can be seen that when an Fe crucible and Fe stirring paddle are used during melting of Mg alloys, a protective coating (e.g., BN coating [66]) should be pre-coated to inhibit the dissolution of Fe into the Mg melt. However, it has been reported that, by taking advantage of the effective removal of Fe by Zr, high-purity Mg alloys containing less than 1 ppm Fe can be produced [78].

4.3.3. Be

Cao et al. [67] investigated the effect of trace addition of Be on the grain refinement of Mg-0.5Zr alloy, as shown in Figure 11, where a significant grain coarsening effect can be observed. However, since Be was added in the form of Al-5Be master alloy, the influence of the consumption of Zr by Al cannot be neglected. Thus, the alloys were designed in two groups, i.e., a low-Zr-content group (ZM1~ZM4) and a high-Zr-content group (ZM5). It can be seen that with increasing addition Be from 10 to 100 ppm, the grain size of ZM1~ZM4 increased gradually, which was mainly because of the continuous consumption of Zr by Al. Nevertheless, with the further addition of 0.5% Zr (ZM5), the grain size continued to coarsen, rather than recovering. Therefore, a poisoning effect of Be was deduced, whereby Be, as a surface-active 'surfactant', readily coats the surfaces of heterogeneous nucleant particles such as Zr particles or native particles (Fe-rich, carbon-rich nuclei) destroying the nucleating potency.

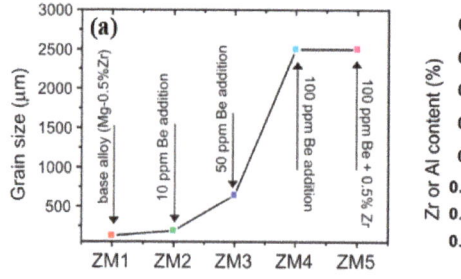 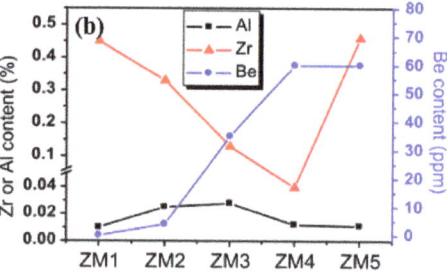

Figure 11. (a) Grain size as a function of Be addition. (b) Be, Al and Zr contents in alloys. ZM1: base alloy Mg-0.5% Zr; ZM2: 10 ppm Be addition; ZM3: 50 ppm Be addition; ZM4: 100 ppm Be addition; ZM5: 100 ppm Be + 0.5% Zr. [67]. Reprinted with permission from ref. [67]. Copyright 2004 Elsevier.

4.3.4. Ca

Chang et al. [68] showed that the grain size of squeeze-cast Mg-xCa-0.6Zr (x = 0.2~1.2) alloy was finer than that of squeeze-cast Mg-0.6Zr alloy. They concluded that Ca was able to increase the effect of grain refinement with Zr. The reason for this was explained from a thermodynamics perspective. Before the addition of Ca, the Mg melt reacts preferentially with O_2, forming MgO inclusions that envelope the Zr particles. This effect increases the difficulty of the dissolution of Zr. After the addition of Ca, the formation of MgO is suppressed due to CaO possessing a lower standard free energy than MgO. Accordingly, the dissolution of Zr particles becomes more efficient, and the interface wettability between Zr particles and the Mg matrix improved. Therefore, the combined addition of Zr and Ca shows a better grain refinement effect.

However, the effect of CS was neglected in [68]. According to StJohn et al. [4], the GRF value for Ca is 11.94, which is the third highest among all solute elements in Mg alloy (the first is Fe, at 52.56, and the second, Zr, at 38.29). Thus, the high *GRF* of Ca may be another reason for the increase in the grain refinement effect of Mg-Ca-Zr alloy.

4.3.5. Zn

Zn has been reported to show contradictory influences on the effect of grain refinement with Zr in Mg-Zn-Zr alloy, depending on the Zn content. According to Hildebrand et al. [69] and Li et al. [70], there exists a critical Zn content (3~4%) below which the grain size is gradually refined, but beyond which the grain size coarsened. The reasons for this have been well explained. When Zn content does not exceed 3~4%, the soluble Zr content increases with increasing Zn content, resulting in a finer grain size [69]. When Zn content exceeds 4%, Zn and Zr form the stable compounds $ZnZr_2$ or Zn_2Zr_3 [79], decreasing the soluble Zr content and resulting in a coarser grain size. However, in partial divergence from Hildebrand et al. [69], Li et al. [70] attributed the grain coarsening effect at higher Zn contents to the "dendrite coherency theory". A large constitutional undercooling caused by higher Zn content leads to instability of the *S-L* interface. The sharp tip of the dendrite increases the growth rate of the dendritic grains, thus leading to coarser grains.

In addition, higher Zn contents have been reported to relieve the grain poisoning effect of Zr on AZ91 alloy. Jafari et al. [71] showed that the grain size of AZ93 alloy was not coarsened when 0.25~0.9% Zr was added, which was ascribed to the higher content of Zn (3%). Compared with lower Zn content (1%), higher Zr content can improve the Zr solubility in Mg alloy, thus improving the grain refinement effect.

5. Methods of Improving the Grain Refinement Efficiency of Zr

As is well known, Mg-Zr master alloy with the largest number density of fine Zr particles (1~5 μm) exhibits the best refinement ability [26,31,61]. Therefore, the Zr particle size distribution in the Mg-Zr master alloy needs to be controlled to be as fine and as uniform as possible. To achieve this aim, a new Mg-Zr master alloy with fine Zr particles [14,80] has been developed, and pre-treatments [27,81–87] have been conducted to modify the Mg-Zr master alloys already available. In addition, melt treatment can be conducted to reduce the settling of Zr during the melting of the Mg alloy, thereby improving the Zr grain refinement efficiency [41,49,88–94]. This section briefly summarizes these methods.

5.1. Pretreatments of the Mg-Zr Master Alloy

The severe plastic deformation (SPD) method can generate massive force, breaking the Zr particles into smaller ones, and can been conducted via hot rolling [27,83], equal channel angular extrusion (ECAE) [47,82,83], friction stirring processing (FSP) [84], or crush [85].

Qian [81] first showed that hot rolling effectively fragmented the Zr particle clusters in the Mg-33Zr master alloy into smaller clusters. Sun [27] further verified that eight passes hot rolling effectively crushed the large Zr particles of the Mg-30Zr master alloy into smaller ones. Saha [82] showed that ECAE was able to effectively break up the Zr particles of the Mg-15Zr master alloy. Wang et al. [84] showed that FSP can be used to modify the Zr particle distribution of the Mg-30Zr master alloy. Recently, Wang et al. [85] employed a new method for preparing "powder Mg-Zr master alloy" that consisted of three steps, i.e., mechanically crushing the original Mg-30Zr master alloy block, ball-milling the pieces, and finally sieving to keep particles smaller than 20 μm.

These methods mainly take advantage of mechanical force to refine the Zr particle size. All of these studies proved that the Zr particle size distribution can be modified to become narrow, achieving better distribution. As a result of modification, the settling of Zr particles during melting can be retarded, and a higher soluble Zr content can be obtained. Consequently, the grain refinement efficiency of pretreated Mg-Zr master alloys on pure Mg [81–83], Mg-10Gd-3Y [27], Mg-3Nd-0.2Zn [84] and Mg-14Li-Zn [85] alloys was effectively improved. However, the pretreatment methods have some obvious disadvantages, such as their complicated processes, low productivity, and the SPD molds and tools (FSP probes) being easily broken.

In addition to the SPD methods, a non-equilibrium re-precipitation method via tungsten inert gas arc remelting with ultra-high frequency pulses (UHFP-TIGR) was recently trialed by Xin et al. [86,87]. Figure 12 shows the Zr particle distribution in the UHFP-TIGR-treated Mg-30Zr master alloy. UHFP-TIGR treatment promoted the precipitation of a considerable number of nano-sized Zr particles (Zr_{np}) from the supersaturated Mg matrix due to the effect of high temperature, strong stirring, and rapid equilibrium cooling. As a result, the UHFP-TIGR-treated Mg-30Zr master alloy exhibited superior grain refinement efficiency on Mg-9Gd-3Y alloy compared to its untreated counterpart. A similar method was also reported by Zhang et al. [95], who improved the refining efficiency of Al-5Ti-1B master alloy on Al alloys by re-precipitating tiny TiB_2 particles.

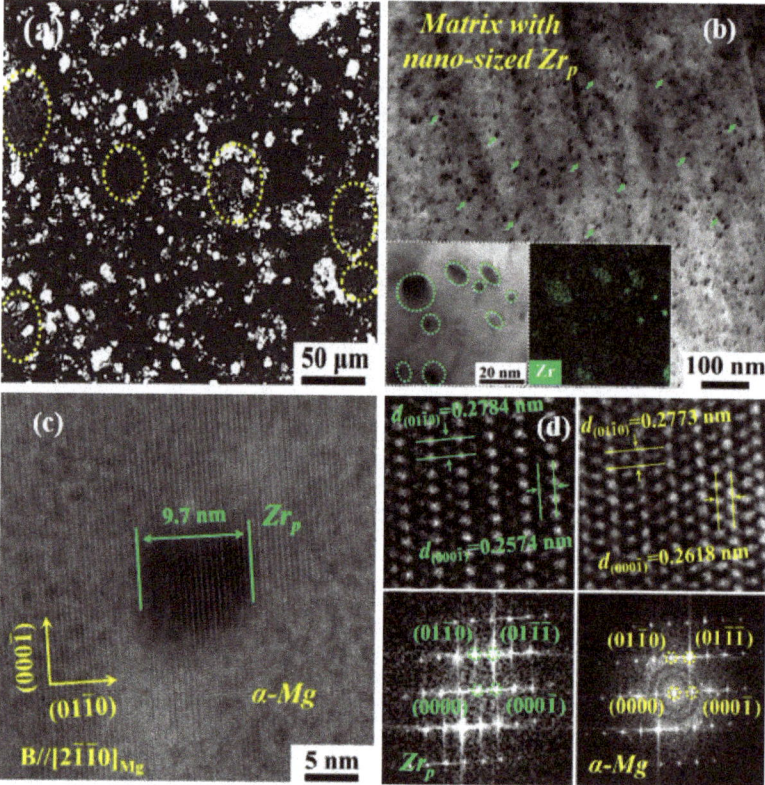

Figure 12. Characterization of UHFP-TIGR-treated Mg-30Zr master alloy: (**a**) SEM BSE image under low magnification; (**b**) TEM dark-field image; (**c,d**) HRTEM image with diffraction patterns of nanoscale Zr_{np} [86]. Reprinted with permission from ref. [86]. Copyright 2022 Elsevier.

5.2. Melt Treatments

Melt treatments such as ultrasonic treatment (UST) [88–91], intensive melt shearing (IMS) [41,49,92,93] or low-frequency electro-magnetic stirring (LFEMS) [94] have been tried after the addition of Zr into the Mg alloy melt. Ramirez et al. [88] showed that UTS enhances the effect of grain refinement with Zr (0.5, 1.0 or 1.5Zr) on Mg-3Zn alloy (Figure 13). For instance, the grain size of Mg-3Zn-0.5Zr alloy was about 91.5 µm, which was further refined to 71.9 µm by means of UST. Nagasivamuni et al. [89,90] showed that UST enhances the grain refinement effect of Mg-Zr alloys (0.2, 0.5 or 1.0Zr). The main reason for this is that UST increases the soluble Zr, activates more Zr nucleation particles, and decreases the Zr settling [89,90]. Additionally, another possible reason for this may be that the surfaces of some contaminated Zr particles were cleaned and wetted by UST, facilitating an increase in the number of nucleation sites.

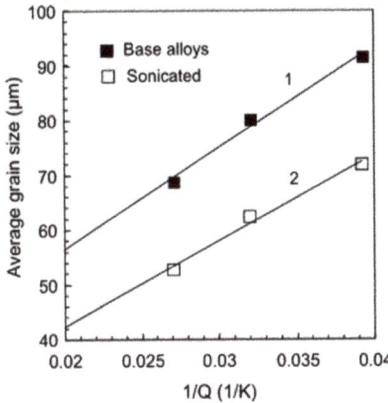

Figure 13. Grain size vs. 1/Q for Mg-3Zn-Zr alloys [88]. "Sonicated" = UST; addition of Zr = 0.5, 10.0 and 1.5, respectively. Reprinted with permission from ref. [88]. Copyright 2008 Elsevier.

Das et al. [92] showed that IMS enhances the effect of grain refinement with Zr on Mg-6Zn alloy. The mechanism for this is that IMS deagglomerates and disperses the Zr particle clusters uniformly in the melt, increasing the number density of Zr nucleation particles. However, Peng et al. [39,49,93] showed that IMS makes the effect of grain refinement with Zr more complicated, as a result of the competition between Zr and MgO particles formed in-situ [49]. On the one hand, when Zr content was as low as 0.1%, IMS led to a significant grain refinement of the Mg-0.1Zr alloy. This was because the MgO particles formed in-situ were able to be well dispersed by IMS, resulting in an increase in the number density of MgO nucleating particles. At the same time, the adsorption of the Zr layer on the surface of the MgO particles was enhanced by IMS, leading to an improvement in the nucleating potency of the MgO particles [49]. Thus, the grain refinement effect could be achieved at a relatively lower Zr content (0.1%). On the other hand, when Zr content was just beyond the peritectic point, grain coarsening occurred. This was because the Zr particles underwent a coarsening growth process with the effect of IMS, resulting in the formation of larger Zr particles and a reduction in Zr number density [49]. Thus, grain coarsening was observed. In addition, when Zr content was as high as 2%, the grain was well refined again, because more Zr nucleating particles were supplied.

6. Conclusions and Remarks

Zr is the most effective and important grain refiner for Al-free Mg alloys, especially for many high-strength Mg alloys. This review paper summarizes the recent advances in the effect of grain refinement with Zr on Mg alloys in detail, including the alloying process of Zr, the grain refinement mechanism of Zr, the grain refinement behavior of Zr, and improvements in the grain refinement efficiency of Zr. This review provides a full understanding of the effect of grain refinement with Zr on Mg alloys. The main points of Zr refinement are summarized in Figure 14.

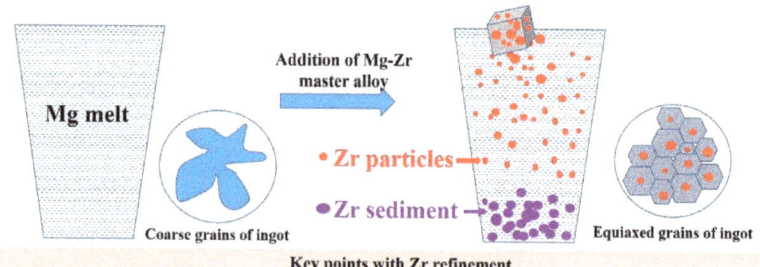

Figure 14. The main features of the effect of grain refinement with Zr on Mg alloys.

The main conclusions are as follows:

(1) Among various methods of introducing Zr into Mg melt, only the Mg-Zr master alloy shows high efficiency. This is because the Mg-Zr master alloy ensures that the interface between the Zr particles and the Mg melt is clean and active, facilitating the diffusion of Zr elements and increased nucleation utilization.

(2) The grain refinement mechanism is attributed to both heterogeneous nucleation and the constitutional supercooling effect. The perfect crystal match between Zr and Mg, and the high GRF value contribute to the powerful effect of grain refinement with Zr.

(3) Many factors influence the effect of grain refinement with Zr, including Zr particle settling, the particle size distribution of the Mg-Zr master alloy, cooling rate, and the alloying elements. In particular, the size distribution of the Mg-Zr master alloy has a great influence on the grain refinement efficiency of Zr. The spontaneous settling of Zr particles increases the alloying cost.

(4) To achieve a better refinement effect and a higher utilization rate of Zr, two methods have been investigated, i.e., pre-treatment of the Mg-Zr master alloy and melt treatment. The newly developed UHFP-TIGR pre-treatment [86] shows remarkable modification effects and remarkable grain refinement efficiency.

However, due to the complicated process of melt treatment or pretreatment of the Mg-Zr master alloy, some work needs to be done in future on tackling the problem of Zr waste at the root and saving expensive Zr resources. The key point is the development of Mg-Zr master alloy with super-fine Zr particles. To achieve this aim, the typical method for preparing the Mg-Zr master alloy, i.e., Mg thermal reduction reaction with Zr-rich halides, should be investigated again in more detail with respect to its thermal dynamic aspects. If the Zr particle size can be controlled to be as small as possible during the reaction, a more economical production of Mg-Zr master alloy will be achieved.

Additionally, if the in-situ method of mixing Zr-rich salts is employed, a suitable purification technology should be developed for removing the salt inclusions in the Mg melt. There are two points worthy of investigation. One is that new fluxes effective in removing inclusions can be developed based on the molten salt system. Another is that complex purification methods such as gas bubbling plus external energy field are deserving of trials.

Author Contributions: Conceptualization, M.S. and Y.Z.; methodology, M.S.; investigation, M.S. and D.Y.; resources, L.M., X.L. and S.P.; data curation, M.S. and D.Y.; writing—original draft preparation, M.S.; writing—review and editing, D.Y., L.M. and Y.Z.; supervision, M.S. and Y.Z.; project administration, M.S.; funding acquisition, M.S., Y.Z. and L.M. All authors have read and agreed to the published version of the manuscript.

Funding: This research was funded by National Natural Science Foundation of China (NSFC), grant number 51701124, 51901027 and 51901137. The APC was funded by 51701124.

Data Availability Statement: Not applicable.

Conflicts of Interest: The authors declare no conflict of interest.

References

1. Karakulak, E. A review: Past, present and future of grain refining of magnesium castings. *J. Magnes. Alloy.* **2019**, *7*, 355–369. [CrossRef]
2. Song, J.; Chen, J.; Xiong, X.; Peng, X.; Chen, D.; Pan, F. Research advances of magnesium and magnesium alloys worldwide in 2021. *J. Magnes. Alloy.* **2022**, *10*, 863–898. [CrossRef]
3. Ali, Y.; Qiu, D.; Jiang, B.; Pan, F.; Zhang, M. Current research progress in grain refinement of cast magnesium alloys: A review article. *J. Alloy. Compd.* **2015**, *619*, 639–651. [CrossRef]
4. StJohn, D.H.; Ma, Q.; Easton, M.A.; Cao, P.; Hildebrand, Z. Grain refinement of magnesium alloys. *Metall. Mater. Trans. A* **2005**, *36*, 1669–1679. [CrossRef]
5. Sauerwald, F. Das Zustandsdiagramm Magnesium-Zirkonium. *Z. Für Anorg. Chemie.* **1947**, *255*, 212–220. [CrossRef]
6. Friedrich, H.E.; Mordike, B.L. *Magnesium Technology: Metallurgy, Design Data, Applications*; Springer: Berlin/Heidelberg, Germany, 2006; pp. 128–143.
7. Mo, N.; Tan, Q.; Bermingham, M.; Huang, Y.; Dieringa, H.; Hort, N.; Zhang, M. Current development of creep-resistant magnesium cast alloys: A review. *Mater. Des.* **2018**, *155*, 422–442. [CrossRef]
8. Sravya, T.; Sankaranarayanan, S.; Abdulhakim, A.; Manoj, G. Mechanical properties of magnesium-rare earth alloy systems: A review. *Metals* **2015**, *5*, 1–39. [CrossRef]
9. Wu, G.; Wang, C.; Sun, M.; Ding, W. Recent developments and applications on high-performance cast magnesium rare-earth alloys. *J. Magnes. Alloy.* **2021**, *9*, 1–20. [CrossRef]
10. Stjohn, D.H.; Easton, M.A.; Ma, Q.; Taylor, J.A. Grain refinement of magnesium alloys: A review of recent research, theoretical developments, and their application. *Metall. Mater. Trans. A* **2013**, *44*, 2935–2949. [CrossRef]
11. Easton, M.A.; Ma, Q.; Prasad, A.; StJohn, D.H. Recent advances in grain refinement of light metals and alloys. *Curr. Opin. Solid. State Mater.* **2016**, *20*, 13–24. [CrossRef]
12. Yu, W.; He, H.; Li, C.; Li, Q.; Liu, Z.; Qin, B. Existing form and effect of zirconium in pure Mg, Mg-Yb, and Mg-Zn-Yb alloys. *Rare Metals* **2009**, *28*, 289–296. [CrossRef]
13. Emley, E.F. *Principles of Magnesium Technology*; Pergamon Press: Oxford, UK, 1966.
14. Ma, Q.; StJohn, D.H.; Frost, M.T. Magnesium-Zirconium Alloying. U.S. Patent 20050161121A1, 2005.
15. Sun, M.; Wu, G.; Dai, J.; Pang, S.; Ding, W. Current research status of grain refinement effect of Zr on magnesium alloy. *Foundry* **2010**, *59*, 255–259. (In Chinese)
16. Sauerwald, F. Process for the Production of Magnesium-Zirconium Alloys. U.S. Patent 2228781, 14 January 1941.
17. Ball, C.J.P.; Jessup, A.C.; Emley, E.F.; Fisher, P.A. Alloying Composition for Introducing Zirconium into Magnesium. U.S. Patent 2497531, 14 February 1950.
18. Michael, D.W. Magnesium Base Alloys Containing Zirconium. U.S. Patent 2698230, 28 December 1954.
19. Whitehead, D.J.; Frederick, E.E. Alloying of Manganese and Zirconium to Magnesium. U.S. Patent 2919190, 29 December 1959.
20. Sun, M.; Wu, G.; Dai, J.; Wang, W.; Ding, W. Grain refinement behavior of potassium fluozirconate (K_2ZrF_6) salts mixture introduced into Mg-10Gd-3Y magnesium alloy. *J. Alloy. Compd.* **2010**, *494*, 426–433. [CrossRef]
21. Wang, W.; Zhang, M.; Wang, W.; Wang, A.; Li, M. Effect of potassium fluorozirconate salts mixture on the microstructure and mechanical properties of Mg-3Y-3.5Sm-2Zn alloy. *Rare Metal. Mat. Eng.* **2020**, *49*, 1151–1158.
22. Le, Q.; Zhang, Z.; Cui, J. The Preparation Method of Zr-rich Salts Mixture Used for Zr Alloying in Mg Alloy. Chinese Patent 102605200A, 25 July 2012. (In Chinese)
23. Tong, X.; Wu, G.; Zhang, L.; Liu, W.; Ding, W. Achieving low-temperature Zr alloying for microstructural refinement of sand-cast Mg-Gd-Y alloy by employing zirconium tetrachloride. *Mater. Charact.* **2021**, *171*, 110727. [CrossRef]
24. Viggers, R.F. The Magnesium-Zirconium Alloys. Ph.D. Thesis, Oregon State College, Corvallis, OR, USA, 1950.
25. Sun, M.; Wu, G.; Easton, M.A.; StJohn, D.H.; Abbott, T.; Ding, W. A comparison of the microstructure of three Mg-Zr master alloys and their grain refinement efficiency. In Proceedings of the 9th International Conference on Magnesium Alloys and Their Applications, Vancouver, Canada, 8–12 July 2012; pp. 873–880.
26. Sun, M.; Easton, M.A.; StJohn, D.H.; Wu, G.; Abbott, T.; Ding, W. Grain refinement of magnesium alloys by Mg-Zr master alloys: The role of alloy chemistry and Zr particle number density. *Adv. Eng. Mater.* **2013**, *15*, 373–378. [CrossRef]

27. Sun, M. Study on Grain Refinement Behavior of Mg-Gd-Y Magnesium Alloy by Zirconium. Ph.D. Thesis, Shanghai Jiao Tong University, Shanghai, China, 2012. (In Chinese)
28. Liu, H.; Ning, Z.; Cao, F.; Zhang, Y.; Sun, J. Effect of melting process on Zr content and grain refinement in ZE41A alloy. *Adv. Mat. Res.* **2011**, *284–286*, 1651–1655. [CrossRef]
29. Ma, Q.; Graham, D.; Zheng, L.; StJohn, D.H.; Frost, M.T. Alloying of pure magnesium with Mg-33.3 wt.% Zr master alloy. *Mater. Sci. Technol.* **2003**, *19*, 156–162. [CrossRef]
30. Ma, Q.; Zheng, L.; Graham, D.; Frost, M.T.; StJohn, D.H. Settling of undissolved zirconium particles in pure magnesium melts. *J. Light Met.* **2001**, *1*, 157–165.
31. Ma, Q.; Das, A. Grain refinement of magnesium alloys by zirconium: Formation of equiaxed grains. *Scr. Mater.* **2006**, *54*, 881–886. [CrossRef]
32. Tamura, Y.; Kono, N.; Motegi, T.; Sato, E. Grain refining of pure magnesium by adding Mg-Zr master alloy. *J. Jpn. Inst. Met.* **1997**, *47*, 679–684. (In Japanese)
33. Qiu, D.; Zhang, M.; Taylor, J.A.; Kelly, P.M. A new approach to designing a grain refiner for Mg casting alloys and its use in Mg-Y-based alloys. *Acta Mater.* **2009**, *57*, 3052–3059. [CrossRef]
34. Dai, J.; Easton, M.A.; Zhang, M.; Qiu, D.; Xiong, X.; Liu, W.; Wu, G. Effects of cooling rate and solute content on the grain refinement of Mg-Gd-Y alloys by aluminum. *Metall. Mater. Trans. A* **2014**, *45*, 4665–4678. [CrossRef]
35. Wang, C.; Dai, J.; Liu, W.; Zhang, L.; Wu, G. Effect of Al additions on grain refinement and mechanical properties of Mg-Sm alloys. *J. Alloy. Compd.* **2015**, *620*, 172–179. [CrossRef]
36. Ma, Q.; StJohn, D.H.; Frost, M.T. Effect of soluble and insoluble zirconium on the grain refinement of magnesium alloys. *Mater. Sci. Forum* **2003**, *419–422*, 593–598. [CrossRef]
37. Ma, Q.; StJohn, D.H.; Frost, M.T. Characteristic zirconium-rich coring structures in Mg-Zr alloys. *Scr. Mater.* **2002**, *46*, 649–654. [CrossRef]
38. Ma, Q.; StJohn, D.H. Grain nucleation and formation in Mg-Zr alloys. *Int. J. Cast. Metal. Res.* **2009**, *22*, 256–259. [CrossRef]
39. Ma., Q.; StJohn, D.H.; Frost, M.T. Zirconium alloying and grain refinement of magnesium alloys. In Proceedings of the Magnesium Technology 2003, San Diego, CA, USA, 2–6 March 2003; pp. 209–214, Edited by Howard I. Kaplan.
40. Ramachandran, T.R.; Sharma, P.K.; Balasubramanian, K. Grain refinement of light alloys. In Proceedings of the 68th WFC-World Foundry Congress, Chennai, India, 7–10 February 2008; pp. 189–193.
41. Peng, G.; Wang, Y.; Chen, K.; Chen, S. Improved Zr grain refining efficiency for commercial purity Mg via intensive melt shearing. *Int. J. Cast. Metal. Res.* **2017**, *30*, 374–378. [CrossRef]
42. Ma, Q. Heterogeneous nucleation on potent spherical substrates during solidification. *Acta Mater.* **2007**, *55*, 943–953. [CrossRef]
43. StJohn, D.H.; Ma, Q.; Easton, M.A.; Cao, P. The Interdependence Theory: The relationship between grain formation and nucleant selection. *Acta Mater.* **2011**, *59*, 4907–4921. [CrossRef]
44. Li, J.; Chen, R.; Ma, Y.; Ke, W. Effect of Zr modification on solidification behavior and mechanical properties of Mg-Y-RE (WE54) alloy. *J. Magnes. Alloy.* **2013**, *1*, 346–351. [CrossRef]
45. Zhao, Y.; Pu, Z.; Wang, L.; Liu, D. Modeling of grain refinement and nucleation behavior of Mg-4Y-0.5Zr (wt.%) alloy via cellular automaton model. *Int. J. Metalcast.* **2022**, *16*, 945–961. [CrossRef]
46. Tamura, Y.; Kono, N.; Motegi, T.; Sato, E. Grain refining mechanism and casting structure of Mg-Zr alloy. *J. Jpn. Inst. Met.* **1998**, *48*, 185–189. (In Japanese)
47. Saha, P. An Analysis of the Grain Refinement of Magnesium by Zirconium. Ph.D. Thesis, The University of Alabama, Tuscaloosa, AL, USA, 2010. ProQuest Dissertations Publishing.
48. Saha, P.; Viswanathan, S. Grain refinement of magnesium by zirconium: Characterization and analysis. In Proceedings of the American Foundry Society (AFS) Proceedings 2011, Dayton, IL, USA, 10–12 October 2011; pp. 175–180.
49. Peng, G.S.; Wang, Y.; Fan, Z. Competitive heterogeneous nucleation between Zr and MgO particles in commercial purity magnesium. *Metall. Mater. Trans. A* **2018**, *49*, 2182–2192. [CrossRef]
50. Yang, W.; Ji, S.; Zhang, R.; Zhang, J.; Liu, L. Abnormal grain refinement behavior in high-pressure die casting of pure Mg with addition of Zr as grain refiner. *JOM* **2018**, *70*, 2555–2560. [CrossRef]
51. Yang, W.; Liu, L.; Zhang, J.; Ji, S.; Fan, Z. Heterogeneous nucleation in Mg-Zr alloy under die casting condition. *Mater. Lett.* **2015**, *160*, 263–267. [CrossRef]
52. Abdallah, E.; Francesco, D.; Comondore, R.; Dimitry, S. Observing the effect of grain refinement on crystal growth of Al and Mg alloys during solidification using in-situ neutron diffraction. *Metals* **2022**, *12*, 793. [CrossRef]
53. Han, Q. The role of solutes in grain refinement of hypoeutectic magnesium and aluminum alloys. *J. Magnes. Alloy* 2022, in press. [CrossRef]
54. StJohn, D.H.; Cao, P.; Ma, Q.; Easton, M.A. A new analytical approach to reveal the mechanisms of grain refinement. *Adv. Eng. Mater.* **2007**, *9*, 739–749. [CrossRef]
55. Easton, M.A.; StJohn, D.H. A model of grain refinement incorporating alloy constitution and potency of heterogeneous nucleant particles. *Acta Mater.* **2001**, *49*, 1867–1878. [CrossRef]
56. Fan, Z.; Gao, F.; Wang, Y.; Men, H.; Zhou, L. Effect of solutes on grain refinement. *Prog. Mater. Sci.* **2022**, *123*, 100809. [CrossRef]
57. Peng, Z.K.; Zhang, X.M.; Chen, J.-M.; Xiao, Y.; Jiang, H. Grain refining mechanism in Mg-9Gd-4Y alloys by zirconium. *Mater. Sci. Technol.* **2005**, *21*, 722–726. [CrossRef]

58. Zhang, D.; Qiu, D.; Zhu, S.; Dargusch, M.; StJohn, D.H.; Easton, M.A. Grain refinement in laser remelted Mg-3Nd-1Gd-0.5Zr alloy. *Scr. Mater.* **2020**, *183*, 12–16. [CrossRef]
59. Ma, Q. Creation of semisolid slurries containing fine and spherical particles by grain refinement based on the Mullins-Sekerka stability criterion. *Acta Mater.* **2006**, *54*, 2241–2252. [CrossRef]
60. Chen, Y.; Zhang, L.; Liu, W.; Wu, G.; Ding, W. Preparation of Mg-Nd-Zn-(Zr) alloys semisolid slurry by electromagnetic stirring. *Mater. Des.* **2016**, *95*, 398–409. [CrossRef]
61. Ma, Q.; StJohn, D.H.; Frost, M.T. Heterogeneous nuclei size in magnesium- zirconium alloys. *Scr. Mater.* **2004**, *50*, 1115–1119. [CrossRef]
62. Liu, H.; Ning, Z.; Cao, F.; Meng, Z.; Sun, J. Effect of cooling condition on Zr-rich core formation and grain size in Mg alloy. *Adv. Mat. Res.* **2011**, *189–193*, 3920–3924. [CrossRef]
63. Sun, M.; Stjohn, D.H.; Easton, M.A.; Wang, K.; Ni, J. Effect of cooling rate on the grain refinement of Mg-Y-Zr alloys. *Metall. Mater. Trans. A* **2020**, *51*, 482–496. [CrossRef]
64. Kabirian, F.; Mahmudi, R. Effects of zirconium additions on the microstructure of as-cast and aged AZ91 magnesium alloy. *Adv. Eng. Mater.* **2009**, *11*, 189–193. [CrossRef]
65. Balasubramani, N.; Wang, G.; StJohn, D.H.; Dargusch, M.S. The poisoning effect of Al and Be on Mg-1 wt.% Zr alloy and the role of ultrasonic treatment on grain refinement. *Front. Mater.* **2019**, *10*, 322. [CrossRef]
66. Cao, P.; Ma, Q.; StJohn, D.H.; Frost, M.T. Uptake of iron and its effect on grain refinement of pure magnesium by zirconium. *Mater. Sci. Technol.* **2004**, *20*, 585–592. [CrossRef]
67. Cao, P.; Qian, M.; StJohn, D.H. Grain coarsening of magnesium alloys by beryllium. *Scr. Mater.* **2004**, *51*, 647–651. [CrossRef]
68. Chang, S.Y.; Tezuka, H.; Kamio, A. Mechanical properties and structure of ignition-proof Mg-Ca-Zr alloys produced by squeeze casting. *Mater. Trans. JIM* **1997**, *38*, 526–535. [CrossRef]
69. Hildebrand, Z.; Qian, M.; StJohn, D.H.; Frost, M.T. Influence of zinc on the soluble zirconium content in magnesium and the subsequent grain refinement by zirconium. In Proceedings of the Magnesium Technology 2004, Warrendale, PA, USA, 14–18 March 2004; pp. 241–245.
70. Li, P.; Hou, D.; Han, E.; Chen, R.; Shan, Z. Solidification of Mg-Zn-Zr alloys: Grain growth restriction, dendrite coherency and grain size. *Acta Metall. Sin.* **2020**, *33*, 1477–1486. [CrossRef]
71. Jafari, H.; Amiryavari, P. The effects of zirconium and beryllium on microstructure evolution, mechanical properties and corrosion behaviour of as-cast AZ63 alloy. *Mat. Sci. Eng. A* **2016**, *654*, 161–168. [CrossRef]
72. Tamura, Y.; Motegi, T.; Kono, N.; Sato, E. Effect of minor elements on grain size of Mg-9%Al alloy. *Mater. Sci. Forum* **2000**, *350–351*, 199–204. [CrossRef]
73. Cao, P.; Ma, Q.; StJohn, D.H. Native grain refinement of magnesium alloys. *Scr. Mater.* **2005**, *53*, 841–844. [CrossRef]
74. Miyahara, Y.; Matsubara, K.; Horita, Z.; Langdon, T.G. Grain refinement and superplasticity in a magnesium alloy processed by equal-channel angular pressing. *Metall. Mater. Trans. A* **2005**, *36*, 1705–1711. [CrossRef]
75. Kabirian, F.; Mahmudi, R. Effects of Zr additions on the microstructure and impression creep behavior of AZ91 magnesium alloy. *Metall. Mater. Trans. A* **2010**, *41*, 3488–3498. [CrossRef]
76. Ma, Q.; Hildebrand, Z.C.G.; StJohn, D.H. The loss of dissolved zirconium in zirconium-refined magnesium alloys after remelting. *Metall. Mater. Trans. A* **2009**, *40*, 2470–2479. [CrossRef]
77. Davis, B.; O'Reilly, K. Electron probe micro analysis of sedimented zirconium particles in magnesium. In *Proceedings of the 6th International Conference Magnesium Alloys and Their Applications*; Wiley: New York, NY, USA, 2003; pp. 242–247. [CrossRef]
78. Prasad, A.; Uggowitzer, P.J.; Shi, Z.; Atrens, A. Production of high purity Mg-X rare earth binary alloys using Zr. *Mater. Sci. Forum* **2013**, *765*, 301–305. [CrossRef]
79. Xing, F.; Guo, F.; Su, J.; Zhao, X.; Cai, H. The existing forms of Zr in Mg-Zn-Zr magnesium alloys and its grain refinement mechanism. *Mater. Res. Express* **2021**, *8*, 066516. [CrossRef]
80. Ma, Q.; StJohn, D.H.; Frost, M.T. A new zirconium-rich master alloy for the grain refinement of magnesium alloys. In *Magnesium: Proceedings of the 6th International Conference Magnesium Alloys and Their Applications*; Wiley: New York, NY, USA, 2003; pp. 706–712. [CrossRef]
81. Ma, Q.; StJohn, D.H.; Frost, M.T.; Barnett, M.R. Grain refinement of pure magnesium using rolled Zirmax®master alloy (Mg-33.3Zr). In Proceedings of the Magnesium: Proceedings of Magnesium Technology 2003, San Diego, CA, USA, 2–6 March 2003; pp. 215–220.
82. Viswanathan, S.; Saha, P.; Foley, D.; Hartwig, K.T. Engineering a more efficient zirconium grain refiner for magnesium. In *Magnesium Technology 2011*; Springer: Berlin/Heidelberg, Germany; pp. 559–564.
83. Viswanathan, S.; Saha, P.; Foley, D.; Hartwig, K.T. Developing an improved zirconium grain refiner for magnesium. In Proceedings of the American Foundry Society (AFS) Proceedings 2011, Dayton, IL, USA, 10–12 October 2011; p. 11-067.
84. Wang, C.; Sun, M.; Zheng, F.; Peng, L.; Ding, W. Improvement in grain refinement efficiency of Mg-Zr master alloy for magnesium alloy by friction stir processing. *J. Magnes. Alloy.* **2014**, *2*, 239–244. [CrossRef]
85. Wang, L.; Guan, H.; Wu, G.; Pan, Y.; Sun, S. A new approach to improving the macrosegregation in zirconium-containing magnesium-lithium alloy. *Materialwiss. Werkstofftechnol.* **2018**, *49*, 1125–1134. [CrossRef]
86. Tong, X.; Wu, G.; Easton, M.A.; Sun, M.; StJohn, D.H.; Jiang, R.; Qi, F. Exceptional grain refinement of Mg-Zr master alloy treated by tungsten inert gas arc re-melting with ultra-high frequency pulses. *Scr. Mater.* **2022**, *215*, 114700. [CrossRef]

87. Tong, X.; Wu, G.; Zhang, L. A Pretreatment Method to Improve the Refining Effect of Mg-Zr Master Alloy on Mg alloy. Chinese Patent 111872517B, 20 July 2021. (In Chinese)
88. Ramirez, A.; Ma, Q.; Davis, B.; Wilks, T.; StJohn, D.H. Potency of high-intensity ultrasonic treatment for grain refinement of magnesium alloys. *Scr. Mater.* **2008**, *59*, 19–22. [CrossRef]
89. Nagasivamuni, B.; Wang, G.; StJohn, D.H.; Dargusch, M.S. Effect of ultrasonic treatment on the alloying and grain refinement efficiency of a Mg-Zr master alloy added to magnesium at hypo- and hyper-peritectic compositions. *J. Cryst. Growth* **2019**, *512*, 20–32. [CrossRef]
90. Nagasivamuni, B.; Wang, G.; Easton, M.A.; StJohn, D.H.; Dargusch, M.S. A comparative study of the role of solute, potent particles and ultrasonic treatment during solidification of pure Mg, Mg-Zn and Mg-Zr alloys. *J. Magnes. Alloy.* **2021**, *9*, 829–839. [CrossRef]
91. Xiao, L.; Tian, Y.; Li, Z.; Zou, W.; Yuan, Y.; Li, B.; Wang, X.; Zhang, X. A Method of Preparing Magnesium Zirconium Master Alloy by Ultrasonic Treatment. CN105385863A, 31 May 2017. (In Chinese)
92. Das, A.; Liu, G.; Fan, Z. Investigation on the microstructural refinement of an Mg-6 wt.% Zn alloy. *Mater. Sci. Eng. A* **2006**, *419*, 349–356. [CrossRef]
93. Peng, G.; Wang, Y.; Fan, Z. Effect of intensive melt shearing and Zr content on grain refinement of Mg-0.5Ca-xZr alloys. *Mater. Sci. Forum* **2013**, *765*, 336–340. [CrossRef]
94. Wang, C.; Dong, Z.; Li, K.; Sun, M.; Wu, J.; Wang, K.; Wu, G.; Ding, W. A novel process for grain refinement of Mg-RE alloys by low frequency electro-magnetic stirring assisted near-liquidus squeeze casting. *J. Mater. Process. Technol.* **2022**, *303*, 117537. [CrossRef]
95. Zhang, L.; Jiang, H.; He, J.; Zhao, J. Improved grain refinement in aluminium alloys by re-precipitated TiB_2 particles. *Mater. Lett.* **2022**, *312*, 131657. [CrossRef]

MDPI
St. Alban-Anlage 66
4052 Basel
Switzerland
www.mdpi.com

Metals Editorial Office
E-mail: metals@mdpi.com
www.mdpi.com/journal/metals

Disclaimer/Publisher's Note: The statements, opinions and data contained in all publications are solely those of the individual author(s) and contributor(s) and not of MDPI and/or the editor(s). MDPI and/or the editor(s) disclaim responsibility for any injury to people or property resulting from any ideas, methods, instructions or products referred to in the content.

www.ingramcontent.com/pod-product-compliance
Lightning Source LLC
LaVergne TN
LVHW070703100526
838202LV00013B/1019